# Statistics
# for the
# Utterly Confused

Other books in the **Utterly Confused** Series include:

*Calculus for the Utterly Confused*

*Financial Planning for the Utterly Confused*, Fifth Edition

*Job Hunting for the Utterly Confused*

*Physics for the Utterly Confused*

# Statistics for the Utterly Confused

## Lloyd R. Jaisingh

**McGraw-Hill**

New York   Chicago   San Francisco   Lisbon   London
Madrid   Mexico City   Milan   New Delhi   San Juan
Seoul   Singapore   Sydney   Toronto

**Library of Congress Cataloging-in-Publication Data**

Jaisingh, Lloyd R.
    Statistics for the utterly confused / Lloyd R. Jaisingh.
      p. cm.
    ISBN 0-07-135005-5
    1. Statistics—Study and teaching.   I. Title.

  QA276.18.J35  2000
  519.5—dc21

                                00-022982

# McGraw-Hill

*A Division of The McGraw·Hill Companies*

7 8 9 10 11  DOC/DOC  0 9 8 7 6 5 4 3

ISBN 0-07-135005-5

*The sponsoring editor for this book was Barbara Gilson, the editing supervisor was Maureen B. Walker, and the production supervisor was Tina Cameron. It was set in Times Ten by North Market Street Graphics.*

*Printed and bound by R. R. Donnelley & Sons Company.*

McGraw-Hill books are available at special quantity discounts to use as premiums and sales promotions, or for use in corporate training programs. For more information, please write to the Director of Special Sales, McGraw-Hill, Two Penn Plaza, New York, NY 10121. Or contact your local bookstore.

 This book is printed on recycled, acid-free paper containing a minimum of 50% recycled, de-inked fiber.

# Dedication

This book is dedicated to my wife, Pam, for allowing me to spend countless hours at the office working on successive projects. Also, to my son Nathan, for all his love, my mother, for all her nurturing, and to the memory of a dedicated father.

# Acknowledgments

I would like to thank my colleague, Terry Irons, for reading the manuscript before it was sent to the desk of the editor. I would like to thank him immensely for all his literary suggestions that helped to make the book a better product. Also, I would like to thank the very capable staff at McGraw-Hill, especially Maureen Walker, Editing Supervisor, for her tremendous help. In addition, I would like to thank my editor, Barbara Gilson, for allowing me the opportunity to write this book. I thank her for her guidance in helping this book to become a reality.

# Contents

Preface                                                                    xiii

Technology Integration                                                      xiv

Organization of the Text                                                     xv

Part I          DESCRIPTIVE STATISTICS                                        1

Chapter 1       *Graphical Displays of Univariate Data*                       3
    •   Do I Need to Read This Chapter?                                        3
    •   1-1  Introduction                                                      4
    •   1-2  Frequency Distributions                                          6
    •   1-3  Dot Plots                                                        10
    •   1-4  Bar Charts or Bar Graphs                                         11
    •   1-5  Histograms                                                       12
    •   1-6  Frequency Polygons                                              13
    •   1-7  Stem-and-Leaf Displays or Plots                                 14
    •   1-8  Time Series Graphs                                              15
    •   1-9  Pie Graphs or Pie Charts                                        16
    •   1-10  Pareto Charts                                                  17
    •   Technology Corner                                                    17
    •   It's a Wrap                                                          18
    •   Test Yourself                                                        18

**Chapter 2**  *Data Description—Numerical Measures of Central Tendency*
          *for Ungrouped Univariate Data*                                27
- Do I Need to Read This Chapter?                                     27
- Get Started                                                         27
- 2-1  The Mean                                                       28
- 2-2  The Median                                                     30
- 2-3  The Mode                                                       32
- 2-4  Shapes (Skewness)                                              33
- Technology Corner                                                   35
- It's a Wrap                                                         36
- Test Yourself                                                       36

**Chapter 3**  *Data Description—Numerical Measures of Variability*
          *for Ungrouped Univariate Data*                             43
- Do I Need to Read This Chapter?                                     43
- Get Started                                                         43
- 3-1  The Range                                                      44
- 3-2  The Interquartile Range                                        45
- 3-3  The Mean Absolute Deviation                                    46
- 3-4  The Variance and Standard Deviation                            47
- 3-5  The Coefficient of Variation                                   51
- 3-6  The Empirical Rule                                             51
- Technology Corner                                                   53
- It's a Wrap                                                         54
- Test Yourself                                                       54

**Chapter 4**  *Data Description—Numerical Measures of Position*
          *for Ungrouped Univariate Data*                             63
- Do I Need to Read This Chapter?                                     63
- Get Started                                                         63
- 4-1  The $z$ Score or Standard Score                                64
- 4-2  Percentiles                                                    65
- Technology Corner                                                   72
- It's a Wrap                                                         73
- Test Yourself                                                       73

**Chapter 5**  *Exploring Bivariate Data*                             82
- Do I Need to Read This Chapter?                                     82
- Get Started                                                         82
- 5-1  Scatter Plots                                                  82
- 5-2  Looking for Patterns in the Data                               84
- 5-3  Correlation                                                    86
- 5-4  Correlation and Causation                                      87
- 5-5  Least-Squares Regression Line                                  88
- 5-6  The Coefficient of Determination                              90

| | | |
|---|---|---:|
| • | 5-7  Residual Plots | 91 |
| • | 5-8  Outliers and Influential Points | 92 |
| • | Technology Corner | 92 |
| • | It's a Wrap | 93 |
| • | Test Yourself | 94 |

**Chapter 6  *Exploring Categorical Data*  103**

| | | |
|---|---|---:|
| • | Do I Need to Read This Chapter? | 103 |
| • | Get Started | 103 |
| • | 6-1  Marginal Distributions | 104 |
| • | 6-2  Conditional Distributions | 105 |
| • | 6-3  Using Bar Charts to Display Contingency Tables | 107 |
| • | 6-4  Independence in Categorical Variables | 109 |
| • | 6-5  Simpson's Paradox | 111 |
| • | Technology Corner | 113 |
| • | It's a Wrap | 113 |
| • | Test Yourself | 113 |

**Part II   PROBABILITY   119**

**Chapter 7  *Randomness, Uncertainty, and Probability*   121**

| | | |
|---|---|---:|
| • | Do I Need to Read This Chapter? | 121 |
| • | Get Started | 121 |
| • | 7-1  Randomness and Uncertainty | 122 |
| • | 7-2  Random Experiments, Sample Space, and Events | 122 |
| • | 7-3  Classical Probability | 123 |
| • | 7-4  Relative Frequency or Empirical Probability | 125 |
| • | 7-5  The Law of Large Numbers | 125 |
| • | 7-6  Subjective Probability | 127 |
| • | 7-7  Some Basic Laws of Probability | 127 |
| • | 7-8  Other Probability Rules | 128 |
| • | 7-9  Conditional Probability | 133 |
| • | 7-10  Independence | 133 |
| • | Technology Corner | 135 |
| • | It's a Wrap | 135 |
| • | Test Yourself | 135 |

**Chapter 8  *Discrete Probability Distributions*   144**

| | | |
|---|---|---:|
| • | Do I Need to Read This Chapter? | 144 |
| • | Get Started | 144 |
| • | 8-1  Random Variables | 145 |
| • | 8-2  Probability Distributions for Discrete Random Variables | 146 |
| • | 8-3  Expected Value | 148 |
| • | 8-4  Variance and Standard Deviation of a Discrete Random Variable | 151 |
| • | 8-5  Bernoulli Trials and the Binomial Probability Distribution | 153 |

- Technology Corner                                                                157
- It's a Wrap                                                                      157
- Test Yourself                                                                    158

Chapter 9    *The Normal Probability Distribution*                                166
- Do I Need to Read This Chapter?                                                  166
- Get Started                                                                      166
- 9-1  The Normal Distribution                                                     167
- 9-2  Properties of the Normal Distribution                                       170
- 9-3  The Standard Normal Distribution                                            173
- 9-4  Applications of the Normal Distribution                                     177
- Technology Corner                                                                179
- It's a Wrap                                                                      181
- Test Yourself                                                                    181

Chapter 10  *Sampling Distributions and the Central Limit Theorem*                189
- Do I Need to Read This Chapter?                                                  189
- Get Started                                                                      189
- 10-1  Sampling Distribution of a Sample Proportion                               190
- 10-2  Sampling Distribution of a Sample Mean                                     193
- 10-3  Sampling Distribution of a Difference between Two Independent
        Sample Proportions                                                        197
- 10-4  Sampling Distribution of a Difference between Two Independent Sample Means  200
- Technology Corner                                                                203
- It's a Wrap                                                                      203
- Test Yourself                                                                    203

Part III    STATISTICAL INFERENCE                                                 213

Chapter 11  *Confidence Intervals—Large Samples*                                  215
- Do I Need to Read This Chapter?                                                  215
- Get Started                                                                      215
- 11-1  Large-Sample Confidence Interval for a Proportion                          215
- 11-2  Large-Sample Confidence Interval for a Mean                                218
- 11-3  Large-Sample Confidence Interval for the Difference
        between Two Population Proportions                                         220
- 11-4  Large-Sample Confidence Interval for the Difference
        between Two Population Means                                               221
- Technology Corner                                                                223
- It's a Wrap                                                                      224
- Test Yourself                                                                    224

Chapter 12  *Hypothesis Tests—Large Samples*                                      235
- Do I Need to Read This Chapter?                                                  235
- Get Started                                                                      235

# Contents

- 12-1 Some Terms Associated with Hypothesis Testing 235
- 12-2 Large-Sample Test for a Proportion 237
- 12-3 Large-Sample Test for a Mean 241
- 12-4 Large-Sample Test for the Difference between Two Population Proportions 243
- 12-5 Large-Sample Test for the Difference between Two Population Means 245
- 12-6 *P*-Value Approach to Hypothesis Testing 248
- Technology Corner 249
- It's a Wrap 251
- Test Yourself 251

Chapter 13 *Confidence Intervals and Hypothesis Tests—Small Samples* 265
- Do I Need to Read This Chapter? 265
- Get Started 265
- 13-1 The *t* Distribution 266
- 13-2 Small-Sample Confidence Interval for a Mean 267
- 13-3 Small-Sample Test for a Mean 268
- 13-4 Independent Small-Sample Confidence Interval for the Difference between Two Population Means 270
- 13-5 Independent Small-Sample Tests for the Difference between Two Population Means 272
- 13-6 Dependent Small-Sample Confidence Interval for the Difference between Two Population Means 274
- 13-7 Dependent Small-Sample Tests for the Difference between Two Population Means 275
- Technology Corner 277
- It's a Wrap 278
- Test Yourself 278

Chapter 14 *Chi-Square Procedures* 289
- Do I Need to Read This Chapter? 289
- Get Started 289
- 14-1 The Chi-Square Distribution 289
- 14-2 The Chi-Square Test for Goodness of Fit 291
- 14-3 The Chi-Square Test for Independence 293
- Technology Corner 295
- It's a Wrap 296
- Test Yourself 296

Appendix 305
- Table 1—The Binomial Distribution 305
- Table 2—The Standard Normal Distribution 312
- Table 3—The *t* Distribution 313
- Table 4—The Chi-Square Distribution 314

Index 315

# Preface

The main goal of this book is to present basic concepts in elementary statistics and to illustrate how to tackle some of the most common problems encountered in any elementary, noncalculus statistics course.

Statistics is a frightful subject for most students. This book provides a friendly, logical, step-by-step approach to any introductory college-level noncalculus statistics course to help students overcome this barrier. It is designed as a supplement to the main text for college students enrolled in any elementary noncalculus course. It is also ideal for the nontraditional student who is returning to school and needs to review or who needs a nontechnical reference book on the subject of statistics. In addition, professionals who need a quick reference guide and high school students who need a quick review of topics in the AP statistics curriculum may use this book. The book is written for nonstatisticians and can be used by students in all disciplines. It takes a "Dummies" or "Idiot's Guide" approach to the learning of concepts. Such an approach offers the student an effortless way to the understanding of statistical concepts and, as such, furnishes him or her with a better chance at doing well in a noncalculus statistics course. In addition, this approach to presenting statistical concepts eases the stress of students who are enrolled in noncalculus statistics courses or who are reviewing before returning to college.

The "Test Yourself" sections at the end of each chapter allow students to build confidence by working problems related to the relevant concepts. Nonthreatening explanations of terms and symbols rather than definitions are given throughout the book. Examples are taken from a wide variety of disciplines that emphasize the concepts and are to the point.

It is the honest desire of the author that this book will help students to have a better understanding of concepts in elementary statistics. It is also the sincere hope of the author that this book will help them to lessen the stress brought about by the subject of statistics.

—Lloyd R. Jaisingh
Morehead State University

# Technology Integration

Because of the rapid changes in technology, the study of elementary statistics has undergone significant changes. Our teaching methods must be redesigned to accommodate these changes and incorporate technology to help students investigate, discover, and understand the needed concepts. In the "Technology Corner" sections of the book, the MINITAB software and the TI-83 calculator are used to illustrate how to alleviate much of the computational drudgery and manipulation within the text, enabling the student to concentrate on the discovery, application, and reinforcement of the concepts. Keep in mind that "Technology Corner" is not intended as a tutorial guide for the technology. It is anticipated that the use of the technology will encourage students to discover and further clarify key concepts within elementary statistics in a relaxed environment.

# Organization of the Text

This book is arranged in 14 chapters. These chapters cover a wide range of topics found in any elementary statistics course. The material is such that it can be used as a stand-alone text or as a supplement to any of the traditional texts. The book is divided into three main themes or parts. Part I deals with the descriptive nature of statistics; Part II deals with probability; and Part III deals with statistical inference. The "Technology Corner" sections illustrate how the MINITAB software and the TI-83 calculator can be used to overcome some of the math anxiety and number crunching when data are used. Each chapter ends with a "Test Yourself" section where students can attempt true/false, fill-in-the-blanks, or multiple-choice questions.

## PART I

Descriptive
Statistics

# CHAPTER 1

# Graphical Displays of Univariate Data

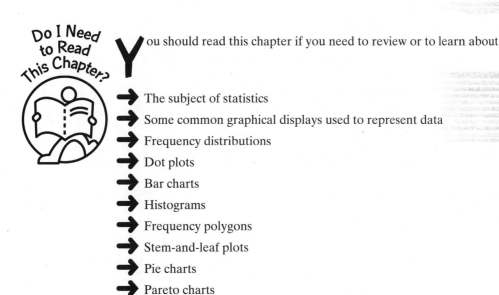

You should read this chapter if you need to review or to learn about

➡ The subject of statistics

➡ Some common graphical displays used to represent data

➡ Frequency distributions

➡ Dot plots

➡ Bar charts

➡ Histograms

➡ Frequency polygons

➡ Stem-and-leaf plots

➡ Pie charts

➡ Pareto charts

In later chapters, you will be introduced to other graphical displays, and you will recognize that these graphical displays can be combined with other measures to describe the data distribution.

## 1-1 Introduction

For us to have an understanding of what the subject of **statistics** is all about, we need to introduce some terminology. First we will explain what we mean by the subject of statistics.

**Explanation of the term—statistics:** Statistics is the science of collecting, organizing, summarizing, analyzing, and making inferences from data.

The subject of statistics is divided into two broad areas that incorporate the collecting, organizing, summarizing, analyzing, and making inferences from data. These categories are *descriptive statistics* and *inferential statistics*. These classifications are shown in **Fig. 1-1.**

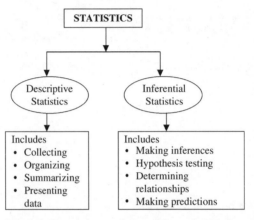

**Fig. 1-1:** Breakdown of the subject of statistics

In order to obtain information, **data** are collected from variables used to describe an event.

**Explanation of the term—data:** Data are the values or measurements that variables describing an event can assume.

Variables whose values are determined by chance are called *random variables.* There are two types of variables: **qualitative variables** and **quantitative variables.** Qualitative variables are nonnumeric in nature. Quantitative variables can assume numeric values and can be classified into two groups: **discrete variables** and **continuous variables.** A collection of values is called a *data set,* and each value is called a *data value.* **Figure 1-2** shows these relationships.

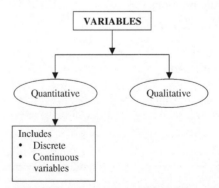

**Fig. 1-2:** Breakdown of the types of variables

**Explanation of the term—quantitative data:** Quantitative data are data values that are numeric. For example, the heights of female basketball players are quantitative data values.

**Explanation of the term—qualitative data:** Qualitative data are data values that can be placed into distinct categories, according to some characteristic or attribute. For example, the eye color of female basketball players is classified as qualitative data.

**Explanation of the term—discrete variables:** Discrete variables are variables that assume values that can be counted—for example, the number of days it rained in your neighborhood for the month of March.

**Explanation of the term—continuous variables:** Continuous variables are variables that can assume all values between any two given values—for example, the time it takes for you to do your Christmas shopping.

In order for statisticians to do any analysis, data must be collected. One of the things statisticians may want to do is to make some inference about a characteristic of a **population.** Sometimes it is impractical or too expensive to collect data from the entire population. In such instances, the statistician may select a representative portion of the population, called a **sample.** This is depicted in **Fig. 1-3.**

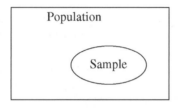

**Fig. 1-3:** The relationship between sample and population

**Explanation of the term—population:** A population consists of all elements that are being studied. For example, we may be interested in studying the distribution of ACT math scores of freshmen at a college campus. In this case, the population will be the ACT math scores of all the freshmen on that particular campus.

**Explanation of the term—sample:** A sample is a subset of the population. For example, we may be interested in studying the distribution of ACT math scores of freshmen at a college campus. In this case, we may select the ACT math score of every tenth freshman from an alphabetical list of the students' last names.

**Explanation of the term—census:** A **census** is a sample of the entire population. For example, we may be interested in studying the distribution of ACT math scores of freshmen at a college campus. In this case, we may list the ACT math scores for all freshmen on that particular campus.

Both populations and samples have characteristics that are associated with them. These are called **parameters** and **statistics,** respectively.

**Explanation of the term—parameter:** A parameter is a characteristic of or a fact about a population. For example, we may be interested in studying the distribution of ACT math scores of freshmen at a college campus. In this case, the average ACT math score for all freshmen on this particular campus may be 25.

**Explanation of the term—statistic:** A statistic is a characteristic of or a fact about a sample. For example, we may be interested in studying the distribution of ACT math scores of freshmen at a college campus. In this case, the average ACT math score for every tenth freshman from an alphabetical list of their last names may be 22.

Since parameters are descriptions of the population, a population can have many parameters. Similarly, a sample can have many statistics. These associations are shown in **Fig. 1-4.**

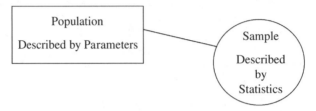

**Fig. 1-4:** The difference between parameters and statistics

When selecting a sample, statisticians would like to select values in such a way that there is no inherent bias. One way of doing this is by selecting a **random sample.**

**Explanation of the term—random sample:** A random sample of a particular size is a sample selected in such a way that each group of the same size has an equal chance of being selected. For example, in a lottery game in which six numbers are selected, this will be a random sample of size six, since each group of size six will have an equal chance of being selected.

- - - - - - - - - - - - - - - - - - - - - - - - - - - - - - - - - - - -

**Quick Tip**

There are other types of samples that will not be discussed in this text. These include systematic, stratified, cluster, and convenience samples.

- - - - - - - - - - - - - - - - - - - - - - - - - - - - - - - - - - - -

## 1-2  Frequency Distributions

In this section, we will deal with **frequency distributions.**

**Explanation of the term—frequency distribution:**  A frequency distribution is an organization of raw data in tabular form, using classes (or intervals) and frequencies.

The types of frequency distributions that will be considered in this section are *categorical, ungrouped,* and *grouped* frequency distributions.

**Explanation of the term—frequency count:** The **frequency** or the **frequency count** for a data value is the number of times the value occurs in the data set.

### Categorical or Qualitative Frequency Distributions

**Explanation of the term—categorical frequency distributions: Categorical frequency distributions** represent data that can be placed in specific categories, such as gender, hair color, or religious affiliation.

***Example 1-1:*** The blood types of 25 blood donors are given below. Summarize the data using a frequency distribution.

| | | | | |
|---|---|---|---|---|
| AB | B | A | O | B |
| O | B | O | A | O |
| B | O | B | B | B |
| A | O | AB | AB | O |
| A | B | AB | O | A |

**Solution:** We will represent the blood types as classes and the number of occurrences for each blood type as frequencies. The frequency table (distribution) in **Table 1-1** summarizes the data.

**Table 1-1:** Frequency Table for Example 1-1

| CLASS (BLOOD TYPE) | FREQUENCY $f$ |
|:---:|:---:|
| A | 5 |
| B | 8 |
| O | 8 |
| AB | 4 |
| Total | 25 |

## Quantitative Frequency Distributions—Ungrouped

**Explanation of the term—ungrouped frequency distribution:** An **ungrouped frequency distribution** simply lists the data values with the corresponding number of times or frequency count with which each value occurs.

***Example 1-2:*** The following data represent the number of defectives observed each day over a 25-day period for a manufacturing process. Summarize the information with a frequency distribution.

| DAY | 1 | 2 | 3 | 4 | 5 | 6 | 7 | 8 | 9 | 10 | 11 | 12 | 13 |
|:---|:---:|:---:|:---:|:---:|:---:|:---:|:---:|:---:|:---:|:---:|:---:|:---:|:---:|
| Defects | 10 | 10 | 6 | 12 | 6 | 9 | 16 | 20 | 11 | 10 | 11 | 11 | 9 |

| DAY | 14 | 15 | 16 | 17 | 18 | 19 | 20 | 21 | 22 | 23 | 24 | 25 |
|:---|:---:|:---:|:---:|:---:|:---:|:---:|:---:|:---:|:---:|:---:|:---:|:---:|
| Defects | 12 | 11 | 7 | 10 | 11 | 14 | 21 | 12 | 6 | 10 | 11 | 6 |

**Solution:** The frequency distribution for the number of defects is shown in **Table 1-2.**

**Table 1-2:** Frequency Table for Example 1-2

| CLASS (DEFECTS) | FREQUENCY $f$ |
|:---:|:---:|
| 6 | 4 |
| 7 | 1 |
| 9 | 2 |
| 10 | 5 |
| 11 | 6 |
| 12 | 3 |
| 14 | 1 |
| 16 | 1 |
| 20 | 1 |
| 21 | 1 |
| Total | 25 |

## Quick Tip

Sometimes frequency distributions are displayed with the relative frequencies as well.

**Explanation of the term—relative frequency:** The relative frequency for any class is obtained by dividing the frequency for that class by the total number of observations.

$$\text{Relative frequency} = \frac{\text{frequency for class}}{\text{total number of observations}}$$

The frequency distribution in **Table 1-3** uses the data in **Example 1-2** and displays the relative frequencies and the corresponding percentages.

**Table 1-3:**    Frequency Distribution Along with Relative Frequencies for Example 1-2

| CLASS (DEFECTS) | FREQUENCY $f$ | RELATIVE FREQUENCY | PERCENTAGE % |
|:---:|:---:|:---:|:---:|
| 6 | 4 | $\frac{4}{25} = 0.16$ | 16 |
| 7 | 1 | $\frac{1}{25} = 0.04$ | 4 |
| 9 | 2 | $\frac{2}{25} = 0.08$ | 8 |
| 10 | 5 | $\frac{5}{25} = 0.20$ | 20 |
| 11 | 6 | $\frac{6}{25} = 0.24$ | 24 |
| 12 | 3 | $\frac{3}{25} = 0.12$ | 12 |
| 14 | 1 | $\frac{1}{25} = 0.04$ | 4 |
| 16 | 1 | $\frac{1}{25} = 0.04$ | 4 |
| 20 | 1 | $\frac{1}{25} = 0.04$ | 4 |
| 21 | 1 | $\frac{1}{25} = 0.04$ | 4 |
| Total | 25 | 1 | 100 |

## Quick Tip

Sometimes frequency distributions are displayed with the *cumulative frequencies* and *cumulative relative frequencies* as well.

**Explanation of the term—cumulative frequency:** The cumulative frequency for a specific value in a frequency table is the sum of the frequencies for all values at or below the given value.

**Explanation of the term—cumulative relative frequency:** The cumulative relative frequency for a specific value in a frequency table is the sum of the relative frequencies for all values at or below the given value.

**Note:** The explanations given for the cumulative frequency and the cumulative relative frequency assume that the values (or classes) are arranged in ascending order from top to bottom.

The frequency distribution in **Table 1-4** uses the data in **Example 1-2** and also displays the cumulative frequencies and the cumulative relative frequencies.

**Table 1-4:** Frequency Distribution Along with Relative Frequencies, Cumulative Frequencies, and Cumulative Relative Frequencies for Example 1-2

| CLASS (DEFECTS) | FREQUENCY | RELATIVE FREQUENCY | CUMULATIVE FREQUENCY | CUMULATIVE RELATIVE FREQUENCY |
|:---:|:---:|:---:|:---:|:---:|
| 6 | 4 | 0.16 | 4 | 0.16 |
| 7 | 1 | 0.04 | 5 | 0.20 |
| 9 | 2 | 0.08 | 7 | 0.28 |
| 10 | 5 | 0.20 | 12 | 0.48 |
| 11 | 6 | 0.24 | 18 | 0.72 |
| 12 | 3 | 0.12 | 21 | 0.84 |
| 14 | 1 | 0.04 | 22 | 0.88 |
| 16 | 1 | 0.04 | 23 | 0.92 |
| 20 | 1 | 0.04 | 24 | 0.96 |
| 21 | 1 | 0.04 | 25 | 1.00 |

## Quantitative Frequency Distributions—Grouped

Here we will discuss the idea of **grouped frequency distributions.**

**Explanation of the term—grouped frequency distribution:** A grouped frequency distribution is obtained by constructing classes (or intervals) for the data, and then listing the corresponding number of values (frequency count) in each interval.

**Quick Tip**

There are several procedures that one can use to construct a grouped frequency distribution. However, because of the many statistical software packages available today, it is not necessary to try to construct such distributions using pencil and paper. Later in the chapter, we will encounter a graphical display called the *histogram*. We will see that one can directly construct grouped frequency distributions from these graphical displays.

**Quick Tip**

A frequency distribution should have a minimum of 5 classes and a maximum of 20. For small data sets, one can use between 5 and 10 classes. For large data sets, one can use up to 20 classes.

***Example 1-3:*** The weights of 30 female students majoring in Physical Education on a college campus are given below. Summarize the information with a frequency distribution using seven classes.

| 143 | 151 | 136 | 127 | 132 | 132 | 126 | 138 | 119 | 104 |
| 113 | 90 | 126 | 123 | 121 | 133 | 104 | 99 | 112 | 129 |
| 107 | 139 | 122 | 137 | 112 | 121 | 140 | 134 | 133 | 123 |

**Solution:** A grouped frequency distribution for the data using seven classes is presented in **Table 1-5.** Observe, for instance, that the upper limit value for the first class and the lower limit value for the second class have the same value, 95. The value of 95 cannot be included in both classes, so the convention that will be used here is that *the upper limit of each class is not included in the interval of values*; only the lower limit value is included in the interval. Thus, the value of 95 is included only in the interval of values for the second class.

**Table 1-5:** Grouped Frequency Distribution for Example 1-3

| CLASS (WEIGHTS) | FREQUENCY | RELATIVE FREQUENCY | PERCENTAGE % |
|:---:|:---:|:---:|:---:|
| 85–95 | 1 | 0.03 | 3 |
| 95–105 | 3 | 0.10 | 10 |
| 105–115 | 4 | 0.13 | 13 |
| 115–125 | 6 | 0.20 | 20 |
| 125–135 | 9 | 0.30 | 30 |
| 135–145 | 6 | 0.20 | 20 |
| 145–155 | 1 | 0.03 | 3 |
| Total | 30 | ≈1 | ≈100 |

**Note:** The *class width* for this frequency distribution is 10. It is obtained by subtracting the lower class limit for any class from the lower class limit for the next class. For the third class, the class limit = 115 – 105 = 10.

---

**Quick Tip**

In the grouped frequency distribution, observe that the relative frequency column did not add up to exactly 1 and the percentage column did not add up to exactly 100 percent. This is due to the rounding of the relative frequency values to two decimal places.

---

## 1-3 Dot Plots

**Explanation of the term—dot plot:** A **dot plot** is a plot that displays a dot for each value in a data set along a number line. If there are multiple occurrences of a specific value, then the dots will be stacked vertically.

***Example 1-4:*** Construct a dot plot for the information given in **Example 1-2.**

**Solution: Figure 1-5** shows the dot plot for the data set. Observe that since there are multiple occurrences of specific observations, the dots are stacked vertically. The number of dots represents the frequency count for a specific value. For instance, the value of 11 occurred 6 times, since there are 6 dots stacked above the value of 11.

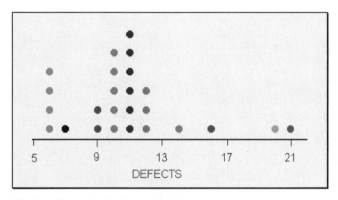

**Fig. 1-5:** Dot plot for Example 1-2

### 1-4 Bar Charts or Bar Graphs

**Explanation of the term—bar chart (graph):** A **bar chart** or a **bar graph** is a graph that uses vertical or horizontal bars to represent the frequencies of the categories in a data set.

---

**Quick Tip**

A bar chart (graph) is a valuable presentation tool, since it is effective at reinforcing differences in magnitude. Bar charts permit the visual comparison of data by displaying the magnitude of each category as a horizontal or vertical bar. Bar charts are useful when the data set has categories (for example, hair color, gender, etc.) and data values that are qualitative in nature. Note that the bars are equally separated.

---

***Example 1-5:*** A sample of 300 college students was asked to indicate their favorite soft drink. The survey results are shown in **Table 1-6.** Display the information using a bar chart.

**Table 1-6:** Frequency Distribution for Example 1-6

| SOFT DRINK | NUMBER OF STUDENTS |
|------------|--------------------|
| Pepsi-Cola | 92 |
| Coca-Cola | 78 |
| Dr. Pepper | 48 |
| 7-Up | 42 |
| Others | 40 |

**Solution:** Observe that these are categorical or qualitative data. The vertical bar chart for this information is shown in **Fig. 1-6.** The number at the top of each category represents the number of values (frequency) for that specific group (soft drink).

A horizontal bar chart for the same soft drink information is shown in **Fig. 1-7.**

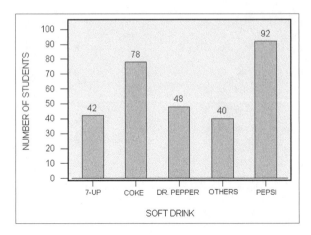

**Fig. 1-6:** Vertical bar chart for Example 1-5

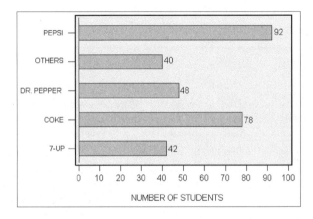

**Fig. 1-7:** Horizontal bar chart for Example 1-5

## 1-5  Histograms

**Explanation of the term—histogram:**  A histogram is a graphical display of a frequency or a relative frequency distribution that uses classes and vertical bars (rectangles) of various heights to represent the frequencies.

------------------------------------------------------------

**Quick Tip**

Histograms are useful when the data values are quantitative. A histogram gives an estimate of the shape of the distribution of the population from which the sample was taken.

------------------------------------------------------------

***Example 1-6:***  Display the data in **Example 1-3** with a histogram using seven classes.

**Solution:**  A histogram with seven classes for the data is shown in **Fig. 1-8.**

The histogram shows the frequency count for each class, with each class having a width of 10.

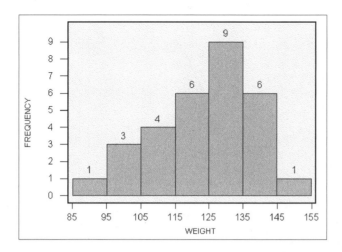

**Fig. 1-8:** Histogram for data in Example 1-3

**Quick Tip**

Observe from the histogram in *Fig. 1-8* that there is a frequency count of 1 for the interval 85–95, a frequency count of 3 for the interval 95–105, etc. From this information, we can construct the grouped frequency distribution given in *Example 1-3.*

**Quick Tip**

If the relative frequencies are plotted along the vertical axis to produce a relative frequency histogram, the shape of the resulting histogram will be the same as that of a histogram in which the frequencies were plotted along the vertical axis. This is true because the relative frequencies are obtained by dividing the frequency values by the total number of values in the data set.

## 1-6 Frequency Polygons

**Explanation of the term—frequency polygon:** A **frequency polygon** is a graph that displays the data using lines to connect points plotted for the frequencies. The frequencies represent the heights of the vertical bars in the histograms.

**Note:** A frequency polygon provides an estimate of the shape of the distribution of the population.

***Example 1-7:*** Display a frequency polygon for the data in **Example 1-3.**

**Solution:** The display given in **Fig. 1-9** shows the frequency polygon superimposed on the histogram for **Example 1-6.**

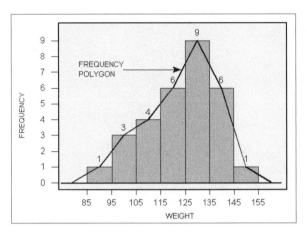

**Fig. 1-9:** Frequency polygon superimposed on the histogram for the data in Example 1-3

---

**Quick Tip**

Observe that the distribution is mound-shaped, with more of the values to the left of the peak. If this were a truly representative sample from the population, then one would expect that the distribution of the *population* of weights would have a similar shape. Also, observe that the line segments pass through the midpoints at the top of the rectangles and that the polygon is "tied down" to the horizontal axis at both ends. The points where the polygon is tied down correspond to the midpoints of the classes with zero frequency. In this case, the midpoints are 80 and 160. Midpoints of classes are called *class marks* or *class midpoints*.

---

### 1-7  Stem-and-Leaf Displays or Plots

**Explanation of the term—stem-and-leaf plot:**  A **stem-and-leaf plot** is a data plot that uses part of a data value as the *stem* to form groups or classes and part of the data value as the *leaf*. A stem-and-leaf plot has an advantage over a grouped frequency distribution, since a stem-and-leaf plot retains the actual data by showing them in graphic form.

The next example will illustrate how a stem-and-leaf plot is constructed.

***Example 1-8:***  Consider the following values: 96, 98, 107, 110, and 112.

(a)  Use the last digit values as the leaves.

**Solution:**  The data and the stems and leaves are shown in **Table 1-7.**

The corresponding stem-and-leaf plot is shown in **Table 1-8.**

(b)  Use the last two digit values as the leaves.

**Solution:**  The data and the stems and leaves are shown in **Table 1-9.**

The corresponding stem-and-leaf plot is shown in **Table 1-10.**

**Table 1-7:** Stems and Leaves for the Data in Example 1-8 with the Last Digit as the Leaves

| DATA | STEM | LEAF |
|------|------|------|
| 96 | 09 | 6 |
| 98 | 09 | 8 |
| 107 | 10 | 7 |
| 110 | 11 | 0 |
| 112 | 11 | 2 |

**Table 1-8:** Stem-and-Leaf Plot with the Last Digit as the Leaves for Example 1-8

| STEM | LEAVES |
|------|--------|
| 09 | 6 8 |
| 10 | 7 |
| 11 | 0 2 |

**Table 1-9:** Stems and Leaves for the Data in Example 1-8 with the Last Two Digits as the Leaves

| DATA | STEM | LEAF |
|------|------|------|
| 96 | 0 | 96 |
| 98 | 0 | 98 |
| 107 | 1 | 07 |
| 110 | 1 | 10 |
| 112 | 1 | 12 |

**Table 1-10:** Stem-and-Leaf Plot with the Last Two Digits as the Leaves for Example 1-8

| STEM | LEAVES |
|------|--------|
| 0 | 96 98 |
| 1 | 07 10 12 |

**Example 1-9:** A sample of the number of admissions to a psychiatric ward at a local hospital during the full phases of the moon is given below. Display the data using a stem-and-leaf plot with the leaves represented by the unit digits.

| 22 | 21 | 31 | 20 | 25 | 21 | 32 | 26 | 43 | 30 | 27 |
|----|----|----|----|----|----|----|----|----|----|----|
| 30 | 27 | 36 | 28 | 33 | 38 | 35 | 19 | 30 | 34 | 41 |

**Solution:** The stem-and-leaf display for the data is given in **Table 1-11.**

**Table 1-11:** Stem-and-Leaf Display for Example 1-9

| STEM | LEAVES |
|------|--------|
| 1 | 9 |
| 2 | 0 1 1 2 5 6 7 7 8 |
| 3 | 0 0 0 1 2 3 4 5 6 8 |
| 4 | 1 3 |

## 1-8 Time Series Graphs

Data collected over a period of time can be displayed using a **time series graph.**

**Explanation of the term—time series graph:** A time series graph displays data that are observed over a given period of time. From the graph, one can analyze the behavior of the data over time.

*Example 1-10:* The data given are the number of hurricanes that occurred each year from 1970 to 1990. Display this information using a time series graph.

| YEAR | 1970 | 1971 | 1972 | 1973 | 1974 | 1975 | 1976 | 1977 | 1978 | 1979 | 1980 |
|------|------|------|------|------|------|------|------|------|------|------|------|
| Number | 5 | 6 | 3 | 4 | 4 | 6 | 6 | 5 | 5 | 5 | 9 |

| YEAR | 1981 | 1982 | 1983 | 1984 | 1985 | 1986 | 1987 | 1988 | 1989 | 1990 |
|------|------|------|------|------|------|------|------|------|------|------|
| Number | 7 | 2 | 3 | 5 | 7 | 4 | 3 | 5 | 7 | 8 |

**Solution:** The time series plot for the data is shown in **Fig. 1-10.**

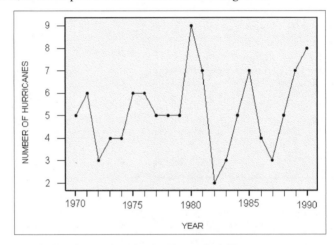

**Fig. 1-10:** Time series plot for Example 1-10

You can see from the graph that the largest number of hurricanes during the 20-year period occurred in 1980 and the smallest number occurred in 1982.

### 1-9  Pie Graphs or Pie Charts

**Explanation of the term—pie graph (chart):** A **pie graph** or **pie chart** is a circle that is divided into slices according to the percentage of the data values in each category.

A pie chart allows us to observe the proportions of sectors relative to the entire data set. It can be used to display either qualitative or quantitative data. However, categorical or qualitative data readily lend themselves to this type of graphical display because of the inherent categories in the data set.

*Example 1-11:* Present a pie chart for the blood type data given in **Example 1-1.**

**Solution:** The pie chart for the data is presented in **Fig. 1-11.**

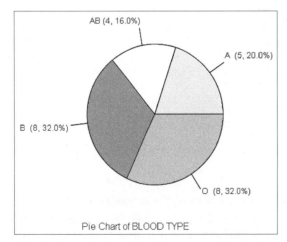

**Fig. 1-11:** Pie chart for the blood type data in **Example 1-1**

Each slice of the pie chart represents a blood type category with its frequency count and the corresponding percentage for the count.

## 1-10 Pareto Charts

**Explanation of the term—Pareto chart:** A **Pareto chart** is a type of bar chart in which the horizontal axis represents categories of interest. When the bars are ordered from largest to smallest in terms of frequency counts for the categories, a Pareto chart can help you determine which of the categories make up the critical few and which are the insignificant many. A cumulative percentage line helps you judge the added contribution of each category.

*Example 1-12:* Display a Pareto chart for the soft drink example in **Example 1-5.**

**Solution:** The Pareto chart for the data is shown in **Fig. 1-12.**

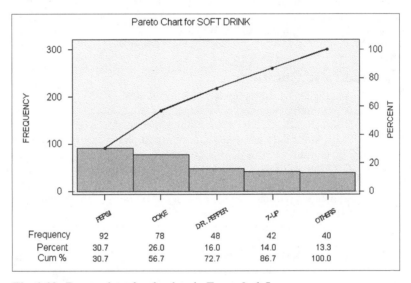

**Fig. 1-12:** Pareto chart for the data in **Example 1-5**

Observe that the categories have been ordered from the highest frequency to the lowest frequency.

## Technology Corner

## Calculators and Computer Software Packages

Statistical calculations usually involve large data sets that cannot be efficiently analyzed using pencil and paper. Modern technology eases this challenge through calculators and computers that can handle large numbers of arithmetic calculations. Most scientific calculators on the market today have statistical features that make statistical computations easy. Many specialized statistical software packages have features that perform quite sophisticated analysis of data.

Calculators are portable and relatively inexpensive. However, calculators have the drawback of not being able to deal with large amounts of data. There are many different types of calculators on the market today, so once you purchase one, in order to get familiar with its statistical features, you will need to consult the owner's manual for the calculator and practice using the appropriate keys. Specialized statistical software packages are usually more expensive than handheld calculators. These packages, however, can handle large amounts of data and usually perform a wider array of statistical calculations. Further, it is much easier to insert screen outputs from the software directly into word-processing documents. Also, it is relatively easy to transport data and analysis of data through electronic mail and floppy disks. For the more modern calculators, screen outputs can be imported into word-processing documents as well. The graphical features of statistical software packages are usually superior to the graphical features of calculators.

✔ All of the above graphical displays can be constructed with many of the software packages on the market today. As a matter of fact, all of the graphs in this chapter were constructed using the MINITAB for Windows software. A few other examples of software packages that can aid in the construction of the graphs in this chapter are Microsoft Excel, SAS, and SPSS for Windows.

## True/False Questions

1. A statistic is a characteristic of a population.
2. A parameter is a characteristic of a population.
3. Discrete data are data values that are measured over a continuous interval.
4. Continuous data are data that are measured over a given interval.
5. The amount of rainfall in your state for the last month is an example of discrete data.
6. The number of days it rained where you live during the last month is an example of discrete data.
7. Statistics is the science of collecting, classifying, presenting, and interpreting numerical data.
8. The subject of statistics can be broadly divided into two areas: descriptive statistics and inferential statistics.
9. A sample is the set of all possible data values for a given subject under consideration.
10. Descriptive statistics involves the collection, organization, and analysis of all data relating to some population or sample under study.
11. Statistics is concerned only with the collection, organization, display, and analysis of data.
12. A population is the set of all possible values for a given subject under consideration.
13. Inferential statistics involves making predictions or decisions about a sample from a population of values.
14. The frequency of a measurement is the number of times that measurement was observed.
15. The lower class limit for a given class is the smallest possible data value for that class.
16. The class mark for a class is the average of the upper and lower class limits for the given class.
17. The cumulative frequency of a class is the total of all class frequencies up to but not including the frequency of the present class.
18. The relative frequency for a given class is the total of all class frequencies before the class divided by the total number of entries.
19. The class midpoint for a class is computed from (upper limit – lower limit)/2, where the upper and lower limits are for the given class.

20. A frequency histogram and a relative frequency histogram for the same (grouped) frequency distribution will always have the same shape.
21. A frequency polygon for a set of data is obtained by connecting the class marks on the histogram displaying the set of data.
22. The choice of a single item from a group is called random if every item in the group has the same chance of being selected as every other item.
23. The class mark of a class is the midpoint between the lower limit of one class and the upper limit of the next class.
24. A population is part of a sample.
25. In stem-and-leaf displays, the trailing digits (digits to the right) are called the leaves.
26. The sum of the relative frequencies in a relative frequency distribution should always equal 1.
27. A population refers to the entire set of data values for a subject under consideration; a sample is a subset of the population.
28. A census is a sample of the entire population.

## Completion Questions

1. A (parameter, statistic) _____ is a characteristic of a population.
2. A (parameter, statistic) _____ is a characteristic of a sample.
3. A sample of the entire population is called a (sample, population, census) _____.
4. Data that are counting numbers are called (discrete, continuous) _____ data.
5. Data that are measured over an interval are called (discrete, continuous) _____ data.
6. Drawing conclusions about a population from a sample is classified as (descriptive, inferential) _____ statistics.
7. (Descriptive, Inferential) _____ statistics is concerned with making predictions about an entire population based on information from a sample that was appropriately chosen from the population.
8. (Descriptive, Inferential) _____ statistics involves the collection, organization, summarization, and presentation of data.
9. A set of all possible data values for a subject under consideration is called a (sample, population) _____.
10. Class marks are the (lower limits, midpoints, upper limits) _____ of each class.
11. A subset of a population is called a (census, sample, small population)_____.
12. The lower class limit is the (smallest, largest) _____ possible data value for a class.
13. The (relative frequency, frequency, cumulative frequency) _____ is the number of occurrences of a measurement or data value.
14. The shape of the frequency distribution and the relative frequency distribution will always be (the same, different, skewed) _____.
15. Name three graphical methods by which you can display a set of data: (a) _____; (b) _____; (c) _____.
16. The (relative, cumulative) _____ frequency of a class is the total of all class frequencies up to and including the present class.

17. Data such as sex, eye color, race, etc., are classified as (quantitative, qualitative) _____ data.

18. The class mark of a class is defined to be the (average, minimum, maximum) _____ of the upper and lower limits of the class.

19. In a histogram there are no (gaps, values) _____ between the classes represented.

20. A pie chart or circle graph can be used to display (qualitative, quantitative, both types of) _____ data.

21. In a stem-and-leaf plot, the trailing digits are called the (leaves, stems)_____ of the plot and the leading digits are called the (leaves, stems)_____ of the plot.

22. The choice of a single item from a group is called (random, biased)_____ if every item from a group has the same chance of being selected as every other item.

## Multiple-Choice Questions

1. The section of statistics which involves the collection, organization, summarizing, and presentation of data relating to some population or sample is
   (a) inferential statistics.
   (b) descriptive statistics.
   (c) an example of a frequency distribution.
   (d) the study of statistics.

2. A subset of the population selected to help make inferences on a population is called
   (a) a population.
   (b) inferential statistics.
   (c) a census.
   (d) a sample.

3. A set of all possible data values for a subject under consideration is called
   (a) descriptive statistics.
   (b) a sample.
   (c) a population.
   (d) statistics.

4. The number of occurrences of a data value is called
   (a) the class limits.
   (b) the frequency.
   (c) the cumulative frequency.
   (d) the relative frequency.

5. A large collection of data may be condensed by constructing
   (a) classes.
   (b) a frequency polygon.
   (c) class limits.
   (d) a frequency distribution.

6. When constructing a frequency distribution for a small data set, it is wise to use
   (a) 5 to 20 classes.
   (b) 5 to 15 classes.

(c) 5 to 10 classes.

(d) less than 10 classes.

7. When constructing a frequency distribution for a large data set, it is wise to use

(a) 5 to 20 classes.

(b) 5 to 15 classes.

(c) 5 to 10 classes.

(d) less than 10 classes.

8. When straight-line segments are connected through the midpoints at the top of the rectangles of a histogram with the two ends tied down to the horizontal axis, the resulting graph is called

(a) a bar chart.

(b) a pie chart.

(c) a frequency polygon.

(d) a frequency distribution.

9. A questionnaire concerning satisfaction with the Financial Aid Office on campus was mailed to 50 students on a university campus. The 50 students in this survey are an example of a

(a) statistic.

(b) parameter.

(c) population.

(d) sample.

The following information relates to **Problems 10 to 15.**

   The Love Your Lawn lawn care company is interested in the distribution of lawns in a certain subdivision with respect to size (in square feet) of the lawn. The following table shows the distribution of the size of the lawns in hundreds of square feet.

| SIZE OF LAWN (100 SQUARE FEET) | NUMBER OF LAWNS |
| --- | --- |
| 10–15 | 2 |
| 15–20 | 12 |
| 20–25 | 27 |
| 25–30 | 19 |
| 30–35 | 6 |
| 35–40 | 3 |

10. The class mark for the class 25–30 is

(a) 24.5.

(b) 29.5.

(c) 4.

(d) 27.5.

11. The relative frequency for the class 15–20 is

(a) 0.2029.

(b) 0.0290.

(c) 0.1739.

(d) 0.4058.

12. The lower class limit for the class 35–40 is
    (a) 34.5.
    (b) 35.
    (c) 37.
    (d) 39.5.

13. The upper class limit for the class 20–25 is
    (a) 24.5.
    (b) 25.
    (c) 24.
    (d) 22.

14. The cumulative frequency for the class 25–30 is
    (a) 41.
    (b) 9.
    (c) 19.
    (d) 60.

15. The cumulative relative frequency for the class 30–35 is
    (a) 0.8696.
    (b) 0.0870.
    (c) 0.1304.
    (d) 0.9565.

16. The graphical display with the relative frequencies along the vertical axis that may be constructed for quantitative data is
    (a) the pie chart.
    (b) the bar chart.
    (c) the histogram.
    (d) all of the above.

17. The cumulative relative frequency for a given class is defined to be
    (a) the proportion of values preceding the given class.
    (b) the proportion of values up to and including the given class.
    (c) the proportion of values for the given class.
    (d) the proportion of values below the given class.

18. A property of a frequency polygon is that
    (a) a histogram is always needed in the construction of the polygon.
    (b) the polygon is made up of line segments.
    (c) the end points of the polygon need not be tied down to the horizontal axis at both ends.
    (d) the polygon can be constructed on a pie chart.

19. You are given that the total number of observed values in a frequency distribution is 50 and the frequency of a given class 25–30 is 10. Also, the cumulative frequency of all classes above this given class is 40. The cumulative frequency for this class is
    (a) 10.
    (b) 50.

(c) 40.

(d) 30.

20. If a class in a frequency distribution for a sample of 50 has a frequency of 5, the cumulative relative frequency for this class
   (a) is 0.1000.
   (b) is 0.9000.
   (c) is 0.1111.
   (d) cannot be determined from the given information.

21. If the first five classes of a frequency distribution have a cumulative frequency of 50 from a sample of 58, the sixth and last class must have a frequency count of
   (a) 58.
   (b) 50.
   (c) 7.
   (d) 8.

The following information relates to **Problems 22 to 28.** *Hint:* Read the examination scores distribution from smallest value to largest value.

The table below shows the distribution of scores on a final elementary statistics examination for a large section of students.

| CLASSES FOR EXAM SCORES | NUMBER OF STUDENTS |
|---|---|
| 90 and over | 5 |
| 80–90 | 12 |
| 70–80 | 40 |
| 60–70 | 18 |
| 50–60 | 13 |
| 40–50 | 6 |
| Under 40 | 6 |

22. The class width is
   (a) 9.
   (b) 10.
   (c) 7.
   (d) 1.

23. The class mark for the class 40–50 is
   (a) 39.5.
   (b) 49.5.
   (c) 45.
   (d) 9.

24. The relative frequency for the class 80–90 is
   (a) 0.1700.
   (b) 0.0500.
   (c) 0.8300.
   (d) 0.1200.

25. The lower class limit for the class 50–60 is
    (a) 49.5.
    (b) 50.
    (c) 59.
    (d) 59.5.

26. The upper class limit for the class 70–80 is
    (a) 69.5.
    (b) 70.
    (c) 80.
    (d) 79.5.

27. The cumulative frequency for the class 60–70 is
    (a) 18.
    (b) 57.
    (c) 43.
    (d) 12.

28. The cumulative relative frequency for the class 50–60 is
    (a) 0.8800.
    (b) 0.1300.
    (c) 0.7500.
    (d) 0.2500.

29. Can a frequency distribution have overlapping classes?
    (a) Sometimes
    (b) No
    (c) Yes
    (d) All of the above

30. An organization of observed data into tabular form in which classes and frequencies are used is called
    (a) a bar chart.
    (b) a pie chart.
    (c) a frequency distribution.
    (d) a frequency polygon.

31. Given the following stem-and-leaf diagram:
    1 | 0 3
    2 | 2 2 4
    3 | 1 2 3 3 3
    4 | 1 1 2 2 2 2 5 6
    5 | 3 3 5 6
    6 | 2 4
    7 | 3

    The number that occurred the most is
    (a) 2.
    (b) 42.

(c) 33.

(d) 3.

## Further Exercises

If possible, you can use any technology available to help you solve the following questions.

1. The at-rest pulse rates for 16 athletes at a meet are

   67   57   56   57   58   56   54   64   53   54   54   55   57   68   60   58

   (a) Construct a relative frequency distribution for this data set using classes 50–55, 55–60, . . . .

   (b) Construct a histogram for this set of data using the distribution in part (a).

2. The speeds (in mph) of 16 cars on a highway were observed to be

   58   56   60   57   52   54   54   59   63   54   53   54   58   56   57   67

   (a) Construct a relative frequency distribution for this data set using classes 52–55, 55–58, . . . .

   (b) Construct a stem-and-leaf plot for the data set.

3. The starting incomes for mathematics majors at a particular university were recorded for five years and are summarized in the following table:

| STARTING SALARY (IN $1000) | FREQUENCY |
| --- | --- |
| 10–15 | 3 |
| 15–20 | 5 |
| 20–25 | 10 |
| 25–30 | 7 |
| 30–35 | 1 |

   (a) Construct a histogram for the data.

   (b) Construct a table with the relative frequencies and the cumulative relative frequencies.

4. The following frequency distribution shows the distances to campus (in miles) traveled by 30 commuter students:

| DISTANCE (IN MILES) | FREQUENCY |
| --- | --- |
| 35–40 | 8 |
| 40–45 | 13 |
| 45–50 | 6 |
| 50–55 | 3 |

   For the class 40–45, find the following:

   (a) Lower class limit

   (b) Upper class limit

   (c) Class width

   (d) Class mark

(e) Cumulative frequency
(f) Relative frequency
(g) Cumulative relative frequency

## ANSWER KEY
### True/False Questions

1. F   2. T   3. F   4. T   5. F   6. T   7. T   8. T   9. F   10. T   11. F
12. T   13. F   14. T   15. T   16. T   17. F   18. F   19. F   20. T   21. T
22. T   23. F   24. F   25. T   26. T   27. T   28. T

## Completion Questions

1. parameter   2. statistic   3. census   4. discrete   5. continuous   6. inferential
7. Inferential   8. Descriptive   9. population   10. midpoints   11. sample
12. smallest   13. frequency   14. the same   15. bar chart, histogram, pie chart,
frequency polygon, stem-and-leaf plot (any three)   16. cumulative   17. qualitative
18. average   19. gaps   20. both types of   21. leaves, stems   22. random

## Multiple-Choice Questions

1. (b)   2. (d)   3. (c)   4. (b)   5. (d)   6. (c)   7. (a)   8. (c)   9. (d)
10. (d)   11. (c)   12. (b)   13. (b)   14. (d)   15. (d)   16. (c)   17. (b)   18. (b)
19. (b)   20. (d)   21. (d)   22. (b)   23. (c)   24. (d)   25. (b)   26. (c)
27. (c)   28. (d)   29. (b)   30. (c)   31. (b)

# Data Description— Numerical Measures of Central Tendency for Ungrouped Univariate Data

**Do I Need to Read This Chapter?**

**Y**ou should read this chapter if you need to review or to learn about

→ Numerical values that measure central tendencies of a numerical data set

→ How to compute these measures and investigate their properties

You will recognize that these measures deal with only one property of the data set: the centralness. Thus, you will need to combine these measures with other properties of the data set in order to fully describe it. Other properties for univariate data will be investigated in future chapters.

**Get Started**

A measure of central tendency for a collection of data values is a number that is meant to convey the idea of centralness for the data set. The most commonly used measures of central tendency for sample data are the mean, the median, and the mode. These measures are discussed in this chapter.

## 2-1 The Mean

**Explanation of the term—mean:** The **mean** of a set of numerical values is the average of the set of values.

**Note:** In the explanation of the mean, the numerical values can be population values or sample values. Hence, we can compute the mean for either population values or sample values.

**Explanation of the term—population mean:** If the values are from an entire population, then the mean of the values is called a **population mean.** It is usually denoted by $\mu$ (read as "mu").

**Explanation of the term—sample mean:** If the values are from a sample, then the mean of the values is called a **sample mean.** It is denoted by $\bar{x}$ (read as "x-bar").

***Example 2-1:*** What is the mean for the following sample values?

3   8   6   14   0   –4   0   12   –7   0   –10

**Solution:** The sample mean is obtained as

$$\bar{x} = \frac{3 + 8 + 6 + 14 + 0 + (-4) + 0 + 12 + (-7) + 0 + (-10)}{11} = 2$$

That is, the value of the sample mean is 2.

**Quick Tip**

When the word *mean* or *average* is used in everyday conversation, it has come to represent a typical value or the center of a set of values. Because of this, the mean is called a measure of central tendency.

***Question:*** **Why do we use the mean as a measure of the center of a set of values?**

The following discussion will give an insight into the question. First, **Fig. 2-1** shows a plot of the data points along with the sample mean.

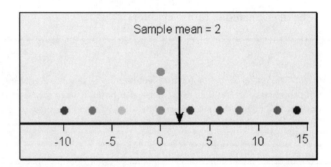

**Fig. 2-1:** Plot of data values for Example 2-1

Next, we compute the deviation from the sample mean for each value in the data set. That is, we compute $(x - \bar{x})$ for each value $x$. These deviations are given in **Table 2-1.**

**Table 2-1:** Deviations from the Mean for Values in Example 2-1

| DATA | DEVIATIONS |
|:---:|:---:|
| 3 | 1 |
| 8 | 6 |
| 6 | 4 |
| 14 | 12 |
| 0 | −2 |
| −4 | −6 |
| 0 | −2 |
| 12 | 10 |
| −7 | −9 |
| 0 | −2 |
| −10 | −12 |

Next, a plot of the deviations from the sample mean is displayed in **Fig. 2-2.**

When the deviations on the left and on the right of the sample means are added, disregarding the sign of the values, we see that when the "balancing point" is the sample mean, then these sums are equal in absolute value. Here in **Example 2-1,** the sum of the deviations to the right of the mean is 33. The sum of the deviations to the left of the mean is −33. However, we use the absolute value of these negative deviations; that is, we use +33. This is depicted in **Fig. 2-3.**

Thus, *the mean is that central point where the sum of the negative deviations (absolute value) from the mean and the sum of the positive deviations from the mean are equal.* This is why the mean is considered a measure of central tendency.

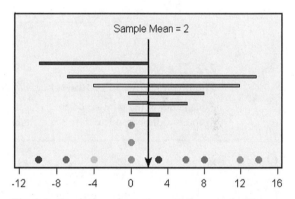

**Fig. 2-2:** Deviations from the sample mean for Example 2-1

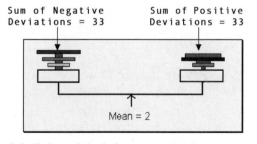

**Fig. 2-3:** Balanced deviations

The next example shows us how to find the mean of a set of values when the data are summarized in a frequency table.

***Example 2-2:*** Find the mean for the following frequency table, using four decimal places.

**Solution:** Observe that the value of 20 has a frequency count of 2, so the total or sum can be written as $20 + 20$ or $2 \times 20$. We can do the same for each of the values and its corresponding frequency count. The total number of values in the table is 15, which is the sum of the frequency values. Thus, we can compute the mean for the frequency distribution as

| *x* VALUES $x_i$ | FREQUENCY $f_i$ |
|---|---|
| 20 | 2 |
| 29 | 4 |
| 30 | 4 |
| 39 | 3 |
| 44 | 2 |

$$\bar{x} = \frac{2 \times 20 + 4 \times 29 + 4 \times 30 + 3 \times 39 + 2 \times 44}{2 + 4 + 4 + 3 + 2} = 32.0667$$

### 2-2  The Median

The next measure of central tendency we will consider is the **median.**

**Explanation of the term—median:**  The median of a numerical data set is the numerical value in the middle when the data set is arranged in order.

***Example 2-3:*** What is the median for the following sample values?

3    8    6    14    0    −4    2    12    −7    −1    −10

**Solution:**  First of all, we need to arrange the data set in order. The ordered set is as follows:

−10    −7    −4    −1    0    2    3    6    8    12    14
                                    ↑
                                   6th

Since the number of values is odd, the median will be the middle value in the ordered set. Thus, the median will be found in the sixth position, since we have a total of 11 values.

That is, the value of the median is 2.

***Question:*** **Why does the middle number in an ordered data set measure central tendency?**

The following discussion will give an insight into the question. **Figure 2-4** shows a plot of the data points and the location of the sample median.

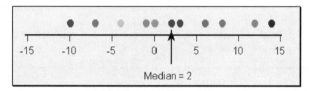

**Fig. 2-4:** Plot of data points for Example 2-3

A list of the values that are above the median and below the median is given in **Table 2-2.**

**Table 2-2:**   List of Values That Are Above or Below the Median for Example 2-3

| DATA | DEVIATIONS | DATA | DEVIATIONS |
|------|-----------|------|-----------|
| 3 | Above | 2 | Neither |
| 8 | Above | 12 | Above |
| 6 | Above | −7 | Below |
| 14 | Above | −1 | Below |
| 0 | Below | −10 | Below |
| −4 | Below | | |

When the values from above and below the median are counted, we see that if the "balancing point" is the sample median, then the number of values above the median balances (equals) the number of values below the median. This is depicted in **Fig. 2-5.**

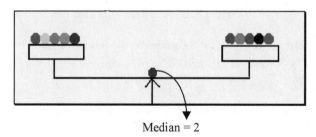

Median = 2

**Fig. 2-5:** Median as a balancing point for data values in Example 2-3

Observe that there are the same number of values above the median as there are below the median. This is why the median is considered a measure of central tendency.

***Example 2-4:*** Find the median for the ages of the following eight college students:

23    19    32    25    26    22    24    20

**Solution:** First order the values. The ordered array is

19 20 22 23 24 25 26 32

Since there is an even number of ages, the median will be the average of the two middle numbers. Since the two middle numbers are located in the fourth and fifth positions, the

$$\text{median} = \frac{23 + 24}{2} = 23.5.$$

### 2-3 The Mode

**Explanation of the term—mode:** The **mode** of a numerical data set is the most frequently occurring value in the data set.

---

**Quick Tips**

1. If all the elements in the data set have the same frequency of occurrence, then the data set is said to have *no mode.*
2. If the data set has one value that occurs more frequently than the rest of the values, then the data set is said to be *unimodal.*
3. If two elements of the data set are tied for the highest frequency of occurrence, then the data set is said to be *bimodal.*

---

***Example 2-5:*** What is the mode for the following sample values?

3 5 1 4 2 9 6 10

**Solution:** We see from **Fig. 2-6** that each value occurs with a frequency of 1. Thus, the data set has no mode.

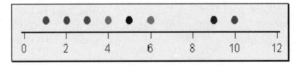

**Fig. 2-6:** Plot of data values for Example 2-5

***Example 2-6:*** What is the mode for the following sample values?

3 5 1 4 2 9 6 10 5 3 4 3 9 3 6 1

**Solution:** **Figure 2-7** shows a plot of the data values.

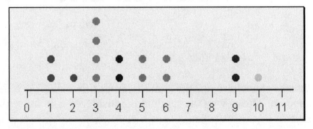

**Fig. 2-7:** Plot of data values for Example 2-6

Observe that the value of 3 occurs with the highest frequency. Thus, the value of the mode is 3, and this data set is unimodal.

***Example 2-7:*** What is the mode for the following sample values?

6   10   5   3   4   3   9   3   6   1   6

**Solution:** **Figure 2-8** shows a plot of the data values.

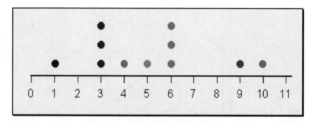

**Fig. 2-8:** Plot of data values for Example 2-7

Observe that the value of 3 and the value of 6 occur with the highest but equal frequency. Thus, the values of the mode are 3 and 6, and this data set is bimodal.

### 2-4  Shapes (Skewness)

The three most important shapes of frequency distributions are positively skewed, negatively skewed, and symmetrical.

#### Positively Skewed Distribution

In a positively skewed distribution, most of the data values fall to the left of the mean, and the "tail" of the distribution is to the right. In addition, the mean is to the right of the median, and the mode is to the left of the median. These properties are depicted in **Fig. 2-9.**

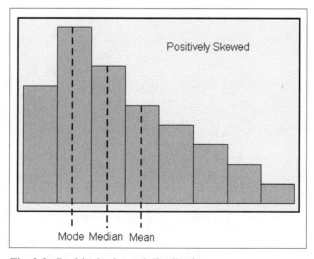

**Fig. 2-9:** Positively skewed distribution

### Negatively Skewed Distribution

In a negatively skewed distribution, most of the data values fall to the right of the mean, and the tail of the distribution is to the left. In addition, the mean is to the left of the median, and the mode is to the right of the median. These properties are depicted in **Fig. 2-10.**

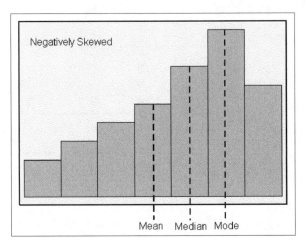

**Fig. 2-10:** Negatively skewed distribution

### Symmetrical Distribution

In a symmetrical distribution, the data values are evenly distributed on both sides of the mean. Also, when the distribution is unimodal, the mean, median, and mode are all equal to one another and are located at the center of the distribution. These properties are depicted in **Fig. 2-11.**

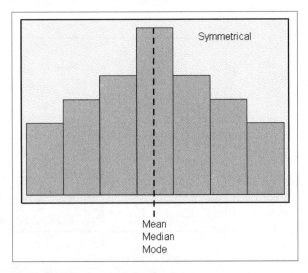

**Fig. 2-11:** Symmetrical distribution

## Technology Corner

All of the concepts discussed in this chapter can be illustrated using most statistical software packages. All scientific and graphical calculators will aid directly in the computations. In addition, some of the newer calculators, such as the TI-83, will allow you to compute the mean and the median directly. If you own a calculator, you should consult the manual to determine what statistical features are included.

**Illustration:** **Figure 2-12** shows the descriptive statistics computed by the MINITAB software. **Figure 2-13** shows the 1-Var Stats (descriptive statistics) computed by the TI-83 calculator. The data used, in both cases, were from **Example 2-3.** Observe that MINITAB has given the value of the mean to two decimal places, while the TI-83 calculator gives the value to nine decimal places. The median using both technologies is 2. Observe that the mode is not displayed in either figure. One will have to use other features of the technologies to obtain the value of the mode. There are other descriptive statistics in the outputs that we will encounter later in the text.

---

### Descriptive Statistics

| Variable | N | Mean | Median | TrMean | StDev | SE Mean |
|----------|-----|------|--------|--------|-------|---------|
| Eg2-3 | 11 | 2.09 | 2.00 | 2.11 | 7.56 | 2.28 |

| Variable | Minimum | Maximum | Q1 | Q3 |
|----------|---------|---------|-------|------|
| Eg2-3 | −10.00 | 14.00 | −4.00 | 8.00 |

---

**Fig. 2-12:** MINITAB descriptive statistics output for Example 2-3

```
1-Var Stats
x̄=2.090909091
Σx=23
Σx²=619
Sx=7.555852638
σx=7.20422282
↓n=11
■
```

```
1-Var Stats
↑n=11
 minX=-10
 Q₁=-4
 Med=2
 Q₃=8
 maxX=14
■
```

**Fig. 2-13:** TI-83 1-Var Stats output for Example 2-3

---

## Quick Tip

Unlike the median and the mode, the mean is sensitive to a change in any of the values of the data set. If the same constant is added to each value in the data set, the mean of the data set will increase by the same amount. If each value in the data set is multiplied by the same constant, the mean of the data set will also be multiplied by that constant.

*It's a Wrap* The three most commonly used measures of central tendency for numeric data are the

✔ Mean

✔ Median

✔ Mode

Care should always be taken when using these measures of central tendency.

**Test Yourself**

## True/False Questions

1. The mean of a set of data always divides the data set such that 50 percent of the values lie above the mean and 50 percent lie below the mean.
2. The mode is a measure of variability.
3. The median of a set of data values is that value that occurs the most.
4. The mean is not equal to the median in a symmetrical distribution.
5. Of the mean, the median, and the mode of a data set, the mean is most influenced by an outlying value in the data set.
6. If the number of observations in a data set is odd, the median cannot be accurately found, but rather is approximated.
7. A data set with more than one mode is said to be bimodal.
8. The sum of the deviations from the mean for any data set is always 0.
9. For a negatively skewed distribution, the tail is to the right of the mean.
10. For a positively skewed distribution, the mode is less than the median, and the median is less than the mean.

## Completion Questions

1. If the number of measurements in a data set is odd, the median is the _____ value when the data set is ordered from the smallest value to the largest value.
2. If the number of measurements in a data set is even, the median is the _____ of the two _____ values when the data set is ordered from the smallest value to the largest value.
3. The (mean, median, mode) _____ for a set of data is the value in the data set that occurs most frequently.
4. Two measures of central tendency are the _____ and the _____ .
5. For a symmetrical distribution, the mean, mode, and median are all (equal to, different from) _____ one another.
6. For a negatively skewed distribution, the mean is (smaller, greater) _____ than the median and the mode.
7. For a positively skewed distribution, the tail of the distribution is to the (right, left) _____ of the distribution.
8. For a positively skewed distribution, the median is (smaller, larger) _____ than the mode.

## Multiple-Choice Questions

1. A student has seven statistics books open in front of him. The page numbers are as follows: 231, 423, 521, 139, 347, 400, 345. The median for this set of numbers is

(a) 139.
(b) 347.
(c) 346.
(d) 373.5.

2. A cyclist recorded the number of miles per day that she cycled for 5 days. The recordings were as follows: 13, 10, 12, 10, 11. The mean number of miles she cycled per day is
   (a) 13.
   (b) 11.
   (c) 10.
   (d) 11.2.

3. An instructor recorded the following quiz scores (out of a possible 10 points) for the 12 students present: 7, 4, 4, 7, 2, 9, 10, 6, 7, 3, 8, 5. The mode for this set of scores is
   (a) 9.5.
   (b) 7.
   (c) 6.
   (d) 3.

4. It is stated that more students are purchasing graphing calculators than any other type of calculator. Which measure is being used here?
   (a) Mean
   (b) Median
   (c) Mode
   (d) None of the above

5. Which of the following is not a measure of central tendency?
   (a) Mode
   (b) Variability
   (c) Median
   (d) Mean

Use the following frequency distribution for **Problems 6 to 8.**

| x VALUES | FREQUENCY |
| --- | --- |
| 20 | 2 |
| 29 | 4 |
| 30 | 4 |
| 39 | 3 |
| 44 | 2 |

6. The mean of the distribution is
   (a) 32.4.
   (b) 30.
   (c) 39.
   (d) 32.07.

7. The median of the distribution is
   (a) 4.

(b) 30.

(c) 29.5.

(d) 34.5.

8. The mode of the distribution is

(a) 29.

(b) 30.

(c) 29 and 30.

(d) none of the above.

9. Given the following data set:

12     32     45     14     24     31

The total deviation from the mean for the data values is

(a) 0.

(b) 26.3333.

(c) 29.5.

(d) 12.

10. The most frequently occurring value in a data set is called the

(a) spread.

(b) mode.

(c) skewness.

(d) maximum value.

11. A single numerical value used to describe a characteristic of a sample data set, such as the sample median, is referred to as a

(a) sample parameter.

(b) sample median.

(c) population parameter.

(d) sample statistic.

12. Which of the following is true for a positively skewed distribution?

(a) Mode = Median = Mean

(b) Mean < Median < Mode

(c) Mode < Median < Mean

(d) Median < Mode < Mean

13. Which of the following would be affected the most if there is an extremely large value in the data set?

(a) The mode

(b) The median

(c) The frequency

(d) The mean

14. If the number of values in a data set is even, and the numbers are ordered, then

(a) the median cannot be found.

(b) the median is the average of the two middle numbers.

(c) the median, mode, and mean are equal.

(d) none of the above answers are correct.

15. What type of distribution is described by the following information?
    Mean = 5.5    median = 5.3    mode = 4.4
    (a) Negatively skewed
    (b) Symmetrical
    (c) Bimodal
    (d) Positively skewed

16. What type of distribution is described by the following information?
    Mean = 56    median = 58.1    mode = 63
    (a) Negatively skewed
    (b) Symmetrical
    (c) Bimodal
    (d) Positively skewed

17. The mean of a set of data is the value that represents
    (a) the middle value of the data set.
    (b) the most frequently observed value.
    (c) the mean of the squared deviations of the values from the mean.
    (d) the arithmetic average of the data values.

18. The median of an ordered set of data is the value that represents
    (a) the middle or the approximate middle value of the data set.
    (b) the most frequently observed value.
    (c) the mean of the squared deviations of the values from the mean.
    (d) the arithmetic average of the data values.

19. Given the following data set: 4, 3, 7, 7, 8, 7, 4, 8, 6. What is the mean value?
    (a) 4
    (b) 5
    (c) 6
    (d) 7

20. Given the following data set: 3, 2, 7, 7, 8, 7, 3, 8, 5. What is the median value?
    (a) 5
    (b) 6
    (c) 7
    (d) 8

21. Given the following data set: 4, 5, 7, 7, 8, 6, 5, 8, 7. What is the mode?
    (a) 4
    (b) 5
    (c) 6
    (d) 7

22. A sample of 10 students was asked by their instructor to record the number of hours they spent studying for a given exam from the time the exam was announced in class. The following data values were the recorded number of hours:
    12    15    8    9    14    8    17    14    8    15

    The median number of hours spent studying for this sample is

(a) 10.

(b) 11.

(c) 12.

(d) 13.

23. The numbers of minutes spent in the computer lab by 20 students working on a project are given below:

*Numbers of Minutes*

30 | 0 2 5 5 6 6 6 8
40 | 0 2 2 5 7 9
50 | 0 1 3 5
60 | 1 3

The median for this data set is

(a) 400.

(b) 402.

(c) 405.

(d) 407.

24. The numbers of minutes spent in the computer lab by 20 students working on a project are given below:

*Numbers of Minutes*

30 | 0 2 5 5 6 6 6 8
40 | 0 2 2 5 7 9
50 | 0 1 3 5
60 | 1 3

The mode for this data set is

(a) 305.

(b) 402.

(c) 306.

(d) 300.

25. The numbers of minutes spent in the computer lab by 20 students working on a project are given below:

*Numbers of Minutes*

30 | 0 2 5 5 6 6 6 8
40 | 0 2 2 5 7 9
50 | 0 1 3 5
60 | 1 3

The mean for this data set is

(a) 306.0.

(b) 402.0.

(c) 403.8.

(d) 450.0.

26. A set of exam scores is given below:

*Exam Scores*

4 | 5 6 8
5 | 3 4 5 6 9

6 | 2 3 5 6 6 9 9
7 | 0 1 1 3 3 4 5 5 5 7 8
8 | 1 2 3 6 9
9 | 3 5 7 8

The mode for this data set is

(a) 75.

(b) 78.

(c) 45.

(d) 98.

## Further Exercises

If possible, you can use any technology available to help you solve the following questions.

1. The at-rest pulse rates for 16 athletes at a meet are

   67   57   56   57   58   56   54   64   53   54   54   55   57   68   60   58

   Find the median, mode, and mean for this set of data.

2. The speeds (in mph) of 16 cars on a highway were observed to be

   58   56   60   57   52   54   54   59   63   54   53   54   58   56   57   67

   Find the mean, mode, and median for this set of data.

3. Estimate the mean for the following frequency distribution. *Hint:* Use the class marks as the actual observed values in each class.

   | CLASS | FREQUENCY |
   | --- | --- |
   | 10–15 | 2 |
   | 15–20 | 4 |
   | 20–25 | 4 |
   | 25–30 | 3 |
   | 30–35 | 2 |

4. Find the mean, median, and mode for the following examination scores.

   *Exam Scores*

   4 | 5 6 8
   5 | 3 4 5 6 9
   6 | 2 3 5 6 6 9 9
   7 | 0 1 1 3 3 4 5 5 5 7 8
   8 | 1 2 3 6 9
   9 | 3 5 7 8

5. The following frequency distribution shows the scores on the exit examination for statistics majors at a four-year college for a given year.

   98   75   85   97   80   87   97   60   83   90

   Find the mean, mode, and median for this set of data.

6. The starting incomes for mathematics majors at a particular university were recorded for five years and are summarized in the following table:

| STARTING SALARY (IN $1000) | FREQUENCY |
|---|---|
| 10–15 | 3 |
| 15–20 | 5 |
| 20–25 | 10 |
| 25–30 | 7 |
| 30–35 | 1 |

  (a)  Construct a histogram for the data.

  (b)  Compute an approximate value for the mean by using the class mark values.

7.  The numbers of 30-second radio advertising spots purchased by each of the 25 members of a local restaurant association last year are given below:

*Numbers of 30-Second Spots*

1 | 1 1 1 2 3 3 3 4 5 6 7
2 | 3 4 4 5 6 6
3 | 1 1 1 2 2 2 3
4 | 0 0 1

  (a)  Find the median.

  (b)  Find the mode.

  (c)  Find the mean.

  (d)  Describe the shape of the distribution.

  (e)  Construct a histogram for the data set.

# ANSWER KEY
## True/False Questions

    1. F    2. F    3. F    4. F    5. T    6. F    7. F    8. T    9. F    10. T

## Completion Questions

    1. middle    2. average, middle    3. mode    4. mean, median, mode (any two)
    5. equal to    6. smaller    7. right    8. larger

## Multiple-Choice Questions

    1. (b)    2. (d)    3. (b)    4. (c)    5. (b)    6. (d)    7. (b)    8. (c)    9. (a)
    10. (b)    11. (d)    12. (c)    13. (d)    14. (b)    15. (d)    16. (a)    17. (d)
    18. (a)    19. (c)    20. (c)    21. (d)    22. (d)    23. (b)    24. (c)    25. (c)
    26. (a)

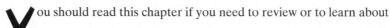

# Data Description—Numerical Measures of Variability for Ungrouped Univariate Data

**Do I Need to Read This Chapter?**

**Y**ou should read this chapter if you need to review or to learn about

➜ Numerical values that measure the spread or variability of a numerical data set

➜ How to compute these measures and investigate some of their properties

You will recognize that these measures deal with only one property of the data set: the spread or variability. Thus, you will need to combine these measures with other properties of the data set in order to fully describe it. Other properties of univariate data will be investigated in future chapters.

**Get Started**

A measure of variability for a collection of data values is a number that is meant to convey the idea of spread for the data set. The most commonly used measures of variability for sample data are the range, the interquartile range, the mean absolute deviation, the variance or standard deviation, and the coefficient of variation. These measures are discussed in this chapter.

## 3-1  The Range

**Explanation of the term—range:**  The **range** is the difference between the largest and smallest values in a data set.

This definition is true for a sample as well as for a finite population of values.

$$\boxed{\text{Range} = R = \text{largest value} - \text{smallest value}}$$

***Example 3-1:***  What is the range for the following sample values?

  3     8     6     14     0     −4     0     12     −7     0     −10

**Solution:**  We can arrange the data values so as to obtain the smallest and the largest values in the data set. The ordered set is as follows:

  −10     −7     −4     0     0     0     3     6     8     12     14

Thus, the sample range = $R = 14 - (-10) = 24$.

***Question:***  **Why does subtracting the smallest value from the largest value in a data set measure spread?**

The following will give an insight into the question. **Figure 3-1** shows a plot of the data points with the locations of the end points. The range measures the distance between the largest and the smallest values and, as such, gives an idea of the spread of the data set.

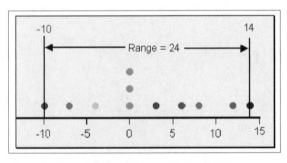

**Fig. 3-1:**  Plot of data values for Example 3-1

---

**Quick Tip**

The range does not use the concept of deviations. It is affected by outliers (large or small values relative to the rest of the data set) and does not utilize all the information in the data set—only the largest and smallest values. Therefore, it is not a very useful measure of variation.

---

***Example 3-2:***  What is the range for the following sample values?

  996     1014     1000     997     1001     1002     999     995     990

**Solution:**  We can arrange the data values so as to obtain the smallest and the largest values in the data set. The ordered set is as follows:

  990     995     996     997     999     1000     1001     1002     1014

Thus, the sample range = $R = 1014 - 990 = 24$.

The range here is also 24, the same as in **Example 3-1.** No information about the values between the extreme data points is involved in the computation. You should note that outlying values in the data set can influence the value of the range.

***Example 3-3:*** What is the range for the following sample values?

9    1014    1000    997    1001    1002    999    995    990

**Solution:** We can arrange the data values so as to obtain the smallest and the largest values in the data set. The ordered set is as follows:

9    995    996    997    999    1000    1001    1002    1014

Thus, the sample range = $R = 1014 - 9 = 1005$.

Here the value of the range is significantly affected by the outlying value of 9. This value is significantly smaller than the rest of the data set.

## 3-2 The Interquartile Range

A measure of spread that is not influenced by any extreme values (outliers) in the data set but still retains the idea of a range is the **interquartile range.**

**Explanation of the term—interquartile range:** The interquartile range measures the spread of the middle 50 percent of an ordered data set.

The interquartile range is obtained by the following steps:

***Step 1:*** Order the data set from the smallest value to the largest value.

***Step 2:*** Find the median of the ordered set. Denote this by $Q_2$.

***Step 3:*** Find the median of the first 50 percent of the data set. The median in Step 2 is not included in this portion of the data set. Let this value be denoted by $Q_1$.

***Step 4:*** Find the median of the second 50 percent of the data set. The median in Step 2 is not included in this portion of the data set. Let this value be denoted by $Q_3$.

***Step 5:*** The interquartile range

$$\boxed{\text{IQR} = Q_3 - Q_1}$$

**Figure 3-2** depicts the idea of the interquartile range.

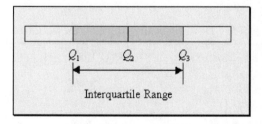

**Fig. 3-2:** General interquartile range

***Example 3-4:*** The following scores for a statistics 10-point quiz were reported. What is the value of the interquartile range for the data?

4    7    8    9    6    8    0    9    9    9    0    0    7    10    9    8    5    7    9

**Solution:** First, let us arrange the data in order from the smallest value to the largest value:

0   0   0   4   5   6   7   7   7   8   8   8   9   9   9   9   9   9   10

From this array, the median $Q_2 = 8$. Next, the medians for the first and second 50 percent of the values, excluding the median, are 5 and 9, respectively.

0   0   0   4   5   6   7   7   7   $\boxed{8}$   8   8   9   9   9   9   9   9   10
               ↑                    $Q_2$               ↑
               $Q_1$                                    $Q_3$

The interquartile range $= IQR = 9 - 5 = 4$. That is, the middle 50 percent of the quiz scores spans a 4-point range.

### 3-3 The Mean Absolute Deviation

The **mean absolute deviation** utilizes deviations of the data values from the mean in its computation.

**Explanation of the term—mean absolute deviation (MAD):** The MAD is the average of the absolute deviation values from the mean. That is, the deviations of the data values from the mean are computed, then absolute (positive) values for these deviations are obtained, and the average of these positive values is calculated.

Generally, if there are $n$ data values in the sample, with $x_i$ being the $i$th value and $\bar{x}$ being the sample mean, then the mean absolute deviation is defined as the average of the absolute deviations from the mean and is given by

$$\text{MAD} = \frac{\sum |x_i - \bar{x}|}{n}$$

The formula says that you subtract the mean from each data value and take the absolute value of the result, then you add these values and divide by the sample size.

***Example 3-5:*** What is the MAD for the following sample values?

3    8    6    12    0    –4    10

**Solution:** First of all, the sample mean $\bar{x} = 5$. (Verify.)

Next, we will construct a table to aid in the computations. **Table 3-1** shows the actual data values, the values for the deviations from the mean, and the absolute values for these deviations.

**Table 3-1:** Deviations and Absolute Deviations for Example 3-5

| DATA VALUES $x_i$ | DEVIATIONS $(x_i - \bar{x})$ | ABSOLUTE DEVIATIONS $|x_i - \bar{x}|$ |
|:---:|:---:|:---:|
| 3 | –2 | 2 |
| 8 | 3 | 3 |
| 6 | 1 | 1 |
| 12 | 7 | 7 |
| 0 | –5 | 5 |
| –4 | –9 | 9 |
| 10 | 5 | 5 |
| Total | 0 | 32 |

The MAD $= \dfrac{32}{7} = 4.57$. That is, the average (absolute) distance of these sample values from the mean is 4.57. **Figure 3-3** displays the values of the absolute deviations, the mean, and the MAD.

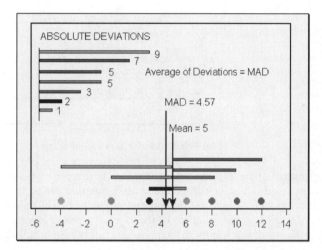

**Fig. 3-3:** Display of absolute deviations for Example 3-5

Observe, for instance, that the absolute deviation of 9 will contribute the most to the total deviation. As a matter of fact, the deviations will contribute to the total in proportion to the size of the deviation. It is desirable to define a variability measure in which each data value contributes in proportion to its distance from the mean, as is the case with the mean absolute deviation.

---

**Quick Tips**

1. If data set A has a larger MAD than data set B, then it is reasonable to believe that the values in data set A are more spread out (variable) than the values in data set B.
2. The MAD is sensitive to values that are very large or very small relative to the rest of the data set.

---

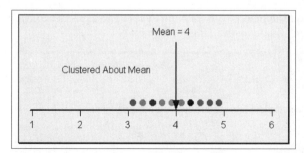

**Fig. 3-4:** Data with small variability

## 3-4 The Variance and Standard Deviation

The variance and the standard deviation are the most common and useful measures of variability. These two measures provide information about how the data vary about the mean.

If the data are clustered around the mean, then the variance and the standard deviation will be somewhat small.

There is small variability when data values are clustered about the mean, as shown in **Fig. 3-4.**

If, however, the data are widely scattered about the mean, the variance and the standard deviation will be somewhat large.

There is large variability when data values are widely scattered about the mean, as shown in **Fig. 3-5.**

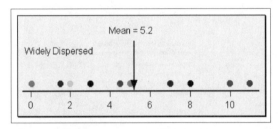

**Fig. 3-5:** Data with large variability

**Explanation of the term—sample variance:** The **sample variance** is an approximate average of the squared deviations from the sample mean. That is, the deviations of the data values from the sample mean are first computed, then the values for these deviations are squared, and the *approximate* average of these square values is found. The average is approximate because we divide not by the sample size but by the sample size minus 1.

**Note:**

1. We divide by the quantity $n - 1$ in order to make the sample variance an unbiased estimator of the population variance.
2. An estimator is unbiased if its average value is equal to the parameter it is estimating. For example, the sample mean $\bar{x}$ is an unbiased estimator of the population mean $\mu$. That is, if we take all possible samples of a fixed size and find the means of these samples, then the average of all these sample means will be equal to the population mean.
3. The sample variance uses the squares of the deviations from the mean, as this will eliminate the effect of the signs (as was also the case when we used the absolute value of the deviations in computing the MAD).

Generally, if there are $n$ data values in the sample, with $x_i$ being the $i$th value and $\bar{x}$ being the sample mean, then the variance of the set of sample values is given by

$$\text{Sample variance} = s^2 = \frac{\sum (x_i - \bar{x})^2}{n - 1}$$

The formula says that you subtract the mean from each data value and square the differences, then you add these values and divide by the sample size minus 1.

---

**Quick Tip**

Do not let the formula frighten you. We will build a table to help compute the variance.

---

**Example 3-6:** What is the variance for the following sample values?

3    8    6    14    0    11

**Solution:** First of all, we need to compute the sample mean:

$$\bar{x} = \frac{3+8+6+14+0+11}{6} = \frac{42}{6} = 7$$

Next, we will build a table to help in the computations. **Table 3-2** displays the data values, the deviations from the sample mean, and the squared deviations.

**Table 3-2:**   Table Used in Helping to Compute the Sample Variance for Example 3-6

| DATA | DEVIATIONS $(x_i - \bar{x})$ | SQUARED DEVIATIONS $(x_i - \bar{x})^2$ |
|:---:|:---:|:---:|
| 3 | $3 - 7 = -4$ | $(-4)^2 = 16$ |
| 8 | $8 - 7 = 1$ | $(1)^2 = 1$ |
| 6 | $6 - 7 = -1$ | $(-1)^2 = 1$ |
| 14 | $14 - 7 = 7$ | $(7)^2 = 49$ |
| 0 | $0 - 7 = -7$ | $(-7)^2 = 49$ |
| 11 | $11 - 7 = 4$ | $(4)^2 = 16$ |
| Total | 0 | 132 |

The table illustrates why we square the deviations. As the second column shows, the sum of the deviations is zero. Squaring the deviations yields a positive value.

From the table, $\sum(x_i - \bar{x})^2 = 132$. Thus the sample variance is

$$s^2 = \frac{\sum(x_i - \bar{x})^2}{n-1} = \frac{132}{6-1} = \frac{132}{5} = 26.4$$

This variance is somewhat large relative to the size of the data values. This can be observed from **Fig. 3-6,** which shows that the data values are very much spread out about the mean value of 7.

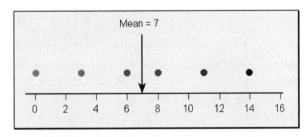

**Fig. 3-6:** Plot of data values for Example 3-6

**Explanation of the term—sample standard deviation:** The **sample standard deviation** is the positive square root of the sample variance. It is given by

$$\text{Sample standard deviation} = s = +\sqrt{s^2} = \sqrt{\frac{\sum (x_i - \bar{x})^2}{n-1}}$$

**Note:** The positive square root of the variance has the same unit as the variable. Since the variance is a square of the variable unit, taking the square root returns the original unit.

***Example 3-7:*** What is the sample standard deviation for the sample data in **Example 3-6?**

**Solution:** The sample standard deviation is $s = \sqrt{26.4} = 5.14$.

***Example 3-8:*** What is the MAD for the sample data in **Example 3-6?**

**Solution:** The sum of the absolute deviations $= 4 + 1 + 1 + 7 - 7 + 4 = 24$. Thus, the MAD $= \frac{24}{6} = 4$. This value can be used as a rough estimate for the value of the standard deviation.

---

## Quick Tips

1. The sample standard deviation is approximately equal to the average distance (MAD) of the observations from their mean.
2. If all of the observations have the same value, the sample standard deviation will be zero. That is, there is no variability in the data set.
3. The variance (standard deviation) is influenced by outliers (very small or very large values) in the data set.
4. The unit for the standard deviation is the same as that for the raw data, so it is preferable to use the standard deviation instead of the variance as a measure of variability.

---

**Explanation of the term—population variance:** The **population variance** is the average of the squared deviations from the population mean. That is, the deviations of the data values from the population mean are first computed, then the values for these deviations are squared, and then the average of these squared values is found. The average will be exact because we divide by the population size.

Generally, if there are $N$ data values in the population, with $x_i$ being the $i$th value and $\mu$ being the population mean, then the population variance, denoted by $\sigma^2$ (read as "sigma squared"), is given by

$$\text{Population variance} = \sigma^2 = \frac{\sum (x_i - \mu)^2}{N}$$

The formula says that you subtract the population mean from each data value and square the result, then you add these values and divide by the population size.

**Explanation of the term—population standard deviation:** The **population standard deviation** is the positive square root of the population variance. It is given by

$$\text{Population standard deviation} = \sigma = +\sqrt{\sigma^2} = \sqrt{\frac{\sum (x_i - \mu)^2}{N}}$$

where the Greek letter $\sigma$ (read as "sigma") is used to represent the population standard deviation. Observe that $\sigma^2$ represents the population variance.

**Note:** Recall that sample values are obtained from a population. Thus the sample mean and the sample variance (standard deviation) will estimate the population mean and the population variance (standard deviation), respectively. Also, observe that the population variance will be the average of the squared deviations from the mean, since we divide by the population size $N$.

### 3-5 The Coefficient of Variation

The coefficient of variation (CV) allows us to compare the variation of two (or more) different variables.

**Explanation of the term—sample coefficient of variation:** The **sample coefficient of variation** is defined as the sample standard deviation divided by the sample mean of the data set. Usually, the result is expressed as a percentage.

$$\text{Sample CV} = \frac{s}{\bar{x}} \times 100\%$$

**Note:** The sample coefficient of variation standardizes the variation by dividing it by the sample mean. The coefficient of variation has no units, since the standard deviation and the mean have the same units, and thus the units cancel each other. Because of this property, we can use this measure to compare variations for different variables with different units.

***Example 3-8:*** The mean number of parking tickets issued in a neighborhood over a four-month period was 90, and the standard deviation was 5. The average revenue generated from the tickets was $5,400, and the standard deviation was $775. Compare the variations of the two variables.

**Solution:**

$$\text{CV (number of tickets)} = \frac{5}{90} \times 100\% = 5.56\%$$

$$\text{CV (ticket revenues)} = \frac{775}{5,400} \times 100\% = 14.35\%$$

Since the CV is larger for the revenues, there is more variability in the recorded revenues than in the number of tickets issued.

**Explanation of the term—population coefficient of variation:** The **population coefficient of variation** is defined as the population standard deviation $\sigma$ divided by the population mean $\mu$. Again, the result is usually expressed as a percentage.

$$\text{Population CV} = \frac{\sigma}{\mu} \times 100\%$$

**Note:** Again, the population coefficient of variation standardizes the variation by dividing it by the population mean. Thus the coefficient of variation has no units, and so we can use this measure to compare variations for different population variables with different units.

### 3-6 The Empirical Rule

Knowing the value of the mean and the value of the standard deviation for a data set can provide a great deal of information about that data set. In particular, if the data set has a single mound and is symmetrical ("bell-shaped"), then one can generalize some properties of the distribution. One such generalization is called the *Empirical Rule*.

## Empirical Rule

The Empirical Rule gives some general statements relating the mean and the standard deviation of a bell-shaped distribution. It relates the mean to one standard deviation, two standard deviations, and three standard deviations. Without the loss of generality, we will use a symmetrical distribution with a mean of zero to discuss the Empirical Rule.

**One-Sigma Rule:** Approximately 68 percent of the data values should lie within one standard deviation of the mean. This is illustrated in **Fig. 3-7.**

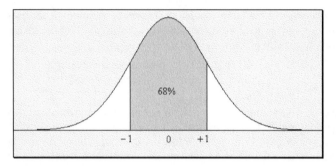

**Fig. 3-7:** One-sigma rule

Thus, one can expect a deviation of more than one sigma from the mean to occur once in every three observations. This is true because approximately 33 percent ≈ 1/3 of the values are outside one standard deviation from the mean.

**Two-Sigma Rule:** Approximately 95 percent of the data values should lie within two standard deviations of the mean. This is illustrated in **Fig. 3-8.**

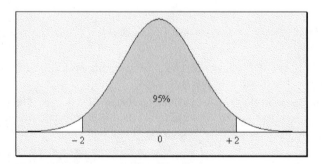

**Fig. 3-8:** Two-sigma rule

Thus, one can expect a deviation of more than two sigma from the mean to occur once in every 20 observations. This is true because approximately 5 percent $= \dfrac{1}{20}$ of the values are outside two standard deviations from the mean.

**Three-Sigma rule:** Approximately 99.7 percent of the data values should lie within three standard deviations of the mean. This is illustrated in **Fig. 3-9.**

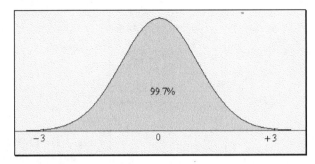

**Fig. 3-9:** Three-sigma rule

Thus, one can expect a deviation of more than three sigma from the mean to occur once in every 333 observations. This is true because approximately 0.3 percent $\approx \dfrac{1}{333}$ of the values are outside three standard deviations from the mean.

***Example 3-9:*** A group of Internet stocks had an average cost per share of $46.20 with a standard deviation of $10.11. If the distribution of the data is bell-shaped, what interval will contain approximately 95 percent of the stock prices?

**Solution:** Using the two-sigma rule, 95 percent of the data will be included in the interval $46.20 ± 2 × $10.11 = $46.20 ± $20.22. The lower limit of the interval is $46.20 – $20.22 = $25.98, and the upper limit of the interval is $46.20 + $20.22 = $66.42. Thus, 95 percent of the stock prices should fall between $25.98 and $66.42.

**Technology Corner**

All of the measures of variability discussed in this chapter can be computed directly or indirectly with most statistical software packages. Most scientific calculators will compute the variance or standard deviation for both sample and population. However, only the newer calculators with extensive statistical capabilities, such as the TI-83, will directly or indirectly compute all of these measures. If you own a calculator, you need to consult the manual to determine what statistical features are included.

**Illustration: Figure 3-10** shows the descriptive statistics computed by the MINITAB software. **Figure 3-11** shows the 1-Var Stats (descriptive statistics) computed by the TI-83 calculator. The data used, in both cases, were from **Example 2-3.** Observe that MINITAB has given the value of the standard deviation (StDev) to two decimal places, while the TI-83 calculator gives the value (Sx) to nine decimal places. Squaring the standard deviation will give

**Descriptive Statistics**

| Variable | N | Mean | Median | TrMean | StDev | SE Mean |
|----------|----|------|--------|--------|-------|---------|
| Eg2-3 | 11 | 2.09 | 2.00 | 2.11 | 7.56 | 2.28 |

| Variable | Minimum | Maximum | Q1 | Q3 | | |
|----------|---------|---------|------|------|--|--|
| Eg2-3 | -10.00 | 14.00 | -4.00 | 8.00 | | |

**Fig. 3-10:** MINITAB descriptive statistics output for Example 2-3

**Fig. 3-11:** TI-83 1-Var Stats
output for Example 2-3

the sample variance. The range can be obtained from both figures by subtracting the minimum value from the maximum value. Also, the interquartile range can be obtained by subtracting $Q_1$ from $Q_3$, which are displayed in both figures. One can use other features of the technologies to illustrate other concepts discussed in the chapter.

## Quick Tip

Unlike the interquartile range, the range, the mean absolute deviation, and the variance (standard deviation) are all sensitive to outlying (very small or very large) values.

**It's a Wrap** The most commonly used measures of variation for numeric data are the
✔ Range
✔ Interquartile range
✔ Mean absolute deviation
✔ Variance or standard deviation
✔ Coefficient of variation
Care should always be taken when using these measures of variation.

## True/False Questions

**Test Yourself**

1. The mean absolute deviation of a set of data always divides the data set in such a way that 50 percent of the values lie above the mean and 50 percent lie below the mean.
2. The median is a measure of variability.
3. The range of a set of data values is the largest value in the data set.
4. If the standard deviation is large, the data are less dispersed.
5. The standard deviation of a data set can be positive or negative.
6. Of a set of data values, 50 percent lie within one standard deviation of the mean of the data set.
7. In an ordered data set, the median of the upper 50 percent of the data set corresponds to a numerical value such that 75 percent of the values are below it.

8. The sum of the absolute deviations from the mean of a data set is always equal to 0.
9. The Empirical Rule states that exactly 95 percent of a data set would lie within two standard deviations of the mean.
10. The unit for the variance of a data set is the square of the unit for the variable associated with the data set.
11. The interquartile range is the average of the medians of the lower and upper 50 percent of an ordered data set.
12. The standard deviation is the square of the variance.
13. The mean absolute deviation is a measure of variability.
14. On a statistics exam, Joe's score was at the median for the lower 50 percent of the scores, and John's score was at the median of the upper 50 percent of the scores. Thus, we can say that John's score was twice Joe's score.
15. The mean value of a data set corresponds to half the value of the range.
16. Of the range, the interquartile range, and the variance of a data set, the interquartile range is most influenced by an outlying value in the data set.
17. Fifty percent of an ordered set of data values constitutes the interquartile range.
18. Quantities that describe populations are called parameters, and quantities that describe samples are called statistics.
19. For a bell-shaped distribution, the range of the values is approximately equal to the length of values within three standard deviations of the mean.
20. The coefficient of variation can be used to compare the variability of data sets that have different units.
21. A data set can have more than one measure of variability.
22. The Empirical Rule applies to all sets of data.
23. The value of the mean absolute deviation represents the number of standard deviations above or below the mean of the data set.
24. The sum of the absolute deviations from the mean for any data set may equal 0.
25. The range is not influenced by an outlying value in a data set.

## Completion Questions

1. The range of a set of data values is the difference between the (largest, smallest) value and the (largest, smallest) _____ value.

2. The mean absolute deviation will always be a (positive, negative) _____ number.

3. The sum of the deviations from the mean in a data set will always be (negative, positive, zero) _____ .

4. The coefficient of variation can be used to compare the (centralness, variability) _____ of data sets.

5. The interquartile range is the middle (25, 50, 75) _____ percent of the ordered data values.

6. The smaller the value of the standard deviation of a data set, the smaller the amount of (range, variability, MAD) _____ in the data set.

7. According to the Empirical Rule, approximately (68, 95, 99.7) _____ percent of the values in a data set will lie within three standard deviations of the (mean, median, mode) _____ if the data set is bell-shaped.

8. The standard deviation of a data set has the same _____ as the variable from which the data was obtained.

9. The mean absolute deviation for a data set measures the (average, median) _____ distance of the values in the set from the mean.

10. It is preferred to use the _____ rather than the variance because it has the same unit as the variable for the data.

11. Two measures of variability are the _____ and the _____ .

12. When computing the mean absolute deviation and the variance, measurements are deviated from the (mean, median, standard deviation) _____ in the computations.

13. If two sets of data values are compared, the larger _____ indicates a larger dispersion about the _____ .

14. If a measurement in a data set is below the mean value for that set, then the deviation from the mean will be (positive, negative) _____ .

15. For a bell curve distribution, approximately (68, 95, 99.7) _____ percent of the data values will lie within one standard deviation of the mean.

## Multiple-Choice Questions

1. A sample of 10 students was asked by the instructor to record the number of hours each spent studying for a given exam from the time the exam was announced in class. The following data values were the recorded numbers of hours:

   12    15    8    9    14    8    17    14    8    15

   The variance for the number of hours spent studying for this sample is
   (a) 10.0000.
   (b) 9.0000.
   (c) 3.4641.
   (d) approximately 12.

2. The numbers of minutes spent in the computer lab by 20 students working on a project are given below:

   *Numbers of Minutes*

   30 | 0, 2, 5, 5, 6, 6, 6, 8
   40 | 0, 2, 2, 5, 7, 9
   50 | 0, 1, 3, 5
   60 | 1, 3

   The range for this data set is
   (a) 300.
   (b) 303.
   (c) 603.
   (d) 600.

3. The numbers of minutes spent in the computer lab by 20 students working on a project are given below:

   *Numbers of Minutes*

   30 | 0, 2, 5, 5, 6, 6, 6, 8
   40 | 0, 2, 2, 5, 7, 9
   50 | 0, 1, 3, 5
   60 | 1, 3

The standard deviation for this data set is
(a) 101.6.
(b) 403.8.
(c) 306.0.
(d) 500.5.

4. The price increases on 5 stocks were $7, $1, $8, $4, and $5. The standard deviation for these price increases is
(a) 2.3.
(b) 2.7.
(c) 3.2.
(d) 4.1.

5. Which of the following is not affected by an extreme value in a data set?
(a) The mean absolute deviation
(b) The median
(c) The range
(d) The standard deviation

6. Given the following set of numbers, what is the variance?
    15    20    40    25    35
(a) 9.27
(b) 86.0
(c) 10.37
(d) 107.5

7. Which of the following is the crudest measure of dispersion?
(a) The mean absolute deviation
(b) The variance
(c) The mode
(d) The range

8. Which of the following is not a measure of central tendency?
(a) Mean
(b) Median
(c) $Q_3$
(d) Mode

9. Given the following data set:
    12    32    45    14    24    31

The total deviation from the mean for the data values is
(a) 0.
(b) 26.3333.
(c) 29.5.
(d) 12.

10. Given that a sample is approximately bell-shaped with a mean of 60 and a standard deviation of 3, the approximate percentage of data values that is expected to fall between 54 and 66 is
(a) 75 percent.

    (b) 95 percent.
    (c) 68 percent.
    (d) 99.7 percent.

11. Given the following frequency distribution:

| $x$ VALUES | FREQUENCY |
|:---:|:---:|
| 20 | 2 |
| 29 | 4 |
| 30 | 4 |
| 39 | 3 |
| 44 | 2 |

    The variance of the distribution is
    (a) 32.07.
    (b) 30.
    (c) 7.44.
    (d) 55.35.

12. Which of the following is a measure of variation?
    (a) Standard deviation
    (b) Midrange
    (c) Mode
    (d) Median

13. The following values are the ages of 15 students in a statistics class:
    18    21    25    21    28    23    21    19    24    26    21    24    18    27    23
    The standard deviation for this set of data is
    (a) 9.
    (b) 9.6857.
    (c) 21.
    (d) 3.1122.

14. An instructor recorded the following quiz scores (out of a possible 10 points) for the 12
    students present:
    7    4    4    7    2    9    10    6    7    3    8    5
    The interquartile range for this set of scores is
    (a) 7.5.
    (b) 6.5.
    (c) 8.
    (d) 3.5.

15. A statement is made that the average distance from the mean in a set of data values is
    10. Which measure is being used here?
    (a) The range
    (b) The interquartile range

    (c) The mean absolute deviation

    (d) The standard deviation

16. Which of the following is not a measure of dispersion?

    (a) Interquartile range

    (b) Range

    (c) Median

    (d) Coefficient of variation

17. For the data set 8, 12, 15, 20, 11, 5, 21, 0, what is the value of the coefficient of variation?

    (a) 62.52 percent.

    (b) 11.5 percent.

    (c) 7.19 percent.

    (d) 159.9 percent.

18. Which of the following statements is correct?

    (a) Two sets of numbers with completely different means and standard deviations may have the same coefficient of variation.

    (b) The most frequently used measure of variation is the standard deviation.

    (c) The range is a crude measure of dispersion, since it involves only the smallest and the largest values in a data set.

    (d) All of the above statements are correct.

19. Given that a sample is approximately bell-shaped with a mean of 25 and a standard deviation of 2, the approximate percentage of data values that are expected to fall between 19 and 31 is

    (a) 75 percent.

    (b) 95 percent.

    (c) 68 percent.

    (d) 99.7 percent.

20. The interquartile range in an ordered data set is the difference between

    (a) the median for the entire data set and the median for the lower 50 percent of the data set.

    (b) the median for the upper 50 percent of the data set and the median for the entire data set.

    (c) the median for the upper 50 percent and the median for the lower 50 percent of the data set.

    (d) the maximum value and the minimum value.

21. A single numerical value used to describe a characteristic of a sample data set, such as the sample median, is referred to as a

    (a) sample parameter.

    (b) sample median.

    (c) population parameter.

    (d) sample statistic.

22. The standard deviation will always be larger than the mean absolute deviation because

    (a) absolute values are not computed for the standard deviation.

    (b) the standard deviation is the square root of the variance.

(c) the larger values in the data set receive stronger emphasis when squared.

(d) of none of the above.

23. Which of the following is not a property of the standard deviation?

(a) It is affected by extreme values in a data set.

(b) It is the most widely used measure of spread.

(c) It uses all the values in the data set in its computation.

(d) It is always a positive number.

24. For which of the following is the coefficient of variation the smallest?

(a) $\bar{x} = 10$ and $s = 2$

(b) $\bar{x} = 14$ and $s = 3$

(c) $\bar{x} = 30$ and $s = 5$

(d) $\bar{x} = 39$ and $s = 8$

25. If a distribution has zero variance, which of the following is true?

(a) All the values are positive.

(b) All the values are negative.

(c) The number of positive values and the number of negative values are equal.

(d) All the values are equal to each other.

26. The following are given for a set of values:

I. The values ranged from 40 to 95.

II. The median value was 79.

III. 25 percent of the values are less than or equal to a value of 62.

IV. 75 percent of the values are less than or equal to 90.

From the above information, the interquartile range for the data set is

(a) 55.

(b) 28.

(c) 50.

(d) 33.

27. A sample of 10 students was asked by the instructor to record the number of hours each spent studying for a given exam from the time the exam was announced in class. The following data values were the recorded number of hours:

12    15    8    9    14    8    17    14    8    15

The mean absolute deviation for the number of hours spent studying for this sample is

(a) 3.0.

(b) 1.41.

(c) 1.33.

(d) 2.5.

## Further Exercises

If possible, you can use any technology available to help you solve the following questions.

1. The at-rest pulse rates for 16 athletes at a meet are

67    57    56    57    58    56    54    64    53    54    54    55    57    68    60    58

Find the range, interquartile range, mean absolute deviation, variance, standard deviation, and coefficient of variation for this set of data.

2. The speeds (in mph) of 16 cars on a highway were observed to be

   58   56   60   57   52   54   54   59   63   54   53   54   58   56   57   67

   Find the range, interquartile range, mean absolute deviation, variance, standard deviation, and coefficient of variation for this set of data.

3. *Estimate* the range, interquartile range, mean absolute deviation, variance, standard deviation, and coefficient of variation for the following frequency distribution. Recall that you can use the class marks for the intervals to approximate the observed values in the distribution.

   | CLASS | FREQUENCY |
   |-------|-----------|
   | 10–15 | 2 |
   | 15–20 | 4 |
   | 20–25 | 4 |
   | 25–30 | 3 |
   | 30–35 | 2 |

4. Find the range, interquartile range, mean absolute deviation, variance, standard deviation, and coefficient of variation for the following examination scores.

   *Exam Scores*

   4 | 5 6 8
   5 | 3 4 5 6 9
   6 | 2 3 5 6 6 9 9
   7 | 0 1 1 3 3 4 5 5 5 7 8
   8 | 1 2 3 6 9
   9 | 3 5 7 8

5. The following frequency distribution shows the scores for the exit examination for statistics majors at a four-year college for a given year.

   98   75   85   97   80   87   97   60   83   90

   Find the range, interquartile range, mean absolute deviation, variance, standard deviation, and coefficient of variation for this set of data.

6. The starting incomes for mathematics majors at a particular university were recorded for five years and are summarized in the following table:

   | STARTING SALARY (IN $1000) | FREQUENCY |
   |----------------------------|-----------|
   | 10–15 | 3 |
   | 15–20 | 5 |
   | 20–25 | 10 |
   | 25–30 | 7 |
   | 30–35 | 1 |

   *Estimate* the range, interquartile range, mean absolute deviation, variance, standard deviation, and coefficient of variation for this frequency distribution. Recall that you can use the class marks for the intervals to approximate the observed values in the distribution.

7. The numbers of minutes spent in the computer lab by 20 students working on a project are given below:

**Numbers of Minutes**

30 | 0, 2, 5, 5, 6, 6, 6, 8
40 | 0, 2, 2, 5, 7, 9
50 | 0, 1, 3, 5
60 | 1, 3

Find the range, interquartile range, mean absolute deviation, variance, standard deviation, and coefficient of variation for this set of data.

8. The following frequency distribution shows the distances traveled to campus (in miles) by 30 commuter students.

| DISTANCE (IN MILES) | FREQUENCY |
|---|---|
| 35–40 | 8 |
| 40–45 | 13 |
| 45–50 | 6 |
| 50–55 | 3 |

*Estimate* the range, interquartile range, mean absolute deviation, variance, standard deviation, and coefficient of variation for this frequency distribution. Recall that you can use the class marks for the intervals to approximate the observed values in the distribution.

# ANSWER KEY
## True/False Questions

1. F   2. F   3. F   4. F   5. F   6. F   7. T   8. F   9. F   10. T   11. F
12. F   13. T   14. F   15. F   16. F   17. T   18. T   19. T   20. T   21. T
22. F   23. F   24. T   25. F

## Completion Questions

1. largest, smallest    2. positive    3. zero    4. variability    5. 50    6. variability
7. 99.7, mean    8. unit    9. average    10. standard deviation    11. range, interquartile range, mean absolute deviation, variance, standard deviation, coefficient of variation (any two)    12. mean    13. variance or coefficient of variation, mean
14. negative    15. 68

## Multiple-Choice Questions

1. (d)   2. (b)   3. (a)   4. (b)   5. (b)   6. (d)   7. (d)   8. (c)   9. (a)
10. (b)   11. (d)   12. (a)   13. (d)   14. (d)   15. (c)   16. (c)   17. (a)
18. (d)   19. (d)   20. (c)   21. (d)   22. (c)   23. (b)   24. (c)   25. (d)
26. (b)   27. (a)

# Data Description— Numerical Measures of Position for Ungrouped Univariate Data

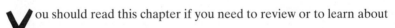

**Do I Need to Read This Chapter?**

**Y**ou should read this chapter if you need to review or to learn about

→ Numerical values that measure the location or position of a data value in a numerical data set

→ How to compute these measures and investigate some of their properties

You will recognize that these measures deal with only one property of the data set: the location or position. Thus, you will need to combine these measures with other properties of the data set in order to fully describe it. Other properties for univariate data will be investigated in future chapters.

## Get Started

A measure of location or position for a collection of data values is a number that is meant to convey the idea of the relative position of a data value in the data set. The most commonly used measures of location for sample data are the *z* score or standard score and percentiles. These measures are discussed in this chapter, as are some special percentiles.

### 4-1 The z Score or Standard Score

**Explanation of the term—z score:** The $z$ *score* for a sample value in a data set is obtained by subtracting the mean of the data set from the value and dividing the result by the standard deviation of the data set. Basically, the $z$ score tells us how many standard deviations a specific value is from the mean value of the data set.

If we let $x$ represent the sample value, $\bar{x}$ the sample mean, and $s$ the sample standard deviation, then the $z$ score can be computed using the following formula:

$$z \text{ score} = \frac{x - \bar{x}}{s}$$

The corresponding $z$ score for a population value $x$ is given by

$$z \text{ score} = \frac{x - \mu}{\sigma}$$

where $\mu$ represents the population mean and $\sigma$ represents the population standard deviation.

---

## Quick Tips

1. The $z$ score is the number of standard deviations the data value falls above (positive $z$ score) or below (negative $z$ score) the mean for the data set.
2. The $z$ score is affected by an outlying value in the data set, since the outlier (very small or very large value) directly affects the value of the mean and the standard deviation.

---

***Example 4-1:*** What is the $z$ score for the value of 14 in the following sample data set?

   3    8    6    14    4    12    7    10

**Solution:** First, compute the sample mean and the sample standard deviation. The sample mean $= \bar{x} = 8$, and the sample standard deviation $= s = 3.8173$. Verify that these values are correct. Thus the $z$ score is

$$z \text{ score} = \frac{14 - 8}{3.8173} = 1.5718 \approx 1.57$$

Thus, the data value of 14 is located 1.57 standard deviations *above* the mean of 8, since the $z$ score is positive.

***Question:*** **Why does a z score measure relative position?**

The following discussion, using the information for **Example 4-1,** will give an insight into the question. **Figure 4-1** shows a plot of the data points with the location of the mean and the data value of 14. The distance between the mean of 8 and the value of 14 is $1.57 \times$ standard deviation $= 5.99 \approx 6$. Observe that if we add the mean of 8 to this value, we will get $8 + 6 = 14$, the data value. Thus, this shows that the value of 14 is 1.57 standard deviations above the mean. That is, the $z$ score gives us an idea of how far away the data value is from the mean, and so it gives us an idea of the position of the data value relative to the mean.

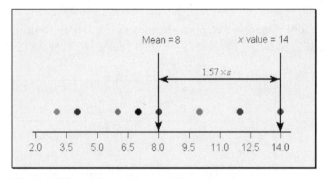

**Fig. 4-1:** Plot of data points for Example 4-1

***Example 4-2:*** What is the $z$ score for the value of 95 in the following sample data set?

| 96 | 114 | 100 | 97 | 101 | 102 | 99 | 95 | 90 |

**Solution:** First, compute the sample mean and the sample standard deviation. The sample mean $= \bar{x} = 99.3333$, and the sample standard deviation $= s = 6.5955$. Verify that these values are correct. Thus, the $z$ score is

$$z \text{ score} = \frac{95 - 99.3333}{6.5955} = -0.6570 \approx -0.66$$

Thus, the data value of 95 is located 0.66 standard deviation *below* the mean of 99.3333, since the $z$ score is negative.

## 4-2 Percentiles

**Explanation of the term—percentiles:** **Percentiles** are numerical values that divide an ordered data set into 100 groups of values with at most 1% of the data values in each group.

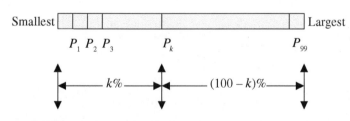

**Fig. 4-2** Illustration of the $k$th percentile

When we discuss percentiles, we generally present the discussion through the $k$th percentile. The $k$th percentile for an ordered array of data is a numerical value $P_k$ (say) such that at most $k$ percent of the data values are smaller than $P_k$, and at most $(100 - k)$ percent of the data values are larger than $P_k$. The idea of the $k$th percentile is illustrated in **Fig. 4-2.**

**Quick Tips**

1. In order for a percentile to be determined, the data set first must be *ordered* from smallest to largest value.
2. There are 99 percentiles in a data set.

*Example 4-3:* Display the 99th percentile pictorially.

**Solution:** **Figure 4-3** displays the idea of the 99th percentile $P_{99}$. Here, at most 99 percent of the data values will be smaller than $P_{99}$, and at most 1 percent will be larger than $P_{99}$.

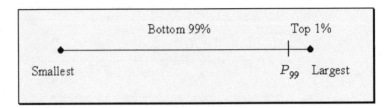

**Fig. 4-3:** Pictorial representation of the 99th percentile

### Percentile Corresponding to a Given Data Value

The percentile corresponding to a given data value, say $x$, in a set is obtained by using the following formula:

$$\text{Percentile} = \frac{\text{Number of values below } x + 0.5}{\text{Number of values in data set}} \times 100\%$$

*Example 4-4:* The shoe sizes, in whole numbers, for a sample of 12 male students in a statistics class were as follows:

13    11    10    13    11    10    8    12    9    9    8    9

What is the percentile rank for a shoe size of 12?

**Solution:** First, we need to arrange the values from smallest to largest. This ordered set is given below:

8    8    9    9    9    10    10    11    11    12    13    13

Observe that the number of values below 12 is 9 and the total number of values in the data set is 12. Thus, using the formula, the corresponding percentile is

$$\frac{9 + 0.5}{12} \times 100\% = 79.17\%$$

That is, the value of 12 corresponds to approximately the 79th percentile.

*Example 4-5:* What is the percentile rank for a shoe size of 10 for the information in **Example 4-4?**

**Solution:** The ordered set from **Example 4-4** is repeated next.

8    8    9    9    9    10    10    11    11    12    13    13

Observe that the number of values below 10 is 5 and the total number of values is 12. Thus, using the formula, the corresponding percentile is

$$\frac{5 + 0.5}{12} \times 100\% = 45.83\%$$

That is, the value of 10 corresponds to approximately the 46th percentile.

### Procedure for Finding a Data Value
### for a General Percentile $P_k$

Assume that we want to determine what data value falls at some general percentile $P_k$. The following steps will enable you to find a general percentile $P_k$ for a data set.

*Step 1:*    Order the data set from smallest to largest.

*Step 2:*    Compute the position $c$ of the percentile.

$$\text{Compute } c = (n \cdot k)/100$$

where $n$ = sample size

   $k$ = the required percentile

*Step 3.1:* If $c$ *is not* a whole number, round up to the next whole number. Locate this position in the ordered set. The value in this location is the required percentile.

*Step 3.2:* If $c$ *is* a whole number, find the average of the values in the $c$ and $c + 1$ positions in the ordered set. The average value will be the required percentile.

*Example 4-6:* The data given below represent the 19 countries with the largest numbers of total Olympic medals—excluding the United States, which had 101 medals—for the 1996 Atlanta Games. Find the 65th percentile for the data set.

63   65   50   37   35   41   25   23   27   21   17   17   20   19   22   15   15   15   15

**Solution:** First, we need to arrange the data set in order. The ordered set is

15   15   15   15   17   17   19   20   21   22   23   25   27   35   37   41   50   63   65

   Next, compute the position of the percentile. Here, $n = 19$ and $k = 65$. Thus, $c = (19 \times 65)/100 = 12.35$, and so we need to round up to 13. Thus, the 65th percentile will be located at the 13th position in the ordered set. The 13th value corresponds to the value of 27. Hence, $P_{65} = 27$.

*Question:* **Why does a percentile measure relative position?**

The following discussion, using the information for **Example 4-6,** will give an insight into the question. **Figure 4-4** shows a plot of the data points with the location of the 65th percentile value of 27. This number is the cutoff point where at most 65 percent of the data values are smaller than the percentile value of 27, and at most 35 percent of the data values are larger than the percentile value of 27. Thus, this shows that the percentile value of 27 is a measure of location. That is, the percentile gives us an idea of the relative position of a value in an ordered data set.

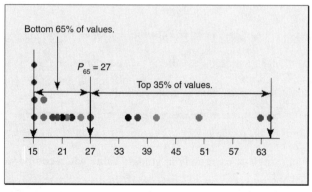

**Fig. 4-4:** Plot to aid in answering the question, Why does a percentile measure relative position?

*Example 4-7:* Find the 25th percentile for the following data set:

6   12   18   12   13   8   13   11   10   16   13   11   10   10   2   14

**Solution:** First, we need to arrange the data set in order. The ordered set is

2   6   8   10   10   10   11   11   12   12   13   13   13   14   16   18

Next, compute the position of the percentile. Here, $n = 16$ and $k = 25$. Thus, $c = (16 \times 25)/100 = 4$. Thus, the 25th percentile will be the average of the values located at the 4th and the 5th positions in the ordered set. The 4th value corresponds to the value of 10, and the 5th value corresponds to the value of 10 as well. Hence, $P_{25} = (10 + 10)/2 = 10$.

### Special Percentiles—Deciles and Quartiles

Deciles and quartiles are special percentiles. Deciles divide an ordered data set into 10 equal parts, and quartiles divide it into four equal parts.

We usually denote deciles by $D_1, D_2, \ldots, D_9$ and quartiles by $Q_1, Q_2,$ and $Q_3$. This is depicted in **Figs. 4-5** and **4-6,** respectively. In the case of **Fig. 4-5,** there is at most 10 percent of the values in each group. In the case of **Fig. 4-6,** there is at most 25 percent of the values in each group.

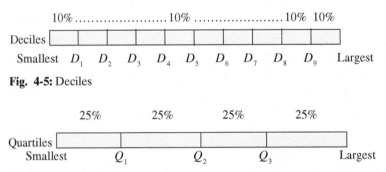

**Fig. 4-5:** Deciles

**Fig. 4-6:** Quartiles

---

### Quick Tips

1.  There are nine deciles and three quartiles.
2.  $Q_1$ is usually referred to as the first quartile, $Q_2$ as the second quartile, and $Q_3$ as the third quartile.
3.  $P_{50} = D_5 = Q_2 = $ median. That is, the 50th percentile, the 5th decile, the 2nd quartile, and the median are all equal to one another.
4.  Finding deciles and quartiles is equivalent to finding the equivalent percentiles.

---

### Outliers

Recall that an outlier is an extremely small or extremely large data value when compared with the rest of the data values.

The procedure below allows us to check whether a data value can be considered an outlier.

### Procedure to Check for Outliers

The following steps will allow us to check whether a given value in a data set can be classified as an outlier.

**Step 1:** Arrange the data in order from smallest to largest.

**Step 2:** Determine the first quartile $Q_1$ and the third quartile $Q_3$. Recall that $Q_1 = P_{25}$ and $Q_3 = P_{75}$.

**Step 3:** Find the interquartile range (IQR) by computing $Q_3 - Q_1$.

**Step 4:** Compute $Q_1 - 1.5 \times \text{IQR}$ and $Q_3 + 1.5 \times \text{IQR}$.

**Step 5:** Let $x$ be the data value that is being checked to determine whether it is an outlier.

    (a) If the value of $x$ is smaller than $Q_1 - 1.5 \times \text{IQR}$, then $x$ is an outlier.

    (b) If the value of $x$ is larger than $Q_3 + 1.5 \times \text{IQR}$, then $x$ is an outlier.

***Example 4-8:*** The data below represent the 20 countries with the largest number of total Olympic medals—including the United States, which had 101 medals—for the 1996 Atlanta Games. Determine whether the number of medals won by the United States is an outlier relative to the numbers for the other 19 countries.

63  65  50  37  35  41  25  23  27  21  17  17  20  19  22  15  15  15  15  101

**Solution:** First, we need to arrange the data set in order. The ordered set is

15  15  15  15  17  17  19  20  21  22  23  25  27  35  37  41  50  63  65  101

Next, we need to determine the first and the third quartiles. That is, we need to determine $P_{25}$ and $P_{75}$. Verify that $Q_1 = P_{25} = 17$ and $Q_3 = P_{75} = 39$. Thus the IQR = 39 − 17 = 22. From these computations, $Q_1 - 1.5 \times \text{IQR} = 17 - 1.5 \times 22 = -16$ and $Q_3 + 1.5 \times \text{IQR} = 39 + 1.5 \times 22 = 72$. Since, 101 > 72, the value of 101 is an outlier relative to the rest of the values in the data set, based on the criterion presented in this section. That is, the number of medals won by the United States is an outlier relative to the numbers won by the other 19 countries for the 1996 Atlanta Olympic Games.

## Box Plots

**Explanation of the term—box plot:** A **box plot** is a graphical display that involves a five-number summary of a distribution of values, consisting of the minimum value, the lower quartile, the median, the upper quartile, and the maximum value.

A horizontal box plot is constructed by drawing a box between the quartiles $Q_1$ and $Q_3$. That is, a box is drawn to indicate the middle 50 percent of the data values. Horizontal lines are then drawn from the middle of the sides of the box to the minimum and maximum values. These horizontal lines are called *whiskers*. A vertical line inside the box marks the median. Outliers are usually indicated by a dot or an asterisk.

***Example 4-9:*** Display the data for **Example 4-8** by a box plot.

**Solution:** The box plot for the data in **Example 4-8** is shown in **Fig. 4-7.**

By observing a box plot, important information concerning the

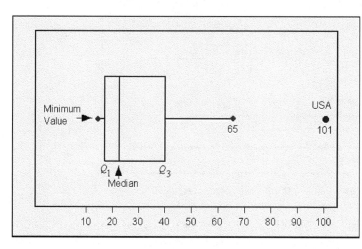

**Fig. 4-7:** Box plot for data in Example 4-8

distribution of the data set can be obtained. A summary of the potential information is given next.

**Quick Tip**

The vertical sides of the horizontal box plot are sometimes called *hinges.*

## Information That Can Be Obtained from a Box Plot

1. If the median is close to the center of the box, the distribution of the data values will be approximately symmetrical.
2. If the median is to the left of the center of the box, the distribution of the data values will be positively skewed.
3. If the median is to the right of the center of the box, the distribution of the data values will be negatively skewed.
4. If the whiskers are approximately the same length, the distribution of the data values will be approximately symmetrical.
5. If the right whisker is longer than the left whisker, the distribution of the data values will be positively skewed.
6. If the left whisker is longer than the right whisker, the distribution of the data values will be negatively skewed.

**Figure 4-8** shows box plots that display these characteristics.

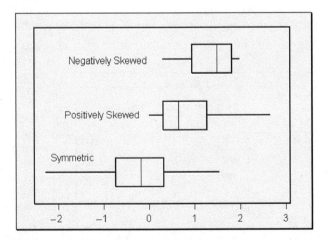

**Fig. 4-8:** Box plots with different characteristics

***Example 4-10:*** Given the following information, determine whether the distribution is negatively skewed, symmetrical, or positively skewed.

(a) Minimum value = 25, $Q_1 = 30$, $Q_2 = 33$, $Q_3 = 52$, maximum value = 70.
(b) Minimum value = 10, $Q_1 = 25$, $Q_2 = 45$, $Q_3 = 65$, maximum value = 80.
(c) Minimum value = 5, $Q_1 = 40$, $Q_2 = 55$, $Q_3 = 60$, maximum value = 65.

**Solution:** (a) Observe that the length of the left whisker = $30 - 25 = 5$ and the length of the right whisker = $70 - 52 = 18$. Also, the median is only $33 - 30 = 3$ units from $Q_1$ and $52 - 33 = 19$ units from $Q_3$. From this information, we can conclude that the distribution is skewed to the right, or positively skewed. The box plot for this information is shown in **Fig. 4-9.**

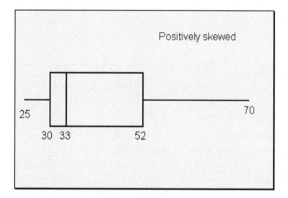

**Fig. 4-9:** Box plot for Example 4-9(a)

(b) Observe that the length of the left whisker = $25 - 10 = 15$ and the length of the right whisker = $80 - 65 = 15$. Also, the median is $45 - 25 = 20$ units from $Q_1$ and $65 - 45 = 20$ units from $Q_3$. From this information, we can conclude that the distribution is symmetrical. The box plot for this information is shown in **Fig. 4-10.**

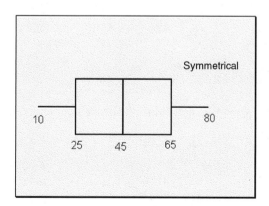

**Fig. 4-10:** Box plot for Example 4-9(b)

(c) Observe that the length of the left whisker = $40 - 5 = 35$ and the length of the right whisker = $65 - 60 = 5$. Also, the median is $55 - 40 = 15$ units from $Q_1$ and only $60 - 55 = 5$ units from $Q_3$. From this information, we can conclude that the distribution is skewed to the left, or negatively skewed. The box plot for this information is shown in **Fig. 4-11.**

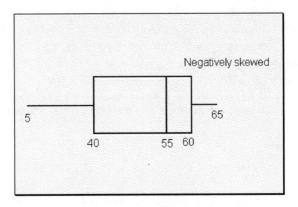

**Fig. 4-11:** Box plot for Example 4-9(c)

### Technology Corner

Usually, only the first, second (median), and third quartiles can be computed directly with most statistical software packages. Box plots can be displayed by most statistical software packages. Most scientific calculators will not compute the measures of position discussed in this chapter. However, the newer calculators with extensive statistical capabilities, such as the TI-83, will directly compute the quartiles. All other percentiles and $z$ scores will have to be computed using the formula or procedures discussed in the chapter. If you own a calculator, you should consult the manual to determine what statistical features are included.

**Illustration: Figure 4-12** shows the descriptive statistics computed by the MINITAB software. **Figure 4-13** shows the 1-Var Stats (descriptive statistics) computed by the TI-83 calcu-

---

### Descriptive Statistics

| Variable | N | Mean | Median | TrMean | StDev | SE Mean |
|---|---|---|---|---|---|---|
| No. Medals | 20 | 32.15 | 22.50 | 29.28 | 22.41 | 5.01 |

| Variable | Minimum | Maximum | Q1 | Q3 |
|---|---|---|---|---|
| No. Medals | 15.00 | 101.00 | 17.00 | 40.00 |

---

**Fig. 4-12:** MINITAB descriptive statistics output for Example 4-8

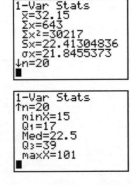

**Fig. 4-13:** TI-83 1-Var Stats output for Example 4-8

lator. The data used, in both cases, were from **Example 4-8.** Observe that MINITAB has given the value of the first, second (median), and third quartiles. These values are also given by the TI-83 calculator. Note that from **Fig. 4-12,** the value for the third quartile is 40. The value given by the TI-83 in **Fig. 4-13** is 39. This discrepancy may be due to rounding or may be due to the procedure used by the calculator to compute the third quartile. One can use other features of the technologies to illustrate other concepts discussed in the chapter.

The most commonly used measures of position for numeric data are the

✔ $z$ score

✔ Percentiles

✔ Quartiles (special percentiles)

Care should always be taken when using these measures of position.

## True/False Questions

1. If a student's exam score corresponds to a negative $z$ score, then the student has a score that is less than the mean of the set of exam scores.
2. At most 50 percent of a set of data values lie between the first and third quartiles of the data set.
3. The third decile corresponds to the 70th percentile.
4. The midrange is the average of the first and third quartiles.
5. On a statistics exam, Joe's score was at the 30th percentile and John's score was at the 60th percentile; thus, we can say that John's score was twice Joe's score.
6. The 50th percentile of a data set corresponds to the mean value of the data set.
7. At most 50 percent of an ordered set of data values lie below the first quartile and above the third quartile.
8. For a symmetrical distribution, exactly 50 percent of the values are below the second quartile and exactly 50 percent of the values are above the second quartile.
9. The $z$ score associated with a data value represents the number of standard deviations that the value lies above or below the mean of the data set.
10. There are 100 percentiles in a data set.
11. In finding percentiles, it is not necessary to first order the data values.
12. If the 75th percentile for scores on a 100-point exam is 75, and the percentile position number $c$ had to be rounded up to a whole number, then a student's score that ranks at the 75th percentile is equal to 75.
13. The 50th percentile is the same as the median.
14. The 9th decile is the same as the 90th percentile.
15. There are four quartiles in any data set.
16. An outlier does not affect the value of the median in a data set.
17. A box plot for a data set can be constructed if one knows the minimum value, the first quartile, the second quartile, the third quartile, and the maximum value of the data set.
18. If the right whisker is significantly longer than the left whisker on a box plot, then the distribution is skewed to the right.
19. In a box plot, if the median is to the right of the center of the box, the distribution of the data values will be positively skewed.
20. In a box plot, if the median is close to the center of the box, the distribution of the data values will be approximately symmetrical.

21. There are 10 deciles in a data set.
22. Given the following: minimum value = 20, $Q_1 = 25$, $Q_2 = 30$, $Q_3 = 35$, maximum value = 40. From this information, we can conclude that the distribution of values from which this information was obtained is symmetrical.
23. A $z$ score always indicates the number of standard deviations a data value is above the mean.
24. $z$ scores can be obtained only for sample values.
25. A large outlying value in a data set may affect the value of the 99th percentile.

## Completion Questions

1. The $z$ value associated with a measurement $x$ represents the number of _____ that $x$ lies above the mean or below the mean.
2. The first quartile is that value at or below which (25, 50, 75) _____ percent of all data entries in the ordered data set fall.
3. Two measures of position are _____ and _____ .
4. If a measurement in a data set is below the mean value for that set, then the $z$ value will be (positive, negative) _____ .
5. Two special types of percentiles are _____ and _____ .
6. There are (two, three, four) _____ quartiles in any data set.
7. The seventh decile corresponds to the (30th, 70th) _____ percentile.
8. There are (99, 100) _____ percentiles in any data set.
9. If the left whisker is significantly longer than the right whisker on a box plot, then the distribution is skewed to the (right, left) _____ .
10. In a box plot, if the median is close to the (left edge, center, right edge) _____ of the box, the distribution of the data values will be approximately symmetrical.
11. The 50th percentile, the median, and the second quartile are all _____ to one another.
12. There are (nine, ten) _____ deciles in a data set.
13. In finding a $z$ score for a given data value, if the $z$ score is zero, then the data value and the mean are (different, equal) _____ .
14. The third quartile corresponds to the (25th, 30th, 75th) _____ percentile.
15. In finding percentiles, it is first necessary to _____ the data values from smallest to largest.

## Multiple-Choice Questions

1. A student has seven statistics books open in front of him. The page numbers are as follows: 231, 423, 521, 139, 347, 400, 345. The second quartile for this set of numbers is
   (a) 231.
   (b) 347.
   (c) 330.
   (d) 423.

2. A cyclist recorded the number of miles per day she cycled for 5 days. The recordings were as follows: 13, 10, 12, 10, 11. The 50th percentile for the number of miles she cycled per day is
   (a) 12.5.
   (b) 11.
   (c) 10.
   (d) 11.5.

3. An instructor recorded the following quiz scores (out of a possible 10 points) for the 12 students present: 7, 4, 4, 7, 2, 9, 10, 6, 7, 3, 8, 5. The 25th percentile for this set of scores is
   (a) 4.
   (b) 6.
   (c) 6.5.
   (d) 7.5.

4. An instructor recorded the following quiz scores (out of a possible 10 points) for the 12 students present: 7, 4, 4, 7, 2, 9, 10, 6, 7, 3, 8, 5. The sixth decile for this set of scores is
   (a) 9.
   (b) 7.5.
   (c) 6.5.
   (d) 7.

5. The following values are the ages of fifteen students in a statistics class:

   18   21   25   21   28   23   21   19   24   26   21   24   18   27   23

   The value of the first quartile for this set of data is
   (a) 25.
   (b) 23.
   (c) 21.
   (d) 22.5.

6. Which of the following is not a measure of position?
   (a) First quartile
   (b) Median
   (c) 4th decile
   (d) Mean

7. A final statistics exam had a mean of 70 and a variance of 25. If Bruce had made an 80 on this exam, what is his $z$ score?
   (a) –2
   (b) 10
   (c) 0.4
   (d) 2

8. When it is necessary to determine whether an observation from a set of data falls in the upper 25 percent or the lower 75 percent of the ordered data set, which measure should be used?
   (a) Third quartile
   (b) Mean

    (c) First quartile

    (d) 70th percentile

9. For the data set 38, 42, 45, 50, 41, 35, 51, 29, the value corresponding to the lower quartile (left vertical side) in a horizontal box plot would be equal to

    (a) 43.5.

    (b) 36.5.

    (c) 41.5.

    (d) 47.5.

10. For the data set 8, 12, 15, 20, 11, 5, 21, 0, what is the value of the seventh decile?

    (a) 6.5

    (b) 11.5

    (c) 17.5

    (d) 15

11. The following values are the ages of 15 students in a statistics class:

    18    21    25    21    28    23    21    19    24    26    21    24    18    27    23

    The percentile rank of 25 would be

    (a) 20th.

    (b) approximately 77th.

    (c) 80th.

    (d) 75th.

12. Which of the following is a measure of position?

    (a) Standard deviation

    (b) Box plot

    (c) Mode

    (d) Quartile

13. Given the following frequency distribution:

| x VALUES | FREQUENCY |
| --- | --- |
| 20 | 2 |
| 29 | 4 |
| 30 | 4 |
| 39 | 3 |
| 44 | 2 |

    The 77th percentile of the distribution is

    (a) 29.

    (b) 30.

    (c) 39.

    (d) 32.

14. Given that a sample is approximately bell-shaped with a mean of 60 and a standard deviation of 3, the approximate value for the 98th percentile for this distribution is

    (a) 63.

    (b) 66.

(c) 69.

(d) 57.

15. Given that a sample is approximately bell-shaped with a mean of 60 and a standard deviation of 3, the approximate value for the 16th percentile for this distribution is

(a) 54.

(b) 66.

(c) 63.

(d) 57.

16. Given the following data set:

12    32    45    14    24    31

The value of the interquartile range is

(a) 14.

(b) 32.

(c) 27.5.

(d) 18.

17. If a sample has a mean of 100 and a standard deviation of 6, what is the value in the data set that corresponds to a $z$ score of 2?

(a) 88

(b) 94

(c) 92

(d) 112

18. The 50th percentile is the same as the

(a) mode.

(b) mean.

(c) median.

(d) midrange.

19. The vertical sides on a horizontal box plot are located at

(a) the minimum value and the first quartile of the data set.

(b) the minimum value and the maximum value of the data set.

(c) the third quartile and the maximum value of the data set.

(d) the first quartile and the third quartile of the data set.

20. The interquartile range is the difference between

(a) the second quartile and the first quartile.

(b) the third quartile and the second quartile.

(c) the third quartile and the first quartile.

(d) the maximum value and the minimum value.

21. For a bell-shaped distribution, the most frequently occurring value in a data set will be the

(a) third quartile.

(b) interquartile range.

(c) second quartile.

(d) first quartile.

22. Which of the following may be affected the most if there is an extremely large value in the data set?
   (a) The first quartile
   (b) The second quartile
   (c) The third quartile
   (d) The 99th percentile

23. If the number of values in any data set is even, then
   (a) the second quartile cannot be found.
   (b) the second quartile will be the average of two numbers.
   (c) the interquartile range will be half the distance between the minimum value and the median.
   (d) the second quartile will be twice the first quartile.

24. The 75th percentile of a data set is equivalent to
   (a) the seventh decile of the data set.
   (b) the first quartile of the data set.
   (c) the second quartile of the data set.
   (d) the third quartile of the data set.

25. Given the following frequency distribution:

| x VALUES | FREQUENCY |
|----------|-----------|
| 20 | 2 |
| 29 | 4 |
| 30 | 4 |
| 39 | 3 |
| 44 | 2 |

The second quartile of the distribution is
   (a) 29.
   (b) 30.
   (c) 39.
   (d) 29.5.

26. In a box plot, if the median is to the left of the center of the box, then the distribution of the data values will be
   (a) positively skewed.
   (b) negatively skewed.
   (c) symmetrical.
   (d) bell-shaped.

27. In a box plot, if the left whisker is longer than the right whisker, the distribution of the data values will be
   (a) positively skewed.
   (b) negatively skewed.
   (c) symmetrical.
   (d) uniform.

28. The following information for a data set is given: minimum value = 105, $Q_1$ = 140, $Q_2$ = 155, $Q_3$ = 160, maximum value = 165. From this information, the distribution is
    (a) symmetrical.
    (b) positively skewed.
    (c) negatively skewed.
    (d) bell-shaped.

29. The number of deciles in a data set is
    (a) 10.
    (b) 12.
    (c) 99.
    (d) 9.

30. A data value is considered to be an outlier if
    (a) it lies between $-1.5 \times$ IQR and $+1.5 \times$ IQR.
    (b) it lies between $Q_1 - 1.5 \times$ IQR and $Q_3 + 1.5 \times$ IQR.
    (c) it lies between $Q_1$ and $Q_3$.
    (d) it is smaller than $Q_1 - 1.5 \times$ IQR or larger than $Q_3 + 1.5 \times$ IQR.

31. The numbers of minutes spent in the computer lab by 20 students working on a project are given below:

    **Numbers of Minutes**

    30 | 0, 2, 5, 5, 6, 6, 6, 8
    40 | 0, 2, 2, 5, 7, 9
    50 | 0, 1, 3, 5
    60 | 1, 3

    The first quartile for this data set is
    (a) 402.
    (b) 306.
    (c) 500.
    (d) 403.5.

32. The numbers of minutes spent in the computer lab by 20 students working on a project are given below:

    **Numbers of Minutes**

    30 | 0, 2, 5, 5, 6, 6, 6, 8
    40 | 0, 2, 2, 5, 7, 9
    50 | 0, 1, 3, 5
    60 | 1, 3

    The interquartile range for this data set is
    (a) 306.
    (b) 402.
    (c) 500.5.
    (d) 194.5.

# Further Exercises

If possible, you can use any technology available to help you solve the following questions.

1. The at-rest pulse rates for 16 athletes at a meet are

   67   57   56   57   58   56   54   64   53   54   54   55   57   68   60   58

   Find the quartiles, interquartile range, deciles, 47th percentile, and 88th percentile for this set of data. Display the data using a box plot.

2. The speeds (in mph) of 16 cars on a highway were observed to be

   58   56   60   57   52   54   54   59   63   54   53   54   58   56   57   67

   Find the quartiles, interquartile range, deciles, 33rd percentile, 66th percentile, and 99th percentile for this set of data. Display the data using a box plot.

3. *Estimate* the quartiles, interquartile range, deciles, 11th percentile, 22nd percentile, 44th percentile, 66th percentile, and 88th percentile for the following frequency distribution. Display the data using a box plot. Recall that you can use the class marks for the intervals to approximate the observed values in the distribution.

   | CLASS | FREQUENCY |
   |-------|-----------|
   | 10–15 | 2 |
   | 15–20 | 4 |
   | 20–25 | 4 |
   | 25–30 | 3 |
   | 30–35 | 2 |

4. Find the quartiles, interquartile range, deciles, 12th percentile, 24th percentile, 36th percentile, 48th percentile, 60th percentile, 72nd percentile, 84th percentile, and 96th percentile for the following examination scores. Display the data using a box plot.

   *Exam Scores*

   4 | 5, 6, 8
   5 | 3, 4, 5, 6, 9
   6 | 2, 3, 5, 6, 6, 9, 9
   7 | 0, 1, 1, 3, 3, 4, 5, 5, 5, 7, 8
   8 | 1, 2, 3, 6, 9
   9 | 3, 5, 7, 8

5. The following frequency distribution shows the scores for the exit examination for statistics majors at a four-year college for a given year.

   98   75   85   97   80   87   97   60   83   90

   Find the quartiles, interquartile range, deciles, 30th percentile, 60th percentile, and 90th percentile for this set of data. Display the data using a box plot.

6. The starting incomes for mathematics majors at a particular university were recorded for five years and are summarized in the following table:

   | STARTING SALARY (IN $1000) | FREQUENCY |
   |----------------------------|-----------|
   | 10–15 | 3 |
   | 15–20 | 5 |
   | 20–25 | 10 |
   | 25–30 | 7 |
   | 30–35 | 1 |

*Estimate* the quartiles, interquartile range, deciles, 33rd percentile, 66th percentile, and 99th percentile for this set of data. Display the data using a box plot. Recall that you can use the class marks for the intervals to approximate the observed values in the distribution.

7. The numbers of minutes spent in the computer lab by 20 students working on a project are given below:

**Numbers of Minutes**

30 | 0, 2, 5, 5, 6, 6, 6, 8
40 | 0, 2, 2, 5, 7, 9
50 | 0, 1, 3, 5
60 | 1, 3

Find the quartiles, interquartile range, deciles, 33rd percentile, 66th percentile, and 99th percentile for this set of data. Display the data using a box plot.

8. The following frequency distribution shows the distances traveled to campus (in miles) by 30 commuter students:

| DISTANCE (IN MILES) | FREQUENCY |
|---|---|
| 35–40 | 8 |
| 40–45 | 13 |
| 45–50 | 6 |
| 50–55 | 3 |

*Estimate* the quartiles, interquartile range, deciles, 30th percentile, 60th percentile, and 99th percentile for this set of data. Display the data using a box plot. Recall that you can use the class marks for the intervals to approximate the observed values in the distribution.

# ANSWER KEY

## True/False Questions

1. T   2. T   3. F   4. F   5. F   6. F   7. T   8. T   9. T   10. F   11. F
12. T   13. T   14. T   15. F   16. T   17. T   18. T   19. F   20. T   21. F
22. T   23. F   24. F   25. T

## Completion Questions

1. standard deviations   2. 25   3. *z* scores, percentiles   4. negative   5. deciles, quartiles   6. three   7. 70th   8. 99   9. left   10. center   11. equal   12. nine
13. equal   14. 75th   15. order

## Multiple-Choice Questions

1. (b)   2. (b)   3. (a)   4. (d)   5. (c)   6. (d)   7. (d)   8. (a)   9. (b)
10. (d)   11. (b)   12. (d)   13. (c)   14. (b)   15. (d)   16. (d)   17. (d)
18. (c)   19. (d)   20. (c)   21. (c)   22. (d)   23. (b)   24. (d)   25. (b)
26. (a)   27. (b)   28. (c)   29. (d)   30. (d)   31. (b)   32. (d)

# CHAPTER 5

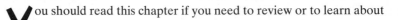

# Exploring Bivariate Data

**Do I Need to Read This Chapter?**

**Y**ou should read this chapter if you need to review or to learn about

➡ Graphical displays used to study the relationship between two variables

➡ The correlation coefficient as a numerical measure of the strength of the relationship between two variables

➡ Least-squares regression as a method for modeling the linear relationship between two variables

So far, you have dealt with univariate (single-variable) data. However, this chapter will introduce you to bivariate (two-variable) data. That is, you will be dealing with data that are associated with two variables. You will study the idea of association through graphical displays as well as through correlation analysis. In addition, you will study how to model the relationship between the two variables through regression analysis.

## Get Started

The most common graphical display used to study the association between two variables is called a *scatter plot*. A measure of association for bivariate data is a number that is meant to convey the idea of the strength of the relationship between the two variables. The most commonly used measure of association is called the *correlation coefficient*. In addition, *regression analysis* will allow us to propose a mathematical model for the association. We can use such models to make predictions for one variable given a value of the other variable. We will, however, restrict our discussions to linear models. Concepts relating to these topics are discussed in this chapter.

## 5-1 Scatter Plots

In simple correlation and regression studies, data are collected on two quantitative variables to determine whether a relationship exists between the two variables.

***Example 5-1:*** The bivariate data given in **Table 5-1** relate the high temperature (°F) reached on a given day and the number of cans of soft drinks sold from a particular vending machine in front of a grocery store. Data were collected for 15 different days.

**Table 5-1:** High-Temperature Data for Example 5-1

| TEMPERATURE | QUANTITY | TEMPERATURE | QUANTITY | TEMPERATURE | QUANTITY |
|---|---|---|---|---|---|
| 70 | 30 | 98 | 59 | 90 | 53 |
| 75 | 31 | 72 | 33 | 95 | 56 |
| 80 | 40 | 75 | 38 | 98 | 62 |
| 90 | 52 | 75 | 32 | 91 | 51 |
| 93 | 57 | 80 | 45 | 98 | 58 |

We would like to study graphically the association between the temperature and the number of cans of soft drinks sold.

To graphically analyze the data, we can display the data on a two-dimensional graph. We can plot the number of cans of soft drinks along the vertical axis, and the temperature along the horizontal axis. Such plots are called **scatter plots.** The variable along the vertical axis is called the *dependent variable,* and the variable along the horizontal axis is called the *independent variable.*

**Notation:** We will let $y$ represent the dependent variable, and we will let $x$ represent the independent variable.

**Explanation of the term—scatter plot:** A scatter plot is a graph of the ordered pairs $(x, y)$ of values for the independent variable $x$ and the dependent variable $y$.

Observe that the number of cans of soft drinks sold from the machine is a *function* of the temperature. Thus, in **Example 5-1,** the dependent variable will be the number of cans of soft drinks sold, and the independent variable will be the temperature.

- - - - - - - - - - - - - - - - - - - - - - - - - - - - - - - - - - - - - - - - - -

## Quick Tip

In a scatter plot, we let the values of the dependent variable be along the vertical (*y*) axis, and the values of the independent variable be along the horizontal (*x*) axis.

- - - - - - - - - - - - - - - - - - - - - - - - - - - - - - - - - - - - - - - - - -

The scatter plot for **Example 5-1** is displayed in **Fig. 5-1.**

We can observe from the plot that the number of cans of soft drinks sold increases as the temperature increases, and that there seems to be a linear trend for this association.

**Fig. 5-1:** Scatter plot for data in Example 5-1

## 5-2  Looking for Patterns in the Data

Detecting an association or a relationship for bivariate data starts with a scatter plot. When examining a scatter plot, one should try to answer the following:

- Is there a straight-line pattern or association?
- Does the pattern or association slope upward or slope downward?
- Are the plotted values tightly clustered together in the pattern or widely separated?
- Are there noticeable deviations from the pattern?

**Quick Tips**

1. Two variables are said to be *positively related* if larger values of one variable tend to be associated with larger values of the other.

2. Two variables are said to be *negatively related* if larger values of one variable tend to be associated with smaller values of the other.

The scatter plots presented in **Figs. 5-2** through **5-7** display different patterns. In **Fig. 5-2**, the variables are positively related, since larger values of the dependent variable are associated with larger values of the independent variable, and vice versa. The values are on a straight line, and therefore one can say that there is a *perfect positive association* between the variables. Perfect association rarely occurs when sample data are collected.

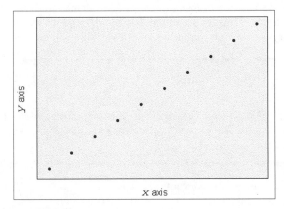

**Fig. 5-2:**  Perfect positive association

**Figure 5-3** shows variables that are negatively related, since larger values of the dependent variable are associated with smaller values of the independent variable, and vice versa. The values are on a straight line, and therefore one can say that there is a *perfect negative association* between the variables. Again, perfect association rarely occurs with sample data.

**Figure 5-4** shows variables that are positively related. The values are not on a straight line, but they are somewhat closely packed together in a linear manner, and so one can say that there is a very strong positive association between the variables.

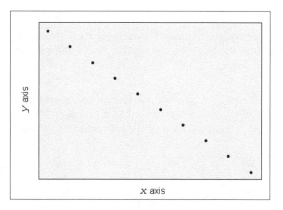

**Fig. 5-3:** Perfect negative association

**Fig. 5-4:** Very strong positive association

    **Figure 5-5** shows variables that are negatively related. The values are not on a straight line, but they are relatively closely packed together in a somewhat linear pattern, and so one can say that there is a very strong negative association between the variables.

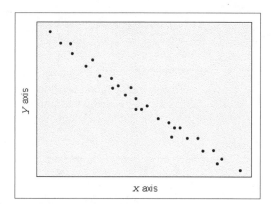

**Fig. 5-5:** Very strong negative association

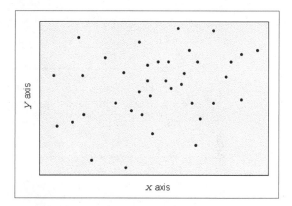

**Fig. 5-6:** No association

    **Figure 5-6** does not show any noticeable pattern. The values are scattered around, and so one can say that there is very little association between the variables.

    **Figure 5-7** does not show any noticeable linear pattern. The scatter plot displays a nonlinear or curvilinear relationship. We will not study such relationships in this text but will concentrate only on linear relationships between two variables.

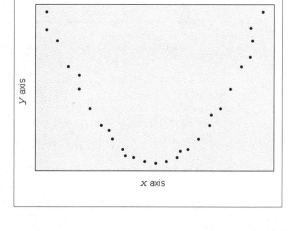

**Fig. 5-7:** Nonlinear association

## 5-3 Correlation

So far, you have seen how a scatter plot can provide a visual of the association between two variables. Here, we will discuss a numerical measure of the association between two variables called the *Pearson product moment correlation coefficient* or simply the *correlation coefficient.*

**Explanation of the term—sample correlation coefficient:** The **sample correlation coefficient** measures the strength and direction of a relationship between two variables using sample data. The sample correlation coefficient is denoted by the letter $r$ and is computed from the following equation:

$$r = \frac{n(\sum xy) - (\sum x)(\sum y)}{\sqrt{[n(\sum x^2) - (\sum x)^2] \times [n(\sum y^2) - (\sum y)^2]}}$$

where $n$ is the number of $(x, y)$ data pairs.

**Note:** This is only one way to compute the value of the correlation coefficient. There are several other ways to find the value of $r$ that will not be discussed in this text.

***Example 5-2:*** Compute the correlation coefficient for the following set of observations for the independent variable $x$ and the dependent variable $y$.

| $x$ | 8 | 4 | 5 | −1 |
|---|---|---|---|---|
| $y$ | −2 | 0 | 2 | 6 |

**Solution:** The formula may look intimidating, but we can construct a table, as shown in **Table 5-2,** to help with the computations. We can use a table such as **Table 5-2** to obtain the different sums in the formula.

**Table 5-2:** Table to Help with the Computation of $r$

| $x$ | $y$ | $x \cdot y$ | $x^2$ | $y^2$ |
|---|---|---|---|---|
| 8 | −2 | −16 | 64 | 4 |
| 4 | 0 | 0 | 16 | 0 |
| 5 | 2 | 10 | 25 | 4 |
| −1 | 6 | −6 | 1 | 36 |
| $\sum x = 16$ | $\sum y = 6$ | $\sum x \cdot y = -12$ | $\sum x^2 = 106$ | $\sum y^2 = 44$ |

Using the values from **Table 5-2** and substituting into the formula, we obtain

$$r = \frac{4(-12) - (16)(6)}{\sqrt{[4(106) - (16)^2] \times [4(44) - (6)^2]}} = -0.939$$

Following is a summary of the properties of the correlation coefficient.

### Properties of the Correlation Coefficient

- The range of the correlation coefficient is from −1 to +1.
- If there is a perfect positive linear relationship between the variables, the value of $r$ will be equal to +1. See **Fig. 5-2.**

- If there is a perfect negative linear relationship between the variables, the value of *r* will be equal to –1. See **Fig. 5-3.**
- If there is a strong positive linear relationship between the variables, the value of *r* will be close to +1. See **Fig. 5-4.**
- If there is a strong negative linear relationship between the variables, the value of *r* will be close to –1. See **Fig. 5-5.**
- If there is little or no linear relationship between the variables, the value of *r* will be close to 0. See **Fig. 5-6.**

**Figure 5-8** gives an idea about the range of the correlation coefficient *r*.

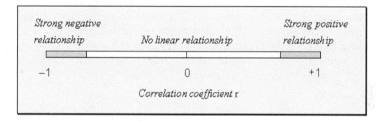

**Fig. 5-8:** Range of the correlation coefficient *r*

**Explanation of the term—population correlation coefficient:** The **population correlation coefficient** measures the strength and direction of a relationship between two variables using population data values. The population correlation coefficient is denoted by the Greek letter $\rho$ (read as "rho") and is computed by using all possible pairs of data values $(x, y)$ taken from the population.

The same formula that is used to compute the sample correlation coefficient is used, except that all possible pairs of values $(x, y)$ from the population are now utilized.

**Quick Tip**

One should always examine the scatter plot and not just rely on the value of the linear correlation coefficient. This measure will not detect curvilinear or other types of complex relationships. That is, there may be a relationship between two variables even though the correlation is close to zero.

## 5-4 Correlation and Causation

It is important to understand the nature of the relationship between the independent variable *x* and the dependent variable *y*. Listed are some possibilities that one should consider.

- There may be a direct cause-and-effect relationship between the two variables. For example, *x* may cause *y*. To illustrate, lack of water causes dehydration, intensive exercise causes thirst, heat causes ice cream to melt, etc.
- There may be a reverse cause-and-effect relationship between the two variables. For example, *y* causes *x*. To illustrate, one may believe that bad grades may be caused by

absences, but one should not fail to also consider the fact that bad grades may cause absences.

- The relationship may be due to chance or coincidence. To illustrate, one may find a relationship between the number of suicides and the increase in the sale of bagels. One can only conclude that any association between these two variables must be due to chance.

- The relationship may be due to confounding. That is, the relationship may be due to the interrelationships among several variables.

The next illustration shows the distinction between association and causation. For example, a large correlation (negative or positive) does not imply causation. Suppose that a high correlation is observed between the weekly sales of hot chocolate and the number of skiing accidents. One can reasonably conclude that hot chocolate sales could not cause skiers to have accidents while skiing, and that more skiing accidents could not cause an increase in sales of hot chocolate. Since the two variables are not actually related, what could explain such a relationship? The apparent relationship between the two variables may be caused by a third variable. In this case, the variables may be related to the weather conditions during the winter months. This is illustrated in **Fig. 5-9.**

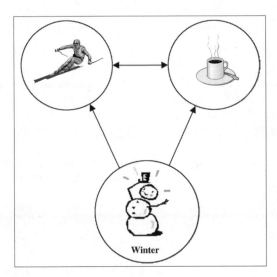

**Fig. 5-9:** Correlation and causation

### 5-5 Least-Squares Regression Line

In investigating the relationship between two variables, the first thing one should do is to prepare a scatter plot after the data are collected. From the plot, one can observe any pattern. If the correlation coefficient is reasonably large (positive or negative), the next step would be to fit the regression line which best fits or models the data. One of the purposes for the model is to help make predictions.

### Line of Best Fit

The scatter plot in **Fig. 5-10** shows two possible straight lines that may be used to model the data. The question is, Which of these lines best represents the association between the two variables?

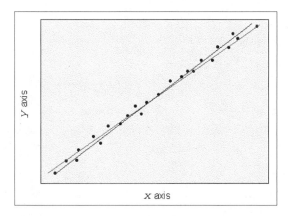

y axis

x axis

**Fig. 5-10:** Line of best fit

Regression analysis allows us to determine which of the two lines best represents the relationship. From algebra, the equation of a straight line is usually given by $y = mx + b$, where $m$ is the slope of the line and $b$ is the $y$ intercept. In elementary statistics, the equation of the regression line is usually written as $\hat{y} = ax + b$, where $a$ is the slope, $b$ is the $y$ intercept, and $\hat{y}$ is read as "$y$ hat," and it gives the predicted $y$ value for a given $x$ value. Least-squares analysis allows us to determine values for $a$ and $b$ such that the equation of the regression line best represents the relationship between the two variables by minimizing the error sum of squares—that is, by minimizing $\sum (y - \hat{y})^2$, where $(y - \hat{y})$ is the error for a given $y$ value. This regression line is usually called the *line of best fit.* We usually refer to this type of regression analysis as *simple regression analysis,* since we are dealing only with straight-line models involving one independent variable. The equations that one can use to compute the values for $a$ and $b$ are

$$a = \frac{n(\sum xy) - (\sum x)(\sum y)}{n(\sum x^2) - (\sum x)^2}$$

$$b = \frac{(\sum y)(\sum x^2) - (\sum x)(\sum xy)}{n(\sum x^2) - (\sum x)^2}$$

Other forms of the equations are possible.

***Example 5-3:*** Determine the equation for the line of best fit for the information in **Example 5-2.**

**Solution:** From **Example 5-2,** we have that $n = 4$, $\sum x = 16$, $\sum y = 6$, $\sum xy = -12$, $\sum x^2 = 106$, and $\sum y^2 = 44$. Substituting into the formulas above gives, to three decimal places,

$$a = \frac{(4)(-12) - (16)(6)}{4(106) - (16)^2} = -0.857 \quad \text{and} \quad b = \frac{(6)(106) - (16)(-12)}{4(106) - (16)^2} = 4.929$$

Thus, the line of best fit is given by $\hat{y} = -0.857x + 4.929$. The line of best fit superimposed on the scatter plot is shown in **Fig. 5-11.** Observe that the slope of the line is negative.

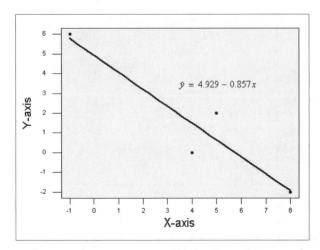

**Fig. 5-11:** Display of the line of best fit for Example 5-3

***Example 5-4:*** Interpret the value for the slope from **Example 5-3.**

**Solution:** Recall that the equation for the line of best fit was $\hat{y} = -0.857x + 4.929$, and so the slope is $-0.857$. We can interpret this value as saying that for a 1-unit *increase* in the $x$ (independent-variable) value, there will be a *decrease* (because of the negative value) of 0.857 unit in the $y$ (dependent-variable) value.

***Example 5-5:*** Predict the value of $y$ when $x = 4$ for the regression model in **Example 5-3.**

**Solution:** All one needs to do is to substitute $x = 4$ in the equation for the line of best fit and compute. Thus, $\hat{y} = -0.857 \times 4 + 4.929 = 1.501$. That is, the predicted value of $y$ is 1.501 when the independent variable $x = 4$. Note that the ordered pair (4, 1.501) will be a point located on the line of best fit. Observe that the value of 4 lies in the experimental range for the $x$ values.

**Quick Tip**

When using the line of best fit to make predictions, care must be taken to use independent values that are within the range of the observed independent variable. Using values outside the range of observed independent values may lead to incorrect predictions because we do not know how the model is behaving outside this range. The model reflects the behavior of the association between the two variables only within the range of the observed values.

### 5-6  The Coefficient of Determination

The **coefficient of determination** is a measure that allows us to determine how certain one can be in making predictions with the line of best fit.

**Explanation of the term—coefficient of determination:** The coefficient of determination measures the proportion of the variability in the dependent variable ($y$ variable) that is explained by the regression model through the independent variable ($x$ variable).

- The coefficient of determination is obtained by squaring the value of the correlation coefficient.
- The symbol used is $r^2$.
- Note that $0 \le r^2 \le 1$.
- $r^2$ values close to 1 would imply that the model is explaining most of the variation in the dependent variable and may be a very useful model.
- $r^2$ values close to 0 would imply that the model is explaining little of the variation in the dependent variable and may not be a useful model.

**Example 5-6:** What is the value of the coefficient of determination for the model in **Example 5-3**?

**Solution:** Recall that the correlation coefficient $r = -0.939$. Thus, the coefficient of determination $r^2 = (-0.939)^2 = 0.882$ or 88.2 percent. That is, the regression model can explain about 88.2 percent of the variation in the $y$ values. This would be a reasonable model to use for prediction because of the large $r^2$ value.

### 5-7 Residual Plots

Residuals are just errors. In particular, a residual is the difference between an actual observed $y$ value and the corresponding predicted $y$ value. Thus, the error for any observation is given by

$$\boxed{\text{Residual} = \text{error} = (\text{observed} - \text{predicted}) = (y - \hat{y})}$$

Plots of residuals may display patterns that would give some idea about the appropriateness of the model. If the functional form of the regression model is incorrect, the residual plots constructed by using the model will often display a pattern. The pattern can then be used to propose a more appropriate model.

The residual plot shown in **Fig. 5-12** displays a linear pattern. Such a residual plot would imply that a linear model is the appropriate model for predicting the dependent $y$ values.

The residual plot shown in **Fig. 5-13** displays a nonlinear pattern. If a linear model were used to generate these residuals, this would imply that a nonlinear model is the appropriate model to use for predicting $y$ instead of the linear model.

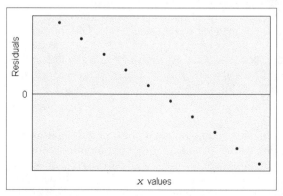

**Fig. 5-12:** Linear residual plot

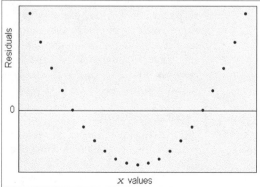

**Fig. 5-13:** Nonlinear residual plot

It is possible to have other curved patterns when residuals are plotted. However, if a linear model is used to generate the residuals, one should reevaluate the model and adjust for the curvature.

## 5-8  Outliers and Influential Points

A value that is well separated from the rest of the data set is called an *outlier*. With respect to the line of best fit, an outlier is an observation with a large absolute residual value. That is, an outlier will fall far from the regression line and will not follow the pattern of the linear relationship expressed by the line of best fit. In **Chap. 4,** we discussed how to test to determine whether a value in a univariate data set can be considered to be an outlier. An observation that causes the values of the slope and the intercept in the line of best fit to be considerably different from what they would be if the observation were removed from the data set is said to be *influential*.

Consider the scatter plot in **Fig. 5-14.** We will use it to discuss the concept of an outlier and an influential point in the context of regression.

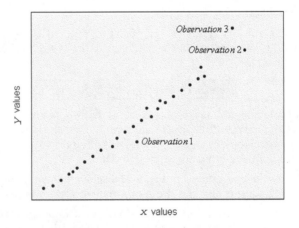

**Fig. 5-14:**  Plot illustrating outliers and influential points

Observations 1 and 2 can be considered to be outliers. Observation 1 is an outlier with respect to its *y* value but not with respect to its *x* value. Its *x* value is near the center of the other *x* values, but its *y* value is not consistent with the linear relationship. Observation 2 is an outlier with respect to its *x* value but not with respect to its *y* value. Its *x* value is outside the range of the majority of the *x* values, but its *y* value is consistent with the linear relationship. Observation 3 is probably an influential point. It is an outlier with respect to its *x* value, and its *y* value is not consistent with the linear relationship.

Technology

## Technology Corner

All the concepts discussed in this chapter can be computed and illustrated using most statistical software packages. Most scientific calculators will not directly compute the correlation coefficient, compute the slope and the intercept for the line of best fit, and do scatter plots. However, some of the newer calculators with extensive statistical and graphical capabilities will directly compute the correlation coefficient and the slope and the intercept for the regression line and display scatter plots. If you own a calculator, you should consult the manual to determine what statistical features are included.

Illustration: **Figure 5-15** shows the regression analysis output computed by the MINITAB software. **Figure 5-16** shows the LinReg output computed by the TI-83 calculator. The data used, in both cases, were from **Example 5-2.** Observe that MINITAB has given the least-squares equation and the coefficient of determination, R-Sq ($r^2$). Also, MINITAB computes other statistics that will not be discussed in this text. The TI-83 calculator output gives the values for the slope, the intercept, the correlation coefficient, and the coefficient of determination. Note: Care should always be taken when using the formulas to compute the values for *a* and *b*. One can use other features of the technologies to illustrate other concepts discussed in the chapter.

```
Regression Analysis


The regression equation is

  y = 4.93 - 0.857 x

Predictor        Coef      SE Coef         T        P
Constant        4.929        1.143      4.31    0.050
X-axis        -0.8571       0.2221     -3.86    0.061


S = 1.439      R-Sq = 88.2%     R-Sq(adj) = 82.2%

Analysis of Variance

Source           DF           SS          MS        F        P
Regression        1       30.857      30.857    14.90    0.061
Residual Error    2        4.143       2.071
Total             3       35.000
```

**Fig. 5-15:** MINITAB regression output for Example 5-2

```
LinReg
y=ax+b
a=-.8571428571
b=4.928571429
r²=.8816326531
r=-.9389529557
```

**Fig. 5-16:** TI-83 LinReg output for Example 5-2

**Note:** MINITAB also uses the form $\hat{y} = ax + b$ for the line of best fit. Other texts and technology may use the form $\hat{y} = a + bx$.

*It's a Wrap*

Linear relationships can be investigated through

✔ Scatter plots
✔ Correlation coefficients
✔ Lines of best fit
✔ Coefficients of determination

Care should always be taken when using the line of best fit to model data. One should be aware of outliers and influential points. One should make use of scatter plots and residual plots to determine whether the model is appropriate. One should be clear as to whether there is an association between two variables and be sure that causation is not being misinterpreted for association.

## True/False Questions

1. In simple regression analysis, if the slope of the line is positive, then there is a positive correlation between the dependent variable $y$ and the independent variable $x$.

2. In simple regression analysis, if the $y$ intercept is negative, then there is a negative correlation between the dependent variable $y$ and the independent variable $x$.

3. The variable that is being predicted in regression analysis is the independent variable.

4. The method of least squares provides the best approximation for the straight-line model that relates the dependent variable $y$ and the independent variable $x$.

5. The correlation coefficient $r$ is always between $-2$ and $+2$.

6. A correlation coefficient of 0.95 indicates that the observations are widely scattered about the regression line.

7. A correlation of almost zero indicates that the strength of the relationship between the dependent and independent variables is very weak.

8. The coefficient of determination can assume negative values.

9. If the least-squares equation relating the independent variable $y$ and the dependent variable $x$ for a given problem is $y = 2x + 5$, then an increase of 1 unit in $x$ is associated with an increase of 2 units in $y$.

10. In regression analysis, the units for the dependent and independent variables will always be the same.

11. A negative correlation between the dependent variable $y$ and the independent variable $x$ indicates that large values of $x$ are associated with small values of $y$.

12. A scatter plot is a graphical display of the dependent and independent variables under study.

13. If the correlation between the dependent and independent variables is $+1$, then the slope of the regression line will also be $+1$.

14. If there is no correlation between the independent and dependent variables, then the value of the correlation coefficient must be $-1$.

15. Causation and correlation explain the same concept in that they both measure the strength of the linear relationship between two variables.

16. The sample correlation coefficient has possible values ranging from $-1$ to $+1$.

17. When data are measured on two variables to determine a possible linear association between these variables, the set of data is called bivariate data.

18. If the least-squares equation between the independent variable $y$ and the dependent variable $x$ for a given set of bivariate data is $y = 2x + 5$, then an increase of 2 units in $x$ is associated with a 1-unit increase in $y$.

19. The coefficient of determination measures the variation in the dependent variable that is explained by the regression model.

20. The least-squares equation minimizes the error sum of squares.

21. If the slope of the regression equation is positive, then the correlation between the dependent variable $y$ and the independent variable $x$ must be 1.

22. If your computed correlation coefficient $r = +1.2$, then you have better than a perfect positive correlation.

23. The variable that is being predicted in regression analysis is the dependent variable.

24. A student might expect that there is a positive correlation between the age of his or her computer and its resale value.

25. Simple regression analysis is used when several independent variables contribute to the variation of the dependent variable.

## Completion Questions

1. A value of the correlation coefficient equal to (0, +1, –1) _____ indicates that there is a perfect negative relationship between the dependent variable $y$ and the independent variable $x$.

2. A negative correlation coefficient between the dependent variable $y$ and the independent variable $x$ indicates that large values of $x$ are associated with (large, small) _____ values of $y$.

3. The sample correlation coefficient has values ranging from _____ to _____ .

4. In simple linear regression analysis, if the $y$ intercept is positive, then the slope of the line of best fit must be (positive, negative, zero, any of the previous three responses) _____ .

5. In regression analysis, the variable that is being predicted is called the (independent, dependent) _____ variable.

6. Correlation analysis is used to determine the _____ of the relationship between the dependent and the independent variables.

7. When there is not a significant relationship between the dependent variable $y$ and the independent variable $x$, the value of the correlation coefficient will be approximately (–2, –1, 0, +1, +2) _____ .

8. You should expect a (positive, negative, zero) _____ correlation between the age of your computer and the resale value of your computer.

9. Which of the following values (–1, 0.66, 0, 1, –1.01, –0.78) cannot be a value of the correlation coefficient? _____

10. If the computed value of the correlation coefficient is 0.71, then the slope of the least-squares line will be (positive, negative, zero) _____ .

11. A correlation coefficient of 0.97 indicates that the observations are (closely, widely) _____ scattered about the regression line.

12. When data are measured on two variables to determine whether there is any association between them, this kind of data is called _____ data.

13. For the least-squares equation $\hat{y} = -x + 1$, the correlation between $x$ and $y$ is (1, –1, positive, negative) _____ .

14. The least-squares equation (minimizes, maximizes) _____ the error sum of squares.

15. Generally speaking, the larger the correlation (either positive or negative) between the independent variable $x$ and the dependent variable $y$ for the simple linear regression model, the (better, worse) _____ will be the predictions of $y$ for given values of $x$.

16. The value of the coefficient of determination lies between (–1 and +1, –1 and 0, 0 and +1, –0.5 and +0.5) _____ .

17. The coefficient of determination can be obtained by squaring the value of the _____ .

18. Regression analysis is used to find the (strength, model) _____ of the linear relationship between the independent and dependent variables.

19. If the slope of the line of best fit is +2, this implies that for a 1-unit increase in the $x$ value, there will be a 2-unit (increase, decrease) _____ in the $y$ value.

20. Without a scatter plot, and with a correlation close to zero, can one say for certain that there is no relationship between the variables? (yes, no) _____

## Multiple-Choice Questions

1. In simple linear regression analysis with $x$ representing the independent variable and $y$ representing the dependent variable, if the $y$ intercept is negative, then
   (a) the correlation between $x$ and $y$ is negative.
   (b) the correlation between $x$ and $y$ is positive.
   (c) the correlation between $x$ and $y$ could be either negative, positive, or zero.
   (d) the value of the predicted $y$ value is always negative.

2. In regression analysis, the input variable that is used to get a predicted value is
   (a) the dependent variable.
   (b) the independent variable.
   (c) the least-squares variable.
   (d) the random variable.

3. In the simple linear regression model with $x$ representing the independent variable and $y$ representing the dependent variable, correlation analysis is used to
   (a) find the least-squares regression line.
   (b) find the slope of the regression line.
   (c) measure the strength of the linear relationship between $x$ and $y$.
   (d) draw a scatter plot.

4. If the correlation coefficient is zero, the slope of a linear regression line will be
   (a) positive.
   (b) negative.
   (c) positive or negative.
   (d) none of the above.

5. In the simple linear regression model, if there is a very strong correlation between the independent and dependent variables, then the correlation coefficient should be
   (a) close to $-1$.
   (b) close to $+1$.
   (c) close to either $-1$ or $+1$.
   (d) close to zero.

6. For the simple linear regression model, if all the points on a scatter plot lie on a straight line with correlation coefficient $r = -1$, then the slope of the regression line is
   (a) $-1$.
   (b) $+1$.
   (c) positive.
   (d) negative.

7. The least-squares equation for the line of best fit
   (a) minimizes the error sum of squares.
   (b) maximizes the error sum of squares.
   (c) does not change the error sum of squares.
   (d) does none of the above.

8. If through some analysis, one can conclude that the slope of the line of best fit is not equal to zero, then the simple linear regression model indicates that there is

(a) a positive relationship between the independent and dependent variables.

(b) a negative relationship between the independent and dependent variables.

(c) a positive or negative relationship between the independent and dependent variables.

(d) no relationship between the independent and dependent variables.

9. Which of the following is not a possible value of the correlation coefficient?

(a) +1

(b) −1

(c) 0.011

(d) 1.11

10. A negative correlation coefficient between the dependent variable $y$ and the independent variable $x$ indicates that

(a) large values of $x$ are associated with small values of $y$.

(b) large values of $x$ are associated with large values of $y$.

(c) small values of $x$ are associated with small values of $y$.

(d) none of the above answers are correct.

11. For the simple linear regression model, if the unit for the dependent variable is square feet, then the unit for the independent variable

(a) must be square feet.

(b) can be some unit of square measurement.

(c) can be any unit.

(d) cannot be a unit of square measurement.

12. In simple linear regression analysis, there

(a) is only one independent variable in the model.

(b) could be several linear independent variables in the model.

(c) is only one nonlinear term in the model.

(d) is at least one nonlinear term in the model.

13. For the regression equation $\hat{y} = 2(1 - x)$, the correlation coefficient

(a) is +2.

(b) is −2.

(c) is −1.

(d) cannot be determined from the information given.

14. In the least-squares regression line, the desired sum of the errors (residuals) should be

(a) positive.

(b) negative.

(c) maximized.

(d) equal to zero.

15. Which of the following is associated with correlation and regression analyses?

(a) Least-squares

(b) Correlation coefficient

(c) Coefficient of determination

(d) All of the above

16. You are given the following set of observations for the independent variable $x$ and the dependent variable $y$:

| $x$ | -3 | -1 | 1 | 3 |
|-----|----|----|----|----|
| $y$ | 8 | 4 | 5 | -1 |

The correlation coefficient is
(a) –1.0.
(b) –0.8971.
(c) +1.
(d) 0.8971.

17. You are given the following set of observations for the independent variable $x$ and the dependent variable $y$:

| $x$ | -3 | -1 | 1 | 3 |
|-----|----|----|----|----|
| $y$ | 8 | 4 | 5 | -1 |

The coefficient of determination is
(a) –1.0.
(b) –0.8048.
(c) +1.
(d) 0.8048.

18. You are given the following set of observations for the independent variable $x$ and the dependent variable $y$:

| $x$ | -3 | -1 | 1 | 3 |
|-----|----|----|----|----|
| $y$ | 8 | 4 | 5 | -1 |

The least-squares estimate of the slope of the regression line is
(a) +4.0.
(b) –1.3.
(c) –0.9.
(d) –4.0.

19. You are given the following set of observations for the independent variable $x$ and the dependent variable $y$:

| $x$ | -3 | -1 | 1 | 3 |
|-----|----|----|----|----|
| $y$ | 8 | 4 | 5 | -1 |

The least-squares estimate for the $y$ intercept of the regression line is
(a) –1.3.
(b) +4.
(c) –0.9.
(d) +1.3.

20. You are given the following set of observations for the independent variable $x$ and the dependent variable $y$:

| $x$ | −3 | −1 | 1 | 3 |
|---|---|---|---|---|
| $y$ | 8 | 4 | 5 | −1 |

The least-squares linear regression equation is
   (a) $\hat{y} = -1.3 + 4x$.
   (b) $\hat{y} = 4.0 - 1.3x$.
   (c) $\hat{y} = -1.3 - 0.8971x$.
   (d) $\hat{y} = -0.897 - 1.3x$.

21. You are given the following set of observations for the independent variable $x$ and the dependent variable $y$:

| $x$ | −3 | −1 | 1 | 3 |
|---|---|---|---|---|
| $y$ | 8 | 4 | 5 | −1 |

The predicted value $\hat{y}$ of the dependent variable $y$ when $x = 2$ is
   (a) 6.7.
   (b) 1.4.
   (c) −3.094.
   (d) −3.497.

22. Given the following information:
$$\sum x = 24, \sum y = 16, \sum x^2 = 180, \sum y^2 = 90, \sum xy = 75, n = 10$$
The correlation coefficient will be
   (a) 0.4122.
   (b) 0.1700.
   (c) 0.2990.
   (d) 0.5683.

23. Given the following information:
$$\sum x = 24, \sum y = 16, \sum x^2 = 180, \sum y^2 = 90, \sum xy = 75, n = 10$$
The coefficient of determination will be
   (a) 0.3230.
   (b) 0.0894.
   (c) 0.0289.
   (d) 0.1699.

24. Given the following information:
$$\sum x = 24, \sum y = 16, \sum x^2 = 180, \sum y^2 = 90, \sum xy = 75, n = 10$$
The least-squares estimate of $a$ is
   (a) 0.4773.
   (b) 0.2990.

(c) 0.2061.

(d) 0.9265.

25. Given the following information:

$$\sum x = 24, \sum y = 16, \sum x^2 = 180, \sum y^2 = 90, \sum xy = 75, n = 10$$

The least-squares estimate of $b$ is

(a) 2.3176.

(b) 0.7176.

(c) 0.8824.

(d) 1.8990.

26. Given the following information:

$$\sum x = 24, \sum y = 16, \sum x^2 = 180, \sum y^2 = 90, \sum xy = 75, n = 10$$

The least-squares regression equation is

(a) $\hat{y} = 0.299 + 0.8824x$.

(b) $\hat{y} = 1.899 + 0.9265x$.

(c) $\hat{y} = 0.7176 + 0.2061x$.

(d) $\hat{y} = 0.8824 + 0.299x$.

27. Given the following information:

$$\sum x = 24, \sum y = 16, \sum x^2 = 180, \sum y^2 = 90, \sum xy = 75, n = 10$$

The predicted value $\hat{y}$ of the dependent variable $y$ when $x = 2$ is

(a) 1.4804.

(b) 1.1296.

(c) 3.752.

(d) 3.2722.

## Further Exercises

If possible, you can use any technology help available to solve the following problems.

1. The scores $x$ on a pretest for a college algebra course and the course grade $y$ were recorded for 10 students. The results are given in **Table 5-3.**

**Table 5-3**

| x | 75 | 81 | 57 | 79 | 68 | 93 | 96 | 84 | 41 | 89 |
|---|----|----|----|----|----|----|----|----|----|----|
| y | 2  | 3  | 1  | 2  | 1  | 4  | 4  | 3  | 1  | 3  |

(a) Present a scatter plot for the data.

(b) Determine the correlation coefficient for the data and interpret the value.

(c) Determine the coefficient of determination and interpret the value.

(d) Compute the least-squares estimate for $a$.

(e) Compute the least-squares estimate for $b$.

(f) State the least-squares regression line.

(g) Find $\hat{y}$ for $x = 60$.

2. Engineers for a car manufacturer wanted to analyze the relationship between the speed $x$ of their new model (The Bullet) and its gas mileage $y$ (in mpg) for regular unleaded gasoline. The car was test driven at different speeds in the laboratory and the data in **Table 5-4** were obtained.

**Table 5-4**

| Speed, $x$ | 30 | 40 | 50 | 60 | 70 | 80 | 90 |
|---|---|---|---|---|---|---|---|
| MPG, $y$ | 39 | 38 | 36 | 32 | 27 | 24 | 22 |

(a) Present a scatter plot for the data.

(b) Determine the correlation coefficient for the data and interpret the value.

(c) Determine the coefficient of determination and interpret the value.

(d) Compute the least-squares estimate for $a$.

(e) Compute the least-squares estimate for $b$.

(f) State the least-squares regression line.

(g) Find $\hat{y}$ for $x = 65$.

3. Twelve people who were advised by their physicians to lose weight for health reasons enrolled in a special weight loss program. **Table 5-5** gives the time in the program ($x$, in days) and the weight lost in the program ($y$, in pounds).

**Table 5-5**

| $x$ | 30 | 41 | 16 | 32 | 54 | 43 | 68 | 91 | 15 | 13 | 59 | 90 |
|---|---|---|---|---|---|---|---|---|---|---|---|---|
| $y$ | 2.9 | 4.5 | 1.8 | 3.6 | 6.8 | 4.7 | 11 | 13.6 | 1.6 | 1.5 | 7.8 | 14.2 |

(a) Present a scatter plot for the data.

(b) Determine the correlation coefficient for the data and interpret the value.

(c) Determine the coefficient of determination and interpret the value.

(d) Compute the least-squares estimate for $a$.

(e) Compute the least-squares estimate for $b$.

(f) State the least-squares regression line.

(g) Find $\hat{y}$ for $x = 50$.

4. In a given community, a survey was conducted to determine whether there is any relationship between the size of one's income $x$ (in thousands of dollars) and the size of one's home $y$ (in square feet). The data in **Table 5-6** were collected for 10 sample points.

**Table 5-6**

| $x$ | 41.2 | 68.3 | 22.4 | 56.7 | 42.2 | 86.1 | 50.3 | 35.7 | 44.4 | 47.5 |
|---|---|---|---|---|---|---|---|---|---|---|
| $y$ | 2.9 | 3.5 | 2.5 | 3.1 | 3.3 | 4 | 3.7 | 2.9 | 3 | 3.1 |

(a) Present a scatter plot for the data.
(b) Determine the correlation coefficient for the data and interpret the value.
(c) Determine the coefficient of determination and interpret the value.
(d) Compute the least-squares estimate for *a*.
(e) Compute the least-squares estimate for *b*.
(f) State the least-squares regression line.
(g) Find $\hat{y}$ for $x = 40$.

## ANSWER KEY
### True/False Questions
1. T   2. F   3. F   4. T   5. F   6. F   7. T   8. F   9. T   10. F   11. T
12. T   13. F   14. F   15. F   16. T   17. T   18. F   19. T   20. T   21. F
22. F   23. T   24. F   25. F

### Completion Questions
1. −1   2. small   3. −1; +1   4. any of the previous three responses   5. dependent
6. strength   7. 0   8. negative   9. −1.01   10. positive   11. closely
12. bivariate   13. negative   14. minimizes   15. better   16. 0 and +1
17. correlation coefficient   18. model   19. increase   20. no

### Multiple-Choice Questions
1. (c)   2. (b)   3. (c)   4. (d)   5. (c)   6. (d)   7. (a)   8. (c)   9. (d)
10. (a)   11. (c)   12. (a)   13. (d)   14. (d)   15. (d)   16. (b)   17. (d)
18. (b)   19. (b)   20. (b)   21. (b)   22. (a)   23. (d)   24. (b)   25. (c)
26. (d)   27. (a)

# CHAPTER 6

# Exploring Categorical Data

**Do I Need to Read This Chapter?**

You should read this chapter if you need to review or to learn about

➡ Two-way tables for a pair of categorical variables

➡ Marginal and conditional distributions of categorical variables

➡ Graphical displays for categorical variables

➡ Independence between categorical variables

➡ Simpson's paradox

In the previous chapter, we dealt with bivariate data for which the variables were quantitative. In this chapter, we will explore the relationship between categorical or qualitative variables.

## Get Started

When we are looking for associations between two qualitative variables, scatter plots will not work to help display any pattern. We use contingency tables to present the association between two or more qualitative or categorical variables. When there are just two qualitative variables, the table is usually called a two-way contingency table or a bivariate frequency table. Examples of categorical variables would be gender, ethnicity, and religious affiliation. These variables can assume values that are qualitative. Examples of values for these variables would be male, Asian, Methodist, etc. A quantitative variable like age can also be a categorical variable when data are classified into age groups. For example, ages 20–25 would be an example of an appropriate category in which to classify a 24 year old. Bar graphs will be used to display the relationship between the qualitative variables.

## 6-1  Marginal Distributions

*Example 6-1:* **Table 6-1** summarizes the information concerning the number of AIDS cases by age (in years) and race/ethnicity for females in the United States, for cases reported through December 1997. Information is presented only for whites, blacks, and Hispanics, and for three broad age groups.

**Table 6-1:**  Summary of the Number of AIDS Cases by Age and Ethnicity

| AGE/ETHNICITY | WHITE | BLACK | HISPANIC | TOTALS |
|---|---|---|---|---|
| 24 and younger | 2,192 | 6,496 | 2,362 | 11,050 |
| 25–49 | 18,200 | 46,614 | 16,726 | 81,540 |
| 50 and older | 2,690 | 4,430 | 1,710 | 8,830 |
| Totals | 23,082 | 57,540 | 20,798 | Grand total 101,420 |

Source: U.S. Department of Health and Human Services.

**Table 6-1** represents a two-way contingency table or a bivariate frequency table, since there are only two qualitative variables. This table is sometimes called a three-by-three (3 × 3) table, since we have three classifications for each of the two variables. The table shows how many observations are allocated to each category. Each row and column combination is called a *cell* in the table. The value of 2,690 in the first column for ethnicity and the third row for the age classification indicates that 2,690 out of the 101,420 females are white and are 50 and older. That is, about 2.65 percent of the total observed values were classified as white females who were 50 and older. Also, there were only 1,710 female Hispanics in the age group 50 and older. This would represent 1.67 percent of the total observed values. However, the 2,690 were out of a total of 23,082 white females, or 11.65 percent, while the 1,710 were out of a total of 20,798 Hispanic females, or 8.22 percent. Thus, the relationship between these two qualitative variables may be better analyzed and understood by using the appropriate percentages. We will explore the relationship between two qualitative variables by using marginal distributions and conditional distributions.

### Marginal Distributions

From the contingency table of frequencies, one can obtain the **marginal distributions** by computing the appropriate percentages.

**Explanation of the term—marginal distribution:**  A marginal distribution for a variable is the percentage of that variable expressed as the row or column totals relative to the grand total for the table.

To obtain the marginal distributions, divide the column or row totals by the grand total. These are usually expressed as percentages.

*Example 6-2:* Compute the marginal distributions for the ethnicity variable.

**Solution:** We need to divide the values of 23,082, 57,540, and 20,798 by the grand total of 101,420. The marginal distributions, in percents, for the classification variable of ethnicity are given in **Table 6-2.**

**Table 6-2:**  Marginal Distributions for Ethnicity

| WHITE | BLACK | HISPANIC | TOTAL |
|---|---|---|---|
| 22.76% | 56.73% | 20.51% | 100% |

Observe that the sum of the marginal percentages equals 100 percent. From the marginal distribution, one can observe that 22.76 percent of the females with AIDS were whites, 56.73 percent were blacks, and 20.51 percent were Hispanics. Also, one can observe that blacks outnumbered whites and Hispanics more than two to one for the data collected.

***Example 6-3:*** Compute the marginal distributions for the age group variable.

**Solution:** We need to divide the values of 11,050, 81,540, and 8,830 by the grand total of 101,420. The marginal distributions, in percentages, for the classification variable of age are given in **Table 6-3.**

Observe that the sum of the marginal percentages equals 100 percent. From the marginal distributions, one can observe that 10.9 percent of the females were 24 years of age and

**Table 6-3:** Marginal Distributions for Age

| | |
|---|---|
| 24 and younger | 10.9% |
| 25–49 | 80.4% |
| 50 and older | 8.7% |
| Total | 100% |

younger, 80.4 percent of the females were between 25 and 49 years of age, and 8.7 percent were 50 years of age and older. Also, one can observe that a very large proportion of the data was from the age classification 25–49. One may want to further subdivide the age groupings to obtain and analyze other properties for the marginal distributions.

## 6-2 Conditional Distributions

From the contingency table, one can obtain the distribution of one variable given the other variable. For example, one may be interested in finding the proportion of female AIDS cases who are 50 or older, given that the female is white. In this case, we are looking for a row classification given a column classification. One can also consider a column classification given a row classification. The proportions computed from such analysis are called **conditional distributions.**

### Conditional Distributions

From the contingency table of frequencies, one can obtain the conditional distributions by computing the appropriate percentages.

**Explanation of the term—conditional distribution:** A conditional distribution for a (first) variable given another (second) variable is the percentage of items for the first variable that is contained in the second variable.

To obtain the conditional distributions for the row classifications given the column classifications, divide the frequency values in the original table by the column totals. These are usually expressed as percentages. The conditional distribution of the column variable given the row variable is obtained by dividing the frequency values in the original table by the row totals and expressing the results as percentages.

***Example 6-4:*** Compute the conditional distributions for the age classifications (row) given the ethnic classifications (column). Use one decimal place.

**Solution:** From the original two-way distribution, we need to compute each frequency entry as a percentage of the respective column totals.

For the class of 24 and younger, the conditional distribution for the entry 2,192, given the white ethnic classification, will be $\dfrac{2,192}{23,082} \times 100$ percent = 9.5 percent. For the class of 50 and older, the conditional distribution for the entry 1,710, given the Hispanic ethnic classification, will be $\dfrac{1,710}{20,798} \times 100$ percent = 8.2 percent. One can continue in this manner to compute the remaining conditional distributions for the rows given the columns. The conditional distributions are given in **Table 6-4.**

**Table 6-4:** Conditional Distributions for Age Given Ethnicity

| AGE/ETHNICITY | WHITE | BLACK | HISPANIC |
|---|---|---|---|
| 24 and younger | 9.5 | 11.3 | 11.4 |
| 25–49 | 78.8 | 81.0 | 80.4 |
| 50 and older | 11.7 | 7.7 | 8.2 |
| Totals | 100% | 100% | 100% |

***Example 6-5:*** Interpret the 81 percent cell value in **Table 6-4** for **Example 6-4.**

**Solution:** The value indicates that 81 percent of females with AIDS who are black are between 25 and 49 years of age.

***Example 6-6:*** Compute the conditional distributions for the ethnic classifications (column) given the age classifications (rows). Use one decimal place.

**Solution:** From the original two-way distribution, **Table 6-1,** we need to compute each frequency cell entry as a percentage of the respective row total.

For the ethnic class of whites, the conditional distribution for the entry 2,192, given the age group 24 and younger, will be $\dfrac{2,192}{11,050} \times 100$ percent = 19.8 percent. For the ethnic class of Hispanics, the conditional distribution for the entry 1,710, given the age group 50 and older, will be $\dfrac{1,710}{8,830} \times 100$ percent = 19.4 percent. One can continue in this manner to compute the remaining conditional distributions for the columns given the rows. The conditional distributions are given in **Table 6-5.**

**Table 6-5:** Conditional Distributions for Ethnicity Given Age

| AGE/ETHNICITY | WHITE | BLACK | HISPANIC | TOTAL |
|---|---|---|---|---|
| 24 and younger | 19.8 | 58.8 | 21.4 | 100% |
| 25–49 | 22.3 | 57.2 | 20.5 | 100% |
| 50 and older | 30.5 | 50.2 | 19.4 | 100% |

**Note:** The sum of the last row of conditional distributions does not add up to exactly 100 percent because of rounding of the percentages to one decimal place.

***Example 6-7:*** Interpret the 19.8 percent cell value in **Table 6-5** for **Example 6-6.**

**Solution:** The value indicates that 19.8 percent of females with AIDS who are 24 years of age and younger are white.

## 6-3  Using Bar Charts to Display Contingency Tables

The following bar charts display the different information given and derived for **Examples 6-1, 6-2, 6-3, 6-4,** and **6-6.** If you need to review the concept of bar charts, see **Chap. 1.**

The bar chart in **Fig. 6-1** displays the information for the original contingency table. It uses the raw frequencies to construct the chart. We can observe from **Fig. 6-1** that for this sample of females with AIDS, the classification 25–49-year-old black females has the highest frequency count. For the white and Hispanic females, whites slightly outnumber Hispanics for the 25–49 age group. For the other two age classifications, black females with AIDS are observed approximately twice as much as either of the other two ethnic groups.

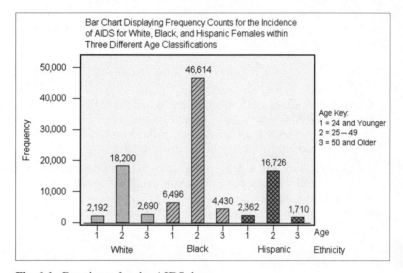

**Fig. 6-1:** Bar charts for the AIDS data

The next chart, **Fig. 6-2,** shows the marginal distributions for the ethnicity classifications. Observe that more than half of the data are related to black females.

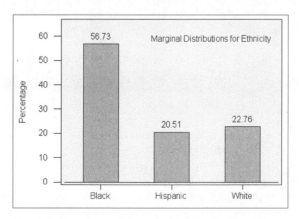

**Fig. 6-2:** Bar chart of the marginal distributions for ethnicity

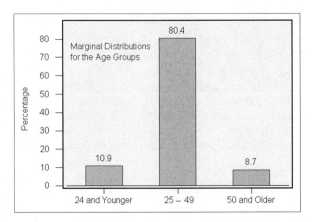

**Fig. 6-3:** Bar chart of the marginal distributions for age

The next chart, **Fig. 6-3,** shows the marginal distributions for the age group classifications for the females. Observe that a significant majority of the females are between the ages of 25 and 49 years. A further breakdown of the age classifications should be investigated; this might reveal other properties of the distributions.

The next chart, **Fig. 6-4,** presents the conditional distributions for the age classifications given the ethnic classifications. From the display, one can observe that there are small deviations within each age classification. Again, one can observe that the majority of the observations are in the 25–49 age range for all three ethnic groups.

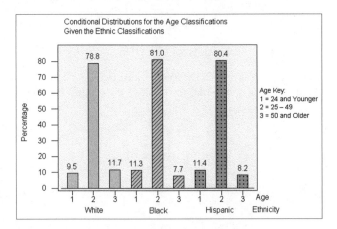

**Fig. 6-4:** Bar chart of the conditional distributions for age given ethnicity

The next chart, **Fig. 6-5,** presents the conditional distributions for the ethnic classifications given the age classifications. From the display, one can observe that within each age group, black females had the highest incidence of AIDS. Observe that the highest incidence of AIDS for white females occurs in the 50 and older age classification. There is not much variation in the incidence of AIDS for Hispanic females for all age classifications.

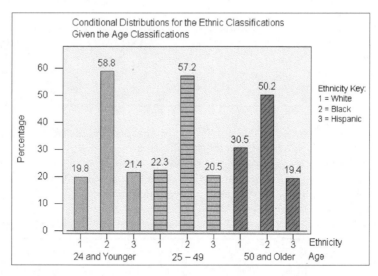

**Fig. 6-5:** Bar chart of the conditional distributions for ethnicity given age

## 6-4 Independence in Categorical Variables

Sometimes it is important to determine whether there is an association between the variables in a contingency table; that is, whether the variables are independent of each other or dependent. We will use the concept of conditional distributions for contingency tables to determine whether there is an association between the variables.

**Explanation of the term—independence:** Two categorical variables are said to be **independent** of each other (have no association) if the conditional distributions of one variable are the same for every category of the other variable.

***Example 6-8:*** A faculty member conducted a survey on a college campus to determine the favorability rating of the college president. One hundred randomly selected students were asked to indicate whether they viewed the president favorably or unfavorably. **Table 6-6** shows a $2 \times 2$ contingency table summarizing the results for both male and female students in the sample.

**Table 6-6:** Favorability Rating of the College President by Gender

| GENDER/RATING | FAVORABLE | UNFAVORABLE | TOTAL |
|---------------|-----------|-------------|-------|
| Male | 54 | 21 | 75 |
| Female | 18 | 7 | 25 |
| Total | 72 | 28 | 100 |

(a) Construct a table of the conditional distributions for the gender classification given the favorable/unfavorable ratings.

**Solution:** Since we are conditioning on the favorable/unfavorable ratings, we need to divide each cell entry by the column totals in the original $2 \times 2$ table. These are expressed as percentages. The conditional distributions are given in **Table 6-7.**

**Table 6-7:** Conditional Distributions of Gender Given Ratings

| GENDER/RATING | FAVORABLE | UNFAVORABLE |
|---|---|---|
| Male | 75 | 75 |
| Female | 25 | 25 |
| Total | 100% | 100% |

(b) Construct a table of the conditional distributions for the favorable/unfavorable ratings given the gender classification.

**Solution:** Since we are conditioning on the gender classification, we need to divide each cell entry by the row totals in the original $2 \times 2$ table. These are expressed as percentages. The conditional distributions are given in **Table 6-8.**

**Table 6-8:** Conditional Distributions of Ratings Given Gender

| GENDER/RATING | FAVORABLE | UNFAVORABLE | TOTAL |
|---|---|---|---|
| Male | 72 | 28 | 100% |
| Female | 72 | 28 | 100% |

(c) Based on the results in parts (a) and (b), can one conclude that the gender variable and the favorable/unfavorable ratings variable are independent? That is, is there no association between the male and female responses of favorable or unfavorable?

**Solution:** From part (a), the distributions of gender of the student were conditioned on the ratings. From the computed conditional distributions for gender, we see that the values are the same for the male classification for both the favorable and unfavorable classifications. Also, for the female classification, the conditional distributions are the same. Based on the definition of independence for contingency tables, one can conclude that there is no association between the gender of the students and their responses of favorable or unfavorable. That is, these variables are independent of each other.

From part (b), the distributions of ratings were conditioned on the gender of the student. From the computed conditional distributions for ratings, we see that the values are the same for the favorable classification for both male and female genders. Also, for the unfavorable classification, the conditional distributions are the same for both genders. Based on the definition of independence for contingency tables, one can again conclude that there is no association between the gender of the students and their responses of favorable or unfavorable. That is, these variables are independent of each other.

In conclusion, if the faculty member knew the gender of the student, he or she would not be at an advantage over someone who did not know the gender of the student in predicting the response.

**Quick Tip**

Contingency tables are not restricted to $2 \times 2$ classifications. Other areas of statistics deal with much more complex tables.

## 6-5 Simpson's Paradox

For a 1973 study on sex bias in admissions to the graduate school at the University of California, Berkeley, **Table 6-9** shows the information obtained for the five largest majors on that campus.

**Table 6-9:** Admissions by Gender and Major

| MAJOR/GENDER | MALES | | FEMALES | |
|---|---|---|---|---|
| | NUMBER OF APPLICANTS | NUMBER ADMITTED | NUMBER OF APPLICANTS | NUMBER ADMITTED |
| Major 1 | 800 | 520 | 120 | 102 |
| Major 2 | 550 | 341 | 32 | 23 |
| Major 3 | 400 | 160 | 410 | 148 |
| Major 4 | 350 | 126 | 347 | 129 |
| Major 5 | 200 | 48 | 387 | 105 |
| Total | 2,300 | 1,195 | 1,296 | 507 |

There was actually a total of 8,442 males and 4,321 females who had applied for admission. Of the males that applied, 3,714 were accepted, and of the females, 1,512 were accepted. That is, $\frac{3,714}{8,442} \times 100$ percent $= 43.99 \approx 44$ percent of the males were accepted, and $\frac{1,512}{4,321} \times 100$ percent $= 34.99 \approx 35$ percent of the females were accepted. These percentages would suggest that there might have been discrimination against females in admission to the graduate school. However, if this was so, then the discrimination should also be apparent in the admission rates for the different majors, since admission was by department.

Let us consider the percentages that were admitted for both males and females. **Table 6-10** displays these data.

**Table 6-10:** Percentages of Males and Females Admitted

| MAJOR/% ADMITTED | % OF MALES ADMITTED | % OF FEMALES ADMITTED |
|---|---|---|
| Major 1 | 65 | 85 |
| Major 2 | 62 | 72 |
| Major 3 | 40 | 36 |
| Major 4 | 36 | 37 |
| Major 5 | 24 | 27 |

For all the majors except Major 3, the admission rate was higher for the female applicants! This reveals that the female applicants were not discriminated against. If anything, it reveals the opposite. How can this reversal be true? By examining **Table 6-9,** one can see that the majors with the highest acceptance rates had a large number of male applicants and fewer females. The majors that had the lowest acceptance rates had fewer males applying and more females applying. That is, the variable of *major* was *confound-*

*ing* the *gender* variable in the computation of the 44 percent and 35 percent. The apparent bias in these percentages is due to the fact that in general, the female applicants were applying to the most difficult majors for acceptance, and not to gender bias. By considering the variable of major, the gender variable was removed from the bias. That is, we say we are controlling for this confounding variable.

To give a different perspective, we will analyze the sample information for the five majors using marginal and conditional distributions for the number of students admitted.

The marginal distributions for *gender* are given in **Table 6-11.**

**Table 6-11:**    Marginal Distributions for Gender

| % OF MALES ADMITTED | % OF FEMALES ADMITTED |
|---|---|
| 70 | 30 |

From this information, one should be alarmed. Here, 70 percent of the persons admitted to these five majors are males and only 30 percent are females.

**Table 6-12** shows the marginal distributions for major.

**Table 6-12:**    Marginal Distributions for Majors

| MAJOR | 1 | 2 | 3 | 4 | 5 |
|---|---|---|---|---|---|
| % | 37 | 21 | 18 | 15 | 9 |

We can observe that *major 1* had the highest acceptance rate of all five majors.

Next, we display the conditional distributions for the *gender* of the applicant given the *major* chosen. These distributions are given in **Table 6-13.**

**Table 6-13:**    Conditional Distributions for Gender Given Major

| MAJOR/GENDER | MALE (%) | FEMALE (%) | TOTALS (%) |
|---|---|---|---|
| Major 1 | 84 | 16 | 100 |
| Major 2 | 94 | 6 | 100 |
| Major 3 | 52 | 48 | 100 |
| Major 4 | 49 | 51 | 100 |
| Major 5 | 31 | 69 | 100 |

From these conditional distributions for gender given the major, one can again observe that for the different majors, the percentages for the males and females are generally going in opposite directions. That is, for majors 1, 2, and 3, there are more males than females, and for majors 4 and 5, there are more females than males. One can make the argument that based on majors 1, 2, and 3, there is gender bias against females. However, one could also argue that there is not a male bias based on the conditional distributions for majors 4 and 5. Again, the variable of *major* was *confounding* the *gender* variable.

The next table, **Table 6-14,** shows the conditional distributions for the majors given the gender of the applicant.

**Table 6-14:** Conditional Distributions for Major Given Gender

| MAJOR/GENDER | MALE (%) | FEMALE (%) |
|:---:|:---:|:---:|
| Major 1 | 44 | 20 |
| Major 2 | 29 | 5 |
| Major 3 | 13 | 29 |
| Major 4 | 11 | 25 |
| Major 5 | 3 | 21 |
| Total | 100 | 100 |

From these conditional distributions for *major* given the *gender* of the applicants, one can observe that for the males, the percentage decreases from major to major, whereas it stays relatively the same for the female applicants except for major 2. For the male applicants, more are accepted in major 1 and major 2, while more females than males are accepted in the other majors. Again, one can make counterarguments for bias. Here again, the variable of *major* was *confounding* the *gender* variable in the computation of the 70 percent of males being accepted compared to the 30 percent of females.

The apparent inconsistency in the example falls into a category of problems known as **Simpson's paradox.** One can claim that there is gender bias in either direction that is not apparent by just looking at the marginal distributions for gender. The moral is to make sure you analyze data thoroughly in order to see what information is hidden within the data.

### Technology Corner

All of the concepts discussed in this chapter can be computed and illustrated using any statistical software package. However, using such software for the computations encountered in this chapter will be technology overkill. All that is required is simple calculations. All scientific and graphical calculators will aid directly in the computations.

Association between two categorical variables can be investigated through

✔ Contingency tables
✔ Marginal distributions
✔ Conditional distributions
✔ Bar charts

Care should always be taken when interpreting the marginal and conditional distributions that are associated with contingency tables. Care should also be taken when interpreting bar charts for contingency tables. One should be clear as to whether there is an association between variables in a contingency table by interpreting the conditional distributions appropriately.

### True/False Questions

1. In finding the marginal distributions for the row variable in a contingency table, one must divide the column totals by the grand total.
2. In finding the marginal distributions for the column variable in a contingency table, one must divide the column totals by the grand total.

3. In finding the marginal distributions for the row variable in a contingency table, one must divide the row totals by the column totals.

4. In finding the conditional distributions for the row variable given the column variable in a contingency table, one must divide the cell frequencies by the grand total.

5. In finding the conditional distributions for the row variable given the column variable in a contingency table, one must divide the cell frequencies by their respective column totals.

6. In finding the conditional distributions for the column variable given the row variable in a contingency table, one must divide the cell frequencies by their respective column totals.

7. Scatter plots can be used to display the association between two qualitative variables.

8. Two categorical variables are said to be independent of each other if the conditional distributions of one variable are the same for every category of the other variable.

9. Bar charts are appropriate graphical displays for marginal and conditional distributions computed for contingency tables.

10. In a contingency table, if the two variables A and B, say, are independent of each other, then knowing the classifications of A will not give you an advantage over someone who does not know the classifications of A in predicting the responses for variable B.

## Completion Questions

1. A two-way contingency table with four classifications for the row variable and three classifications for the column variable will have (how many) _____ cells in the table.

2. Marginal distributions for the row variable in a contingency table are the percentages of that variable expressed as the row totals relative to the (row, column, grand) _____ total(s) for the table.

3. In computing the conditional distributions for the row variable given the column variable in a contingency table, the cell entries are divided by the (row, column) _____ totals.

4. Scatter plots are (appropriate, not appropriate) _____ graphical displays for investigating the association between qualitative variables.

5. Marginal distributions for the column variable in a contingency table are the percentages of that variable expressed as the (row, column) _____ totals relative to the grand total for the table.

6. In computing the conditional distributions for the column variable given the row variable in a contingency table, the cell entries are divided by the (row, column) _____ totals.

7. In a contingency table with variables X and Y, if knowing the classifications of X does not give you an advantage over someone who does not know the classifications of X in predicting the responses for variable Y, then X and Y are said to be _____ .

8. _____ charts can be used to display the marginal and conditional distributions for the variables in a contingency table.

9. Two categorical variables are said to be independent of each other if the conditional distributions of one variable are _____ for every category of the other variable.

10. Frequency tables that are used to describe the association between qualitative variables are called _____ tables.

## Multiple-Choice Questions

1. The number of cells for a $5 \times 7$ contingency table is
   (a) 35.
   (b) 24.
   (c) 48.
   (d) 28.

2. A cross-classification of two categorical variables in tabular form is called a
   (a) frequency distribution table.
   (b) probability distribution table.
   (c) twofold table.
   (d) contingency table.

Consider **Table 6-15,** formed by cross-classifying age group and brand of cola consumed. Use this information for Problems 3 to 8.

**Table 6-15**

| COLA/AGE | UNDER AGE 15 | AGES 15–25 | AGES 25–35 | TOTAL |
|----------|-------------|------------|------------|-------|
| Cola 1   | 150         | 100        | 200        | 450   |
| Cola 2   | 300         | 125        | 200        | 625   |
| Cola 3   | 300         | 200        | 300        | 800   |
| Total    | 750         | 425        | 700        | 1,875 |

3. The marginal distribution for the *under age 15* classification is
   (a) 40 percent.
   (b) 24 percent.
   (c) 20 percent.
   (d) 33.33 percent.

4. The observed cell frequency for *ages 15–25* and *cola 3* consumers is
   (a) 300.
   (b) 200.
   (c) 125.
   (d) 100.

5. The marginal distribution for the *cola 3* classification is
   (a) 40 percent.
   (b) 37.5 percent.
   (c) 42.67 percent.
   (d) 93.75 percent.

6. The conditional distribution of *cola 2* given that the person is classified as having *ages 25–35* is
   (a) 42.67 percent.
   (b) 28.57 percent.

  (c) 32 percent.
  (d) 42.86 percent.

7. The conditional distribution of a person's being classified in the *under age 15* group given cola 1 is
  (a) 20 percent.
  (b) 24 percent.
  (c) 50 percent.
  (d) 33.33 percent.

8. The conditional distributions for the three colas given the age group *ages 15–25* are
  (a) 5.33 percent, 6.67 percent, and 10.67 percent.
  (b) 22.22 percent, 20 percent, and 25 percent.
  (c) 23.25 percent, 29.41 percent, and 47.06 percent.
  (d) 13.33 percent, 29.41 percent, and 28.57 percent.

  A survey was done by a car manufacturer concerning a particular make and model. A group of 500 individuals was asked whether they purchased their cars because of appearance, performance ratings, or fixed price (no negotiating). The results are given in the contingency table, **Table 6-16,** for both male and female owners. Use this information for Problems 9 to 17.

**Table 6-16**

| OWNER/REASON | APPEARANCE | PERFORMANCE | PRICE | TOTAL |
|---|---|---|---|---|
| Male | 100 | 50 | 35 | 185 |
| Female | 80 | 170 | 65 | 315 |
| Total | 180 | 220 | 100 | 500 |

9. The marginal distribution for the *performance* classification is
  (a) 22.73 percent.
  (b) 77.27 percent.
  (c) 44 percent.
  (d) 78.57 percent.

10. The marginal distribution for the *female* classification is
  (a) 63 percent.
  (b) 25.4 percent.
  (c) 53.96 percent.
  (d) 20.63 percent.

11. The contingency table can be classified as
  (a) $4 \times 5$.
  (b) $3 \times 4$.
  (c) $3 \times 3$.
  (d) $2 \times 3$.

12. The observed cell frequency for the number of females who purchased the car because of *appearance* is

    (a) 315.

    (b) 180.

    (c) 80.

    (d) 100.

13. The marginal distribution for the *male* classification is

    (a) 37 percent.

    (b) 54.1 percent.

    (c) 27.03 percent.

    (d) 18.92 percent.

14. The conditional distribution of a *female* given that the reason was *price* is

    (a) 31.75 percent.

    (b) 13 percent.

    (c) 53.85 percent.

    (d) 65 percent.

15. The conditional distribution that the reason is *performance* given that the gender is *male* is

    (a) 27.03 percent.

    (b) 84.09 percent.

    (c) 22.73 percent.

    (d) 29.41 percent.

16. The conditional distributions for the three reasons given the *female* gender are

    (a) 16 percent, 34 percent, and 13 percent.

    (b) 25.4 percent, 53.97 percent, and 20.63 percent.

    (c) 44.44 percent, 77.27 percent, and 65 percent.

    (d) 54.05 percent, 27.027 percent, and 18.92 percent.

17. The conditional distributions for the gender given the *appearance* classification are

    (a) 20 percent and 16 percent.

    (b) 55.56 percent and 44.45 percent.

    (c) 54.05 percent and 25.4 percent.

    (d) 45.45 percent and 80 percent.

## Further Exercises

    If possible, you can use any technology help available to solve the following problems.

1. A sample of four one-pound bags of Skittles was examined, and the different flavors of Skittles in each bag are summarized in **Table 6-17**. Observe that this can be considered as a $4 \times 5$ contingency table.

    (a) Find the marginal distributions for the flavors.

    (b) Find the marginal distributions for the bags.

    (c) Use bar graphs to display the marginal distributions in parts (a) and (b). Discuss any observations.

    (d) Find the conditional distributions for the bags given the flavors. Display the percentages in tabular form.

**Table 6-17**

| BAG/FLAVOR | WILDBERRY | FRUIT PUNCH | RASPBERRY | WILD CHERRY | STRAWBERRY | TOTAL |
|---|---|---|---|---|---|---|
| Bag 1 | 85 | 87 | 84 | 92 | 89 | 437 |
| Bag 2 | 108 | 85 | 88 | 71 | 86 | 438 |
| Bag 3 | 95 | 99 | 61 | 84 | 80 | 419 |
| Bag 4 | 103 | 80 | 82 | 71 | 98 | 434 |
| Total | 391 | 351 | 315 | 318 | 353 | 1,728 |

(e) Use bar charts to display the conditional distributions in part (d). Discuss any observations.

(f) Find the conditional distributions for the flavors given the bags. Display the percentages in tabular form.

(g) Use bar charts to display the conditional distributions in part (f). Discuss any observations.

(h) Determine whether the variables are independent of each other. Discuss your results.

## ANSWER KEY
### True/False Questions

1. F    2. T    3. F    4. F    5. T    6. F    7. F    8. T    9. T    10. T

### Completion Questions

1. 12    2. grand    3. column    4. not appropriate    5. column    6. row
7. independent    8. Bar    9. the same    10. contingency

### Multiple-Choice Questions

1. (a)    2. (d)    3. (a)    4. (b)    5. (c)    6. (b)    7. (d)    8. (c)    9. (c)
10. (a)    11. (d)    12. (c)    13. (a)    14. (d)    15. (a)    16. (b)    17. (b)

## PART II

Probability

# CHAPTER 7

# Randomness, Uncertainty, and Probability

**Do I Need to Read This Chapter?**

Y ou should read this chapter if you need to review or to learn about

➡ Relative frequency probability

➡ The Law of Large Numbers

➡ The Addition Rule in probability

➡ Conditional probability

➡ The Multiplication Rule in probability

➡ Independence in probability

---

**Get Started**

Probability statements are everywhere around us. Examples of probability statements include:

- There is a 60% chance of its raining today.
- The chance of my winning the lottery is one in 80 million.
- There is a 50-50 chance of observing a head when a fair coin is tossed.

Just what is meant by "chance" in the above statements? Chance is a measure of uncertainty, and we call this measure probability. In this chapter, we will study this concept of probability.

---

## 7-1 Randomness and Uncertainty

### Randomness

The term *randomness* suggests unpredictability. A simple example of randomness is the tossing of a coin. Unless someone consults a psychic for a "perfect reading" on the outcome when a coin is tossed, the outcome is uncertain. The outcome can either be an observed head (H) or an observed tail (T). Because the outcome of the toss cannot be predicted for sure, we say it displays randomness. This is an example of an easily describable random process. However, other random processes can be quite intricate; for example, the fluctuating prices of stocks are difficult to explain because there are so many variables and combinations of variables that are influencing the prices.

### Uncertainty

At some time or another, everyone will experience *uncertainty*. For example, if you are playing a game of softball, and the pitch is on its way, you may be uncertain as to whether to take a swing at the ball or not. Or consider the case when you are approaching the traffic signals and the light changes from green to amber. You have to decide whether you can make it through the intersection or not. You may be uncertain as to what the correct decision should be.

### Probability

When you ask yourself the question as to whether you believe that you can make it through the amber light, the answer may be "probably." That is, you believe that you can make it across the intersection, but you still may have some doubt. The concept of *probability* is used to quantify this measure of doubt. If you believe that you have a 0.99 probability of getting across the intersection, you have made a clear statement about your doubt. The probability statement provides a great deal of information, much more than statements such as "Maybe I can make it across," "I should make it across," etc.

## 7-2 Random Experiments, Sample Space, and Events

Before we discuss the concept of probability, we need to introduce some terms that we will encounter later in this chapter.

### Random Experiment

When we toss a coin, as mentioned earlier, we do not know the outcome. Let us refer to this process of tossing the coin as an *experiment*. We will define such an experiment as a random or probability experiment.

**Explanation of the term—random experiment:** A **random experiment** is an experiment in which the outcome on each trial is uncertain and distinct.

Examples of random experiments are rolling a die, selecting items at random from a manufacturing process to examine for defects, selection of numbers by a lottery machine, etc.

### Sample Space

When we toss a coin, we have two possible outcomes, summarized by {H, T}. When a child is born, the child is either a boy (B) or a girl (G), summarized by {B, G}. If we consider a two-child family; the possibilities can be summarized by {BB, BG, GB, GG}. In each case, the outcomes enclosed in { } include all the possible outcomes. Such a list is called a **sample space.**

**Explanation of the term—sample space:** The sample space for an experiment is the list or set of all possible outcomes for the experiment.

***Example 7-1:*** A fair regular six-sided die is rolled. List the sample space for this random experiment.

**Solution:** Let $S$ represent the sample space. Then $S = \{1, 2, 3, 4, 5, 6\}$.

***Example 7-2:*** List the sample space for a two-child family.

**Solution:** Let B represent the outcome of a boy and G for a girl. The diagram in **Fig. 7-1,** called a *tree diagram,* depicts the possibilities.

From **Fig. 7-1,** if you follow along the "branches" of the tree, you will trace out all the possible outcomes as listed on the right-hand side. Thus, the sample space is $S = \{$BB, BG, GB, GG$\}$.

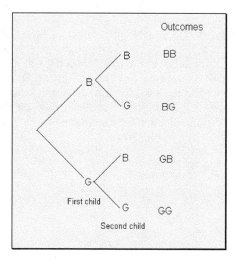

**Fig. 7-1:** Tree diagram for a two-child family

## Events

We may be interested in only part of the sample space. For example, we may only be concerned with one girl in a two-child family; that is, the outcomes BG and GB. These two outcomes constitute a subset of the sample space. Such subsets are called **events.**

**Explanation of the term—event:** An event is a subset of the sample space.

**Note:** Each outcome in a sample space is an event. These events are called *simple events.*

### 7-3 Classical Probability

If we can assume that all the simple events in a sample space have the same chance of occurring, then we can measure the probability of an event as a proportion, relative to the number of points in the sample space. Such a probability measure is referred to as **classical probability.**

**Explanation of the term—classical probability of an event:** If the outcomes in a sample space are equally likely to occur, then the classical probability of an event $A$ is defined to be

$$P(A) = \frac{\text{number of simple events in } A}{\text{total number of simple events in the sample space}}$$

***Example 7-3:*** If a two-child family is selected at random, what is the probability of two boys?

**Solution:** Recall that the sample space was $S = \{BB, BG, GB, GG\}$. In this sample space, the event of two boys occurs once, and there are 4 simple events in the sample space. Thus,

$$P(BB) = \frac{1}{4} = 0.25.$$

***Example 7-4:*** In a manufacturing process, a quality control inspector selected three items at random. Let D represent the event of a defective item, and let N represent the event of a nondefective item. List the possible outcomes for the sample space.

**Solution:** The possible points in the sample space are given in the set $S$.

$$S = \{DDD, DDN, DND, DNN, NDD, NDN, NND, NNN\}.$$

The sample space can be obtained from the tree diagram given in **Fig. 7-2.** There are two possible outcomes when the first item is selected: a defective (D) or a nondefective (N). If a defective is selected, then on the second selection, there are again two possibilities, D or N. If a nondefective was selected on the first selection, then on the second selection, a defective or a nondefective can be selected. Continuing in this manner, you can display the outcomes by the tree diagram given in **Fig. 7-2.**

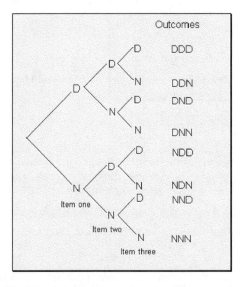

**Fig. 7-2:** Tree diagram for the selection of three items

***Example 7-5:*** For **Example 7-4,** what is the probability of the quality control inspector's observing at least two defective items?

**Solution:** Let $A$ be the event of at least two defectives. Then $A = \{DDD, DDN, DND, NDD\}$. Event $A$ is made up of 4 simple events, and there are 8 simple events in the sample space. Thus, $P(A) = \frac{4}{8} = 0.5.$

## 7-4 Relative Frequency or Empirical Probability

If we flip a fair coin once, we say that the probability of getting a head is $\frac{1}{2} = 0.5$. This is because we have two possible outcomes: a head or a tail. This probability of 0.5 is the theoretical probability of observing a head on a single toss of a coin. In an experiment, however, if we flip the coin 10 times, say, and observe 4 heads, then, based on this information, we say that the chance of observing a head will be $\frac{4}{10} = 0.4$, which is not the same as 0.5. If, however, we flip the coin a large number of times, we would expect about 50 percent of the flips to result in a head.

Observe that

$$\frac{4}{10} = \frac{\text{frequency of occurrence}}{\text{number of trials}}$$

That is, chance or probability can be measured by relative frequency when the trials are exactly repeatable, as in the case of tossing a coin a repeated number of times. Thus, the probability of an event's occurring can be measured by the proportion of times the event occurs if the process is repeated a large number of times. This is called the *long-term relative frequency* of the event.

**Explanation of the term—relative frequency or empirical probability of an event:** The **relative frequency probability** of an event's occurring is the proportion of times the event occurs over a given number of trials.

If $A$ is the event in which we are interested, then the relative frequency probability of $A$'s occurring, denoted by $P(A)$, is computed from

$$P(A) = \frac{\text{frequency of occurrence}}{\text{number of trials}}$$

**Example 7-6:** Of the first 42 presidents of the United States, 26 were lawyers. What is the probability of randomly selecting from these 42 presidents a president who was a lawyer?

**Solution:** Let $A$ represent the event of a president's being a lawyer. Thus, since there are 42 presidents and 26 were lawyers, $P(A) = \frac{26}{42} = 0.619$ (correct to three decimal places).

**Example 7-7:** During a flu season, a campus health clinic observed that on one day, 12 out of 60 students examined had strep throats, while a week later on the same day, 18 out of 75 examined had strep throats. Compute the relative frequencies for the given information.

**Solution:** The relative frequencies are $\frac{12}{60} = 0.2$ and $\frac{18}{75} = 0.24$. Observe that these relative frequencies are different. However, if data are collected over a long period of time, the clinic may be able to conclude that during the flu season, a student who is examined will have strep throat with a probability of 0.22.

## 7-5 The Law of Large Numbers

In any experiment, the relative frequency for an event will change from trial to trial. However, if the experiment is conducted a large number of times, the relative frequency of the event will tend to converge toward a number that is called the *probability* of the event. This concept is called the *Law of Large Numbers.*

### Law of Large Numbers

When an experiment is conducted a large number of times, the relative frequency (empirical) probability of an event can be expected to be close to the theoretical probability of the event. This approximation will improve as the number of replications is increased.

The following example will demonstrate this concept.

***Example 7-8:*** A fair coin is tossed 200 times. Display the graph of the cumulative (running) relative frequency for the number of observed heads.

**Note:** Refer to **Chap. 1** for a review of cumulative relative frequency.

**Solution:** Since we are using a fair coin, the probability of observing a head on a single toss is $P(H) = 0.5$, where H represents the outcome of a head, which is the (theoretical) probability of observing a head. We should expect that as the number of trials increases, the proportion of observed heads will approach 0.5. The experiment is simulated using the MINITAB statistical software package, and the graph of the cumulative relative frequency is displayed in **Fig. 7-3.**

Observe that the cumulative relative frequency varies a great deal at first, but then starts to level off at around 0.5. Recall that $P(H) = 0.5$, and thus one can observe that as the number of trials increases, the relative frequency probability tends toward the probability of observing a head. The display in **Fig. 7-4** shows the simulation for 1,000 tosses of the coin. Observe a distinct convergence of the cumulative relative frequencies to a value of 0.5. This can be expected, since the number of trials was increased to 1,000.

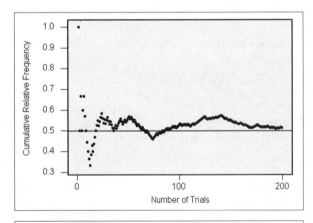

**Fig. 7-3:** Cumulative relative frequency graph for 200 trials

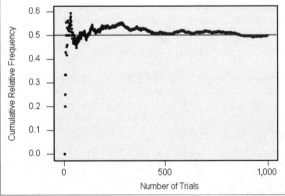

**Fig. 7-4:** Cumulative relative frequency graph for 1,000 trials

**Figure 7-5** shows what is happening as the number of trials gets very large. The proportion flattens out around the 0.5 mark, which is the theoretical probability of observing a head when a coin is tossed.

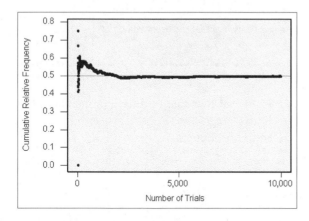

**Fig. 7-5:** Cumulative relative frequency graph for 10,000 trials

## 7-6  Subjective Probability

Subjective probability is a measure of belief. This measure depends on your life experiences. Thus, based on their life experiences, two reasonable persons may have different measures of belief for a particular event's occurring. An example of a subjective probability is a probability value that you assign to your chance of passing an exam. This will be based on your experiences—number of hours you studied, number of classes missed, etc. Subjective probability cannot be uniformly used to define the chance of an event's occurring because the value may be different for different people.

## 7-7  Some Basic Laws of Probability

Following are four basic laws of probability:

**Law 1:**  If the probability of an event is 1, then the event must occur.

For example, the probability of each of us dying is 1. We know that dying is certain to occur.

**Law 2:**  If the probability of an event is 0, then the event will never occur.

For example, the probability of a person who was born outside the United States becoming its president is zero. This is the decree of the U.S. Constitution.

**Law 3:**  The probability of any event must assume a value between 0 and 1, inclusively.

For example, the probability of its raining today is 0.7 = 70 percent. We cannot be more than 100 percent certain that it will rain, nor we cannot be less than 0 percent certain that it will rain.

**Law 4:**  The sum of the probabilities of all the simple events in a sample space must be equal to 1. Another way of saying this is to say that the probability of the sample space in any experiment is always 1.

For example, if we consider the sample space for **Example 7-4,** there are 8 simple events. By the classical approach, each simple event has an equal chance of occurring. That is, each simple event has a $\frac{1}{8}$ chance of occurring. When we sum these probabilities, we have $8 \times \frac{1}{8} = 1$.

### 7-8 Other Probability Rules

Here we will consider some other rules that will enable us to find probabilities between events.

### Compound Events

Sometimes we may have to combine events in order to define another event. Such events are called **compound events.**

**Explanation of the term—compound event:** A compound event is an event that is defined by combining two or more events.

As an illustration, consider the following example.

***Example 7-9:*** Let $A$ be the event that a student owns a portable CD player. Let $B$ be the event that a student owns a laptop computer. Let $C$ be the event that a student owns both a portable CD player and a laptop computer. Discuss the event $C$ that is common to both $A$ and $B$.

**Solution:** From the given information, $C$ is the event that is common to both $A$ and $B$. Since the event $C$ is obtained by combining $A$ and $B$, $C$ is a compound event.

### Union of Events

Consider two events $A$ and $B$. We may be interested in the event that is obtained by considering the elements that are in $A$, or in $B$, or in both $A$ and $B$. Such a compound event is called the **union** of events $A$ and $B$.

**Explanation of the term—union of two events:** The union of two events $A$ and $B$ is the set of outcomes that are included in $A$ or $B$ or both $A$ and $B$.

**Notation:** The union of $A$ and $B$ will be denoted by $A \cup B$.

The diagram given in **Fig. 7-6,** called a *Venn diagram*, depicts the union of events $A$ and $B$. The shaded area represents the event of $A \cup B$.

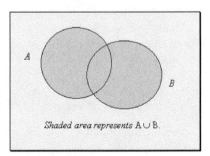

*Shaded area represents* A ∪ B.

**Fig. 7-6:** Venn diagram for $A \cup B$

***Example 7-10:*** Let *A* be the event of rolling a fair six-sided die. Let *B* be the event of an even number between 0 and 9. What is *A* ∪ *B*?

**Solution:** *A* = {1, 2, 3, 4, 5, 6} and *B* = {2, 4, 6, 8}. Thus, *A* ∪ *B* = {1, 2, 3, 4, 5, 6, 8}. Note that elements that are common to both *A* and *B* are not repeated when listing the elements in *A* ∪ *B*. This is shown in **Fig. 7-7.**

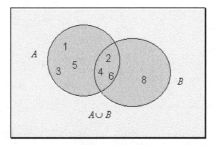

**Fig. 7-7:** Elements in *A* ∪ *B*

**Quick Tip**

In the union of events, elements common to different events are *not* repeated.

***Example 7-11:*** Given that the probability that *only* event *A* will occur is 0.3, the probability that *only* event *B* will occur is 0.4, and the probability that *both* events *A* and *B* will occur is 0.1. Depict this information on a Venn diagram.

**Solution:** The information is displayed in **Fig. 7-8.** Note that the sum of the probabilities in the Venn diagram equals 1.

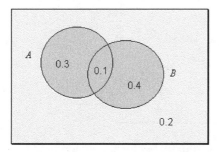

**Fig. 7-8:** Venn diagram displaying information for Example 7-11

### Intersection of Events

Consider two events *A* and *B*. We may be interested in the event that is obtained by considering the elements that are in *both A* and *B*. Such a compound event is called the **intersection** of events *A* and *B*.

**Explanation of the term—intersection of two events:** The intersection of two events *A* and *B* is the set of outcomes that are included in both *A* and *B*.

**Notation:** The intersection of $A$ and $B$ will be denoted by $A \cap B$.

The diagram in **Fig. 7-9** depicts the intersection of the events $A$ and $B$.

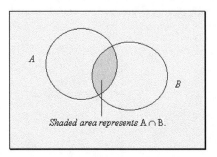

*Shaded area represents* $A \cap B$.

**Fig. 7-9:** Venn diagram for $A \cap B$

***Example 7-12:*** Let $A$ be the event of rolling a fair six-sided die. Let $B$ be the event of an even number between 0 and 9. What is $A \cap B$?

**Solution:** Recall that $A = \{1, 2, 3, 4, 5, 6\}$ and $B = \{2, 4, 6, 8\}$. Thus, $A \cap B = \{2, 4, 6,\}$. $A \cap B$ is shown in **Fig. 7-7** for **Example 7-10.**

***Example 7-13:*** In a sample of 100 college students, 60 said they own a car, 30 said they own a stereo, and 10 said they own both a car and a stereo. Compute probabilities for these events and depict this information on a Venn diagram.

**Solution:** Let $C$ be the event that a student owns a car, and let $D$ be the event that a student owns a stereo. Thus, $P(C) = \dfrac{60}{100} = 0.6$, $P(D) = \dfrac{30}{100} = 0.3$, and $P(C \cap D) = \dfrac{10}{100} = 0.1$. Thus, the probability of *only C* occurring is $0.6 - 0.1 = 0.5$, and the probability of *only D* occurring is $0.3 - 0.1 = 0.2$. Note that we have to subtract the portion that is common to both $C$ and $D$ in order to get *only C* and *only D*. This information is depicted in **Fig. 7-10.**

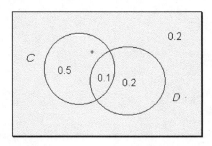

**Fig. 7-10:** Venn diagram for Example 7-13

## Mutually Exclusive Events

Sometimes events may have nothing in common. In such cases, we may deal with the concept known as **mutually exclusive events.**

**Explanation of the term—mutually exclusive events:** Two events $A$ and $B$ are said to be mutually exclusive if they have no elements in common—in other words, if the intersection is empty.

**Figure 7-11** shows two mutually exclusive events. Observe that they do not have any common portion.

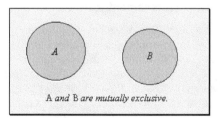

**Fig. 7-11:** Venn diagram depicting two mutually exclusive events $A$ and $B$

**Law 5:** If two events $A$ and $B$ are mutually exclusive, then $P(A \cup B) = P(A) + P(B)$.

## Quick Tips

1. If two events are mutually exclusive, then if one of them occurs, the other cannot occur.

2. Another term used for mutually exclusive is *disjoint*.

***Example 7-14:*** Persons are being selected for a survey. Let M be the event that a male is selected. Let F be the event that a female is selected. Are these mutually exclusive events?

**Solution:** These events are mutually exclusive, since, when a person is selected, that person will be either a male or female. There is no commonality to these two events.

### Complement of an Event

Sometimes it is more convenient to consider what is outside of a given event than to consider what is inside the given event. This deals with the complement of the event.

**Explanation of the term—complement of an event:** The **complement of an event** $A$ is the set of all outcomes that are not in $A$.

**Notation:** We will let $A^c$ represent the complement of the event $A$.

The diagram in **Fig. 7-12** depicts the complement of the event $A$.

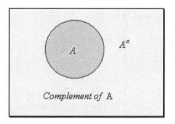

**Fig. 7-12:** Venn diagram depicting the complement of an event

***Example 7-15:*** If a fair six-sided die with faces numbered 1 to 6 is rolled and $A$ is the event of rolling a 2, compute $P(A^c)$.

**Solution:** Now, $P(A) = \dfrac{1}{6}$. Thus, $P(A^c) = P\{1, 3, 4, 5, 6\} = \dfrac{5}{6}$.

## The Complement Rule

Observe that the complement of an event and the event itself are mutually exclusive. If we are dealing with only a single event in a sample space, then the union of the event and its complement will be the same as the event of the sample space. If $A$ is the event, then $A \cup A^c = S$, where $S$ is the sample space. Thus $P(A \cup A^c) = P(S)$. Now, the probability of the sample space for any experiment is 1. That is, $P(S) = 1$. So $P(A \cup A^c) = 1$. Since $A$ and $A^c$ are mutually exclusive, then $P(A \cup A^c) = P(A) + P(A^c)$. This gives us that $P(A) + P(A^c) = 1$. We usually state this as a law.

**Law 6:** The sum of the probability of an event and the probability of its complement equals 1.

$$P(A) + P(A^c) = 1$$
$$or$$
$$P(A^c) = 1 - P(A)$$

***Example 7-16:*** The probability of your favorite college basketball team's winning a game is 0.6. What is the probability of the team's not winning the next game?

**Solution:** Let $A$ be the event that your team does not win the next game. Then $A^c$ is the event of your team's winning the next game and $P(A^c) = 0.6$. Thus,

$$P(A) = 1 - P(A^c)$$
$$= 1 - 0.6$$
$$= 0.4$$

## The Addition Rule

A more generalized rule for the union of two events is given next.

**Law 7:** For any two events $A$ and $B$, the probability of their union is given by

$$P(A \cup B) = P(A) + P(B) - P(A \cap B)$$

This is usually called the **Addition Rule** of probability. The Venn diagram in **Fig. 7-13** depicts this law. The striped area represents $P(A \cup B)$.

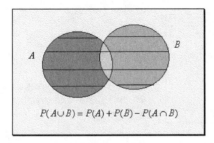

$$P(A \cup B) = P(A) + P(B) - P(A \cap B)$$

**Fig. 7-13:** Venn diagram depicting the Addition Rule

***Example 7-17:*** In **Example 7-13**, what is the probability of a student's having a car, or a stereo, or both a car and a stereo?

**Solution:** We need to find $P(C \cup D)$. From the Venn diagram in **Fig. 7-10** for **Example 7-13**, $P(C) = 0.6$, $P(D) = 0.3$, and $P(C \cap D) = 0.1$. Thus, $P(C \cup D) = 0.6 + 0.3 - 0.1 = 0.8$.

## 7-9 Conditional Probability

Sometimes it is important to find the probability of one event *given* that another event has occurred. Such a probability is called a conditional probability.

**Notation:** We will let $P(A|B)$ represent the conditional probability of the event $A$ given that event $B$ has occurred. It is read as "the probability of $A$ given $B$."

**Law 8:** The conditional probability of an event $A$, given that event $B$ has occurred, is computed from the following formula:

$$P(A|B) = \frac{P(A \cap B)}{P(B)}, \text{ where } P(B) \neq 0$$

***Example 7-18:*** In **Example 7-13,** what is the probability of a student's having a stereo given the student has a car?

**Solution:** We need to compute $P(D|C)$. From the Venn diagram in **Fig. 7-10** for **Example 7-13,** $P(C) = 0.6$, $P(D) = 0.3$, and $P(C \cap D) = 0.1$. Thus, $P(D|C) = \dfrac{P(D \cap C)}{P(C)} = \dfrac{P(C \cap D)}{P(C)} =$ $0.1/0.6 = 0.167$ (correct to three decimal places).

---

**Quick Tip**

In finding a conditional probability, we restrict the sample space to the event on which we condition.

---

In **Example 7-18,** we are restricting the sample space to the event $C$. This is shown in **Fig. 7-14.**

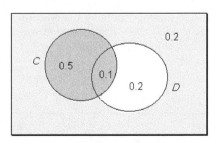

**Figure 7-14**

**Law 9 (multiplication rule for two dependent events):** If events $A$ and $B$ are dependent, then $P(A \cap B) = P(A) \times P(B|A)$ or $P(A \cap B) = P(B) \times P(A|B)$.

## 7-10 Independence

Independence illustrates a special relationship between events. If having knowledge of one event does not affect the probability of occurrence of another event, then these two events are said to be independent. For example, if $P(A|B) = 0.5$ and $P(A) = 0.5$, then having information about event $B$ does not affect the probability of $A$ occurring.

**Explanation of the term—independence in probability:** Two events are **independent** if the occurrence of one does not alter the probability of the other. So, if events are independent, then symbolically, we can express this as

$$P(A|B) = P(A)$$
$$or$$
$$P(B|A) = P(B)$$

If events $A$ and $B$ are not independent, then the events are said to be **dependent.**

***Example 7-19:*** A part-time student is enrolled in a course in geometry (G) and a course in music (M). The probabilities that the student will pass geometry, music, or both subjects are, respectively, $P(G) = 0.8$, $P(M) = 0.7$, and $P(G \cap M) = 0.56$.

(a) What is the probability that the student will pass geometry given that the student passes music?

**Solution:** We need to find $P(G|M)$.

$$P(G|M) = \frac{P(G \cap M)}{P(M)} = \frac{0.56}{0.7} = 0.8$$

(b) Are the events G and M independent?

**Solution:** From part (a), $P(G|M) = 0.8$, and also $P(G) = 0.8$. That is, $P(G|M) = P(G)$. Thus, G and M are independent.

**Law 10 (multiplication rule for two independent events):** If events $A$ and $B$ are independent, then $P(A \cap B) = P(A) \times P(B)$.

***Example 7-20:*** For the information given in **Example 7-19,** use this law to verify that the events G and M are independent.

**Solution:** We need to show that $P(G \cap M) = P(G) \times P(M)$. Given $P(G \cap M) = 0.56$, and $P(G) \times P(M) = 0.8 \times 0.7 = 0.56$. Thus $P(G \cap M) = P(G) \times P(M)$, and so one can conclude that events G and M are independent.

***Example 7-21:*** A consumer group studied the service provided by fast-food restaurants in a given community. One of the things they looked at was the relationship between service and whether the server had a high school diploma or not. The information is summarized in **Table 7-1.**

**Table 7-1:** Table for Example 7-19

| QUALIFICATION/SERVICE | GOOD SERVICE | POOR SERVICE | TOTAL |
|---|---|---|---|
| HS Diploma | 61 | 28 | 89 |
| No HS Diploma | 30 | 81 | 111 |
| Total | 91 | 109 | Grand Total = 200 |

Let

G = the event of good service
B = the event of poor service

H = the event of having a high school diploma
N = the event of not having a high school diploma

(a) Find $P(G)$.

**Solution:** $P(G) = \dfrac{91}{200} = 0.455$. Recall from **Chap. 6** that this is equivalent to the marginal distribution of the variable *good service*.

(b) Find $P(N)$.

**Solution:** $P(N) = \dfrac{111}{200} = 0.555$. Again, recall from **Chap. 6** that this is equivalent to the marginal distribution of the variable *no high school diploma*.

(c) Find $P(B \cap N)$.

**Solution:** The number of observations in $B \cap N$ is 81. Thus, $P(B \cap N) = \dfrac{81}{200} = 0.405$.

(d) Find $P(B|N)$.

**Solution:** Now, $P(B|N) = \dfrac{P(B \cap N)}{P(N)} = \dfrac{0.405}{0.555} = 0.73$. Again, recall from **Chap. 6** that this is equivalent to the conditional distribution of the variable *poor service* given *no high school diploma*.

### Technology Corner

All of the concepts discussed in this chapter can be computed and illustrated using any statistical software package. However, using such software for the computations encountered in this chapter will again be technology overkill. All that is required is simple calculations. All scientific and graphical calculators will aid directly in the computations.

Probability concepts can be investigated through
✔ Relative frequency
✔ The Law of large numbers
✔ Sample spaces
✔ Tree diagrams
✔ Union and intersection of events

Care should always be taken when computing the probability of an event. In certain branches of work in the real world, such as the insurance field, a great deal of emphasis is placed on probabilities.

### True/False Questions

1. The sample space for an experiment is the set of all possible outcomes in that experiment.
2. An event is a subset of the sample space.
3. The probability of an event is a measure of the likelihood that the event will not occur.
4. If we toss a coin 100 times and 50 heads are observed, we can estimate the probability of a head's occurring to be 50 percent. This estimate is known as the Law of Large Numbers.

5. In the classical interpretation of probability, it is not necessary to assume that the outcomes in the sample space are equally likely.
6. If two events $A$ and $B$ are independent, then $P(A|B) \neq P(A)$.
7. Two events are mutually exclusive if the occurrence of one depends on the occurrence of the other.
8. If events $A$ and $B$ are independent, then the probability of both of them occurring, $P(A \cap B)$, is the sum of their respective probabilities.
9. Two events $A$ and $B$ are dependent if $P(A|B) \neq P(A)$.
10. In a Venn diagram, the intersection of two events indicates that the two events are not mutually exclusive.
11. If two events are mutually exclusive, then they are independent.
12. The complement of an event and the event itself are mutually exclusive.
13. The probability of an event is always a number between 0 and 1, inclusive.
14. In selecting a card from a regular deck of cards, the event of "drawing a queen" and the event of "drawing a king" are mutually exclusive.
15. The outcome of an event and the complement of that event together make up the sample space.
16. If a regular six-sided die is rolled, then the complement of "rolling an even number" is the set $\{2, 4, 6\}$.
17. A sample space is a list of all the possible outcomes of the experiment.
18. Two events are mutually exclusive if they cannot both occur at the same time.
19. If two events $A$ and $B$ are independent, then $P(A \cap B) = P(A) \times P(B)$.
20. If $P(A \cap B) = 0$, then the two events must be independent.

## Completion Questions

1. If $A$ and $B$ are mutually exclusive, then $P(A \cap B)$ must equal $(0, 0.5, 1)$ _____ .
2. The probability that an event will occur will assume values between _____ and _____ .
3. An event is a(n) _____ of a sample space.
4. If the probability of an event is 0, then the event will (always, never) _____ occur.
5. The sum of the probabilities of all simple events in a sample space must be equal to _____ .
6. If events $A$ and $B$ are mutually exclusive, then $P(A \cup B) = \{P(A) + P(B); P(A) \times P(B)\}$ _____ .
7. If events $A$ and $B$ are independent, then $P(A \cap B) = \{P(A) + P(B); P(A) \times P(B)\}$ _____ .
8. If the probability of an event's occurring is equal to $(1, 0)$ _____ , then the event must occur.
9. A sample space is the collection of all possible (outcomes, compound events, heads) _____ for an experiment.
10. In classical probability, it is assumed that all outcomes in the sample space are _____ to occur.
11. The complement of an event $A$ is the event that $A$ (will, will not) _____ occur.
12. When any one of the outcomes in an experiment has the same likelihood of occurring as any other, we say the outcomes are _____ .

13. Suppose that we toss a fair coin (with the likelihood of the coin's staying on its edge being zero) a large number of times and we use the observed number of tails to help compute the probability of a tail occurring for this coin. This approach to probability is known as the (relative frequency, classical, subjective) _____ concept of probability.

14. The Addition Rule for probability is helpful when we are computing the probability for (independent, mutually exclusive) _____ events.

15. If $P(A|B) = P(A)$, then $A$ and $B$ must be (independent, dependent) _____ events.

16. If two events $A$ and $B$ have nonzero probabilities and are mutually exclusive, then $P(A \cup B)$ must be (0, 1, neither) _____ .

17. Rolling a regular six-sided die and observing a 6 and then rolling the die again and observing another 6 would be an example of (independent, mutually exclusive) _____ events.

18. If a regular six-sided die is rolled and the outcomes are equally likely, then this is an example of (relative frequency, classical, subjective) _____ probability.

19. The intersection of two events $A$ and $B$ is the set of outcomes that are included in (both $A$ and $B$, only $A$, only $B$, $A$ or $B$) _____ .

20. If $A$ and $B$ are independent, then if $P(A) = 0.7$, $P(B) = 0.8$, and $P(A \cap B) = 0.4$, these assigned probabilities are (valid or invalid) _____ .

## Multiple-Choice Questions

1. If $P(A) = 0.5$, $P(B) = 0.6$, and $P(A \cap B) = 0.3$, then $P(A \cup B)$ is
   (a) 0.8000.
   (b) 0.5000.
   (c) 0.6000.
   (d) 0.0000.

2. If $P(A) = 0.6$, $P(B) = 0.5$, and $P(A \cup B) = 0.9$, then $P(A \cap B)$ is
   (a) impossible.
   (b) 0.2000.
   (c) 0.3000.
   (d) 0.6000.

3. If $P(A) = 0.3$, $P(B) = 0.5$, and $P(A \cup B) = 0.6$, then $P(A|B)$ is
   (a) 0.5000.
   (b) 0.8333.
   (c) 0.4000.
   (d) 0.4500.

4. If $P(A) = 0.5$, $P(B) = 0.4$, and $P(B|A) = 0.3$, then $P(A \cap B)$ is
   (a) 0.7500.
   (b) 0.6000.
   (c) 0.1200.
   (d) 0.1500.

5. If $P(A) = 0.6$, $P(B) = 0.3$, and $P(A|B) = 0.4$, then $P(A \cup B)$ is
   (a) 0.7800.
   (b) 0.1200.
   (c) 0.6667.
   (d) 0.2200.

6. If $A$ and $B$ are mutually exclusive events and $P(A) = 0.5$ and $P(B) = 0.4$, then $P(A \cup B)$ is
   (a) 0.0000.
   (b) 0.9000.
   (c) 0.2000.
   (d) 0.8000.

7. If $A$ and $B$ are independent events and $P(A) = 0.3$ and $P(B) = 0.6$, then $P(A \cap B)$ is
   (a) 0.7200.
   (b) 0.9000.
   (c) 0.1800.
   (d) 0.5000.

8. If $A$ and $B$ are independent events and $P(A) = 0.3$ and $P(B) = 0.6$, then $P(A \cup B)$ is
   (a) 0.9000.
   (b) 0.1800.
   (c) 0.5000.
   (d) 0.7200.

9. If $P(A) = 0.6$, $P(B) = 0.3$, and $P(A|B) = 0.4$, then $P(A^c)$ is
   (a) 0.4000.
   (b) 0.1000.
   (c) 0.6000.
   (d) 0.1200.

10. If $A$ and $B$ are mutually exclusive events and $P(A) = 0.2$ and $P(B) = 0.7$, then $P(A \cap B)$ is
    (a) 0.1400.
    (b) 0.0000.
    (c) 0.9000.
    (d) 0.2857.

Problems 11 to 18 are based on the following information.

In a survey of 120 college students living in the dorms, 60 said that they had *only* a stereo set in their rooms, 40 said that they had *only* a microcomputer in their rooms, and 15 said that they had *both* a stereo and a microcomputer in their rooms. The remaining 5 students had neither.

11. If a student is randomly chosen from this group, the probability that the student has both a stereo and a microcomputer is
    (a) 0.1250.
    (b) 0.2174.
    (c) 0.2143.
    (d) 0.8333.

12. If a student is randomly chosen from this group, the probability that the student has either a stereo or a microcomputer or both is
    (a) 0.9583.
    (b) 0.8333.
    (c) 0.8000.
    (d) 0.7273.

13. If a student is randomly chosen from this group, the probability that the student does not have a stereo is
    (a) 0.4783.
    (b) 0.4583.
    (c) 0.5000.
    (d) 0.3750.

14. If a student is randomly chosen from this group, the probability that the student does not have a microcomputer is
    (a) 0.6522.
    (b) 0.2727.
    (c) 0.5417.
    (d) 0.6667.

15. If a student is randomly selected from this group, the probability that the student has a stereo given that the student does not have a microcomputer is
    (a) 0.9286.
    (b) 0.9231.
    (c) 0.8889.
    (d) 0.8000.

16. If a student is randomly chosen from this group, the probability that the student does have a microcomputer given that the student has a stereo is
    (a) 0.2000.
    (b) 0.6667.
    (c) 0.1500.
    (d) 0.3750.

17. If a student is randomly chosen from this group, the probability that the student does not have either a stereo or a microcomputer is
    (a) 0.0435.
    (b) 0.0417.
    (c) 0.8696.
    (d) 0.8750.

18. If a student is randomly chosen from this group, the probability that the student does have a stereo given that the student has a microcomputer is
    (a) 0.5333.
    (b) 0.6667.
    (c) 0.3750.
    (d) 0.2727.

19. The probability that any one of two engines on an aircraft will fail is 0.001. Assuming that the engines operate independently of each other, the probability that both engines will not fail is

(a) $(0.001)^2$.

(b) 0.002.

(c) $(0.001)(0.999)$.

(d) $(0.999)^2$.

20. If two nontrivial events (probability not equal to zero) $A$ and $B$ are mutually exclusive, which of the following must be true?

(a) $P(A \cap B) = P(A) + P(B)$

(b) $P(A \cup B) = P(A) + P(B)$

(c) $P(A \cup B) = P(A) \times P(B)$

(d) $P(A \cap B) = P(A) \times P(B)$

Problems 21 to 27 are based on the following information.

A clothing store that targets young customers (ages 18 through 22) wishes to determine whether the size of the purchase is related to the method of payment. A sample of 300 customers was analyzed, and the information is given in **Table 7-2.**

**Table 7-2**

| SIZE OF PURCHASE/ METHOD OF PAYMENT | CASH | CREDIT CARD | LAYAWAY PLAN | TOTAL |
|---|---|---|---|---|
| Under $40 | 60 | 30 | 10 | 100 |
| $40 or more | 40 | 100 | 60 | 200 |
| Total | 100 | 130 | 70 | Grand Total = 300 |

21. If a customer is selected at random from this group of customers, the probability that the customer paid cash is

(a) $\frac{1}{3}$.

(b) $\frac{3}{5}$.

(c) $\frac{2}{5}$.

(d) $\frac{2}{3}$.

22. If a customer is selected at random from this group of customers, the probability that the customer paid with a credit card is

(a) $\frac{17}{30}$.

(b) $\frac{13}{30}$.

(c) $\frac{3}{13}$.

(d) $\frac{10}{13}$.

23. If a customer is selected at random from this group of customers, the probability that the customer paid with the layaway plan is

(a) $\frac{6}{7}$.

(b) $\frac{1}{10}$.

(c) $\frac{7}{30}$.

(d) $\frac{1}{7}$.

24. If a customer is selected at random from this group of customers, the probability that the customer purchased under $40 is
    (a) $\frac{3}{5}$.
    (b) $\frac{2}{5}$.
    (c) $\frac{2}{3}$.
    (d) $\frac{1}{3}$.

25. If a customer is selected at random from this group of customers, the probability that the customer purchased $40 or more is
    (a) $\frac{2}{3}$.
    (b) $\frac{2}{5}$.
    (c) $\frac{6}{7}$.
    (d) $\frac{10}{13}$.

26. If a customer is selected at random from this group of customers, the probability that the customer paid with a credit card given that the purchase was under $40 is
    (a) $\frac{3}{10}$.
    (b) $\frac{13}{30}$.
    (c) $\frac{1}{3}$.
    (d) $\frac{2}{3}$.

27. If a customer is selected at random from this group of customers, the probability that the customer paid with a layaway plan given that the purchase was $40 or more is
    (a) $\frac{3}{10}$.
    (b) $\frac{2}{3}$.
    (c) $\frac{7}{30}$.
    (d) $\frac{7}{20}$.

28. Manufactured bolts are collected in a large bin. Suppose one bolt is selected at random and examined to determine whether it is defective or nondefective. The bolt is returned to the bin, and another is selected and examined. If the probability of a defective bolt is 0.01, the probability of selecting two nondefective bolts is
    (a) $2(0.01)$.
    (b) $2(0.99)$.
    (c) $(0.01)^2$.
    (d) $(0.99)^2$.

29. In a particular rural region, 65 percent of the residents are smokers, and research indicates that 15 percent of the smokers have some form of lung cancer. The probability of a resident's having lung cancer given that the resident is a smoker is
    (a) 0.0975.
    (b) 0.2308.
    (c) 0.1500.
    (d) 0.6500.

30. From past experience, an instructor estimates that the probability that a student will cheat on an exam is 0.05. The probability that a student cheats and is caught is 0.01. The probability that a student will be caught, given that the student is cheating, is
    (a) 0.0005.
    (b) 0.0600.

    (c) 0.0400.

    (d) 0.2000.

31. For a certain brand of tire, the probability that a tire will last beyond 40,000 miles is 0.8 and the probability that it will last beyond 50,000 miles is 0.25. Given that a tire lasts beyond 40,000 miles, the probability that it will last beyond 50,000 miles is

    (a) 0.2500.

    (b) 0.0000.

    (c) 0.3125.

    (d) 0.2000.

32. Given $P(A) = 0.5$, $P(B) = 0.6$, and $P(A \cup B) = 0.8$, then $P(A|B)$ is

    (a) 0.5000.

    (b) 0.7500.

    (c) 0.6250.

    (d) 0.0480.

33. If $P(\text{only } A) = 0.4$, $P(\text{only } B) = 0.2$, and $P(A \cup B) = 0.8$, then $P(A \cap B)$ is

    (a) 0.5000.

    (b) 0.2000.

    (c) 0.0800.

    (d) 0.2500.

34. If $P(A) = 0.6$, $P(B) = 0.3$, and $P(A|B) = 0.4$, then $P(A \cup B)$ is

    (a) 0.1200.

    (b) 0.9000.

    (c) 0.7800.

    (d) 0.2400.

35. If $P(A) = 0.5$, $P(B) = 0.4$, and $P(A|B) = 0.9$, then $P(B|A)$ is

    (a) 0.4500.

    (b) 0.3600.

    (c) 0.5556.

    (d) 0.7200.

36. If $A$ and $B$ are two mutually exclusive events with $P(A) = 0.15$ and $P(B) = 0.7$, then $P(A \cup B)$ is

    (a) 0.2143.

    (b) 0.8500.

    (c) 0.0980.

    (d) 0.5500.

37. If $A$ and $B$ are two independent events with $P(A) = 0.15$ and $P(B) = 0.7$, then $P(A \cap B)$ is

    (a) 0.5500.

    (b) 0.2143.

    (c) 0.1050.

    (d) 0.8500.

38. If $A$ and $B$ are two independent events with $P(A) = 0.35$ and $P(B) = 0.6$, then $P(A|B)$ is

    (a) 0.6000.

    (b) 0.3500.

   (c)  0.2100.

   (d)  0.5833.

39.  If $A$ and $B$ are two independent events with $P(A) = 0.15$ and $P(B) = 0.4$, then $P(A \cup B)$ is

   (a)  0.5500.

   (b)  0.3750.

   (c)  0.0600.

   (d)  0.4900.

40.  In a three-child family, the probability that there are at least two girls is

   (a)  $\frac{3}{7}$.

   (b)  $\frac{1}{2}$.

   (c)  $\frac{4}{7}$.

   (d)  $\frac{3}{8}$.

41.  If $A$ and $B$ are two mutually exclusive events with $P(A) = 0.15$ and $P(B) = 0.4$, then $P(A \cap B^c)$ is

   (a)  0.8500.

   (b)  0.1500.

   (c)  0.4000.

   (d)  0.6000.

## ANSWER KEY
### True/False Questions

   1. T    2. T    3. F    4. F    5. F    6. F    7. F    8. F    9. T    10. T    11. F
   12. T    13. T    14. T    15. T    16. F    17. T    18. T    19. T    20. F

### Completion Questions

   1. 0    2. 0, 1    3. subset    4. never    5. 1    6. $P(A) + P(B)$    7. $P(A) \times P(B)$
   8. 1    9. outcomes    10. equally likely    11. will not    12. equally likely
   13. relative frequency    14. mutually exclusive    15. independent    16. neither
   17. independent    18. classical    19. both $A$ and $B$    20. invalid

### Multiple-Choice Questions

   1. (a)    2. (b)    3. (c)    4. (d)    5. (a)    6. (b)    7. (c)    8. (d)    9. (a)
   10. (b)    11. (a)    12. (a)    13. (d)    14. (c)    15. (b)    16. (a)    17. (b)
   18. (d)    19. (d)    20. (b)    21. (a)    22. (b)    23. (c)    24. (d)    25. (a)
   26. (a)    27. (a)    28. (d)    29. (c)    30. (d)    31. (c)    32. (a)    33. (b)
   34. (c)    35. (d)    36. (b)    37. (c)    38. (b)    39. (d)    40. (b)    41. (b)

# CHAPTER 8

# Discrete Probability Distributions

**Do I Need to Read This Chapter?**

Y ou should read this chapter if you need to review or to learn about

→ Probability distributions
→ Expectation or expected values
→ Variance
→ Bernoulli trials
→ The binomial distribution

## Get Started

Here we will discuss ideas relating to randomness for the sample space produced by a random experiment. For example, if a fair coin is tossed, then the resulting sample space is $S = \{H, T\}$. Using the classical definition of probability, we can summarize the outcomes with the associated probabilities as

| OUTCOME | PROBABILITY |
|---------|-------------|
| H | 0.5 |
| T | 0.5 |

To analyze more complex random experiments, we will introduce and use the ideas of random variables and probability distributions.

## 8-1 Random Variables

Let us consider the experiment of tossing a single coin. Recall the sample space $S = \{H, T\}$. Let $X$ represent the number of heads. Thus $X$ can assume the values 0 and 1. We will let $x$ be these values. That is, $x = 0, 1$. What is the relationship between the sample space and the values of $X$? The relationship is shown in **Fig. 8-1.**

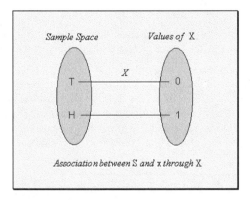

**Fig. 8-1:** Outcomes of tossing a single coin and values of $X$

Next, consider the example of a two-child family. Recall that the sample space is $S = \{$BB, BG, GB, GG$\}$. Let $X$ represent the number of girls. Possible numbers of girls in $S$ are 0, 1, and 2. Thus, $x = 0, 1, 2$. **Figure 8-2** shows the relationship between the sample space $S$ and the values of $X$.

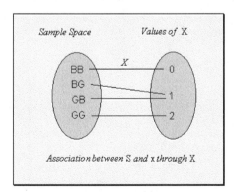

**Fig. 8-2:** Outcomes for a two-child family and values of $X$

By observing **Figs. 8-1** and **8-2,** you will see that the points in the sample space are associated with values on the number line through $X$. Also observe that these numbers are based on the random outcome of the experiment. Because of this, we refer to the variable $X$ as a **random variable.**

**Explanation of the term—random variable:** A random variable assigns one and only one numerical value to each point in the sample space for a random experiment.

**Note:** Random variables are usually denoted by *uppercase* letters near the end of the alphabet, such as $X$, $Y$, and $Z$. We will use *lowercase* letters to represent the values of the random variables, such as $x$, $y$, and $z$.

## Types of Random Variables

We will encounter two types of random variables, *discrete* and *continuous*.

**Explanation of the term—discrete random variable:** A **discrete random variable** is one that can assume a countable number of possible values.

For example, the number of days it rained in your community during the month of March is an example of a discrete random variable. If $X$ is the number of days it rained during the month of March, then the possible values for $X$ are $x = 0, 1, 2, 3, \ldots, 31$.

**Explanation of the term—continuous random variable:** A **continuous random variable** is one that can assume any value in an interval on the real number line.

For example, the amount (in inches) of rainfall in your community during the month of March is an example of a continuous random variable. If $X$ is the amount it rained during the month of March, then the possible values for $X$ will be in the interval $[0, \infty)$. That is, the amount can vary from zero inches to an infinite number of inches. Theoretically, the number of inches of rainfall can go to infinity ($\infty$), but from a practical standpoint, this may never happen. A practical continuous interval may be $[0, 12]$ inches, as shown in **Fig. 8-3.**

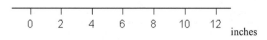

**Fig. 8-3**

**Note:** We will deal only with discrete random variables in this chapter.

## 8-2 Probability Distributions for Discrete Random Variables

Recall the two-child family example from **Chap. 7.** The sample space was given as $S = \{BB, BG, GB, GG\}$, and the tree diagram is repeated in **Fig. 8-4** for convenience.

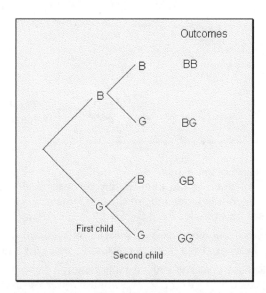

**Fig. 8-4:** Tree diagram for a two-child family

Let $X$ represent the number of girls in the family; then the values for $X$ are $x = 0, 1, 2$. Using the classical definition of probability,

$$P(X = 0) = P(\text{BB}) = \frac{1}{4} = 0.25$$

$$P(X = 1) = P(\text{BG or GB})$$

$$= P(\text{BG} \cup \text{GB})$$

$$= P(\text{BG}) + P(\text{GB}) \qquad \text{since BG and GB are mutually exclusive events}$$

$$= \frac{1}{4} + \frac{1}{4} = \frac{1}{2} = 0.5$$

$$P(X = 2) = P(\text{GG})$$

$$= \frac{1}{4} = 0.25$$

We can arrange the values of the random variable and the associated probabilities in tabular form, as shown in **Table 8-1.**

**Table 8-1**

| $x$ | $P(X = x)$ |
|:---:|:---:|
| 0 | 0.25 |
| 1 | 0.50 |
| 2 | 0.25 |
| | $\sum P(x) = 1$ |

Such a table is called a *probability distribution.* In particular, it is a **discrete probability distribution,** since the random variable is discrete.

**Explanation of the term—discrete probability distribution:** A discrete probability distribution consists of all possible values of a discrete random variable with their corresponding probabilities.

The diagram in **Fig. 8-5** shows a bar graph representation of the probability distribution for the number of girls in a two-child family.

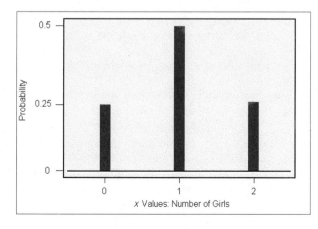

**Fig. 8-5:** Bar graph for the probability distribution for the number of girls in a two-child family

## 8-3  Expected Value

A very important concept in probability is the idea of expected values. The expected value for a random variable is the mean or average value of the random variable. If the random variable is observed over a period of time, the expected value should be close to the average value of the observations generated by the random process. The larger the number of observations, the closer the expected value will be to the average value of the observations.

**Explanation of the term—expected value for a discrete random variable:** The **expected value for a discrete random variable** $X$ is the mean value of the random variable. It is denoted by $E(X)$ or $\mu$ (read as "mu") and is obtained by computing

$$\mu = E(X) = \sum x \times P(x)$$

***Example 8-1:***  Find the expected number of girls in a two-child family.

**Solution:**  Let $X$ represent the number of girls in a two-child family. Using the formula and the information from the probability distribution in **Table 8-1,** we have

$$E(X) = 0 \times 0.25 + 1 \times 0.5 + 2 \times 0.25 = 1$$

That is, if we sample from a large number of two-child families, on average, there will be one girl in each family.

**Figure 8-6** shows the result of a simulation for the average values for the observations generated for the number of girls for a two-child family distribution. Observe that as the number of simulations increases, the average value for the observations fluctuates about the value of 1. Recall that the expected value for the number of girls in a two-child family was 1, so this pattern in the simulation should be expected.

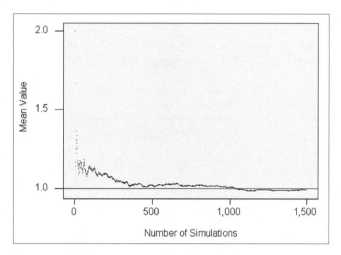

**Fig. 8-6:** Simulation illustrating the expected number of girls in a two-child family

**Quick Tip**

For discrete random variables, the expected value is very seldom one of the possible outcomes of the random variable.

*Example 8-2:* If 1,000 raffle tickets were sold for a bicycle worth $400, what is the expected value of the raffle?

**Solution:** The probability of winning the bicycle is $\dfrac{1}{1,000}$, since there were 1,000 raffle tickets. So the expected value of the raffle will be $\$400 \times \dfrac{1}{1,000} = \$0.40$ or 40 cents. That is, if you purchase a large number of tickets, on average, the return on each ticket will be 40 cents. Thus, it would be unwise to spend more than 40 cents for the ticket. Of course, if the money is going to a worthy cause, you may wish to take that into consideration.

*Example 8-3:* What is the expected value of a raffle with a first prize of $400, a second prize of $300, and a third prize of $200 if 1,000 tickets are sold?

**Solution:** If the raffle is repeated a large number of times, we will lose $\dfrac{997}{1000} \times 100$ percent = 99.7 percent of the time. We will win the first prize $\dfrac{1}{1,000} \times 100$ percent = 0.1 percent of the time, since we can choose from 1,000 tickets. We will win the second prize $\dfrac{1}{999} \times 100$ percent = 0.1 percent of the time, since we only have 999 remaining tickets to choose from for the second prize. We will win the third prize $\frac{1}{998} \times 100$ percent $= 0.1$ percent of the time, since we only have 998 remaining tickets from which to choose the third prize. Converting the percentages to probability, we will have the probability distribution given in **Table 8-2.**

**Table 8-2:** Probability Distribution for Example 8-3

| PRIZE VALUE $x$ (IN $) | $P(x)$ |
|:---:|:---:|
| 0 | 0.997 |
| 200 | 0.001 |
| 300 | 0.001 |
| 400 | 0.001 |

Thus, the expected value of the raffle is $0 \times (0.997) + 200 \times 0.001 + 300 \times 0.001 + 400 \times 0.001 = \$0.90$ or 90 cents. Again, it would be unwise to spend more than 90 cents for a ticket.

***Example 8-4:*** A game is set up such that you have a $\frac{1}{5}$ chance of winning \$350 and a $\frac{4}{5}$ chance of losing \$50. What is your expected gain?

**Solution:** Let $X$ represent the amount of gain. Note that a loss will be considered a negative gain. The probability distribution for $X$ is given in **Table 8-3.**

**Table 8-3:** Probability Distribution for Example 8-4

| $x$ (IN $) | $P(X = x)$ |
|:---:|:---:|
| 350 | 1/5 |
| −50 | 4/5 |

Thus, the expected value of the game is $E(X) = 350 \times \frac{1}{5} + (-50) \times \frac{4}{5} = \$30$. That is, if you play the game a large number of times, on average, you will win \$30 per game.

Sometimes we may be able to use expected values to help make a decision. The following example illustrates this.

***Example 8-5:*** Suppose you are given the option of two investment portfolios, A and B, with the potential profits and associated probabilities displayed in **Table 8-4.** Based on expected profits, which portfolio will you choose?

**Table 8-4:** Probability Distributions for Portfolios A and B

| PORTFOLIO A | | PORTFOLIO B | |
|:---:|:---:|:---:|:---:|
| **PROFIT** | **PROBABILITY** | **PROFIT** | **PROBABILITY** |
| −1,500 | 0.2 | −2,500 | 0.2 |
| −100 | 0.1 | −500 | 0.1 |
| 500 | 0.4 | 1,500 | 0.3 |
| 1,500 | 0.2 | 2,500 | 0.3 |
| 3,500 | 0.1 | 3,500 | 0.1 |

Let $X$ represent the profit for portfolio A, and let $Y$ represent the profit for portfolio B. Then,

$$E(X) = (-1{,}500) \times 0.2 + (-100) \times 0.1 + 500 \times 0.4 + 1{,}500 \times 0.2 + 3{,}500 \times 0.1 = \$540$$

$$E(Y) = (-2{,}500) \times 0.2 + (-500) \times 0.1 + 1{,}500 \times 0.3 + 2{,}500 \times 0.3 + 3{,}500 \times 0.1 = \$1{,}000$$

Since $E(Y) > E(X)$, you should invest in portfolio B based on the expected profit. That is, in the long run, portfolio B will outperform portfolio A. Thus, with repeated investments in portfolio B, you will, on average, gain $\$(1{,}000 - 540) = \$460$ over portfolio A.

### 8-4  Variance and Standard Deviation of a Discrete Random Variable

The following diagrams, **Figs. 8-7** through **8-11,** show the shapes of different distributions about a mean of zero for different random variables. Observe that some are more spread out about zero, while others are more clustered about zero. The spread gives us an idea about the *variability* of the random variable about the mean.

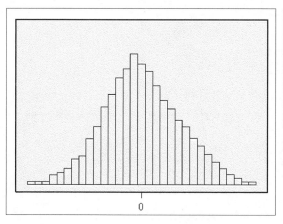

**Fig. 8-7:** Distribution with a fair amount of spread about the mean of zero

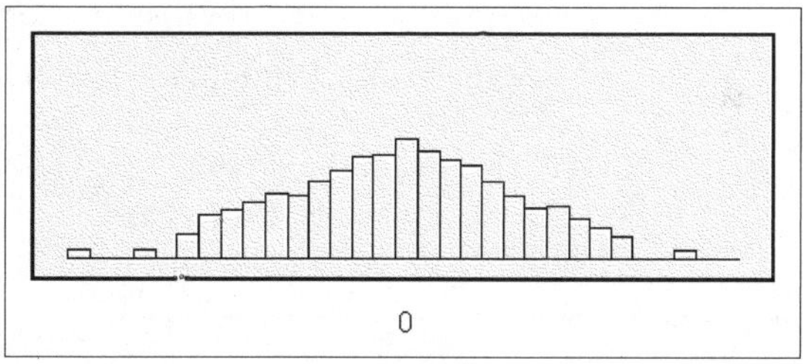

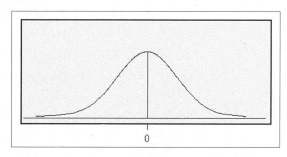

**Fig. 8-8:** Distribution with a large amount of spread about the mean of zero

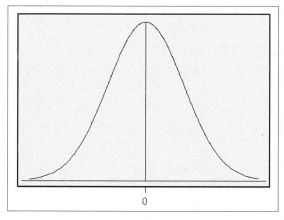

**Fig. 8-9:** Distribution with a fair amount of spread about the mean of zero

**Fig. 8-10:** Distribution with a fair amount of spread about the mean of zero

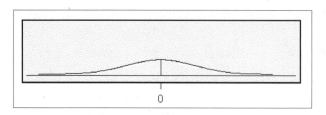

**Fig. 8-11:** Distribution with a large amount of spread about the mean of zero

## Variance

**Explanation of the term—variance for a discrete random variable:** The **variance for a discrete random variable** $X$ measures the spread of the random variable about the expected value (mean) and is computed with the following formula:

$$V(X) = \sum (x - \mu)^2 \times P(x)$$

where $\mu$ is the expected value for the random variable.

An equivalent computational formula for the variance is given as

$$V(X) = \sum [x^2 \times P(x)] - \mu^2$$

**Note:** In the above formula, the summation is only for $[x^2 \times P(x)]$ over all the $x$ values.

---

## Quick Tip

Some texts may use the symbol $\sigma^2$ or $\sigma_X^2$ to represent the variance of the random variable $X$.

---

***Example 8-6:*** Find the variance for the winnings in the raffle for **Example 8-3.**

**Solution:** If we let $X$ represent the winnings, then $V(X) = 0^2 \times 0.997 + 200^2 \times 0.001 + 300^2 \times 0.001 + 400^2 \times 0.001 - 0.90^2 = 289.19$. The units here will be (dollar)$^2$.

**Note:** Variance is measured in square units.

***Example 8-7:*** Find the variance for the gain in **Example 8-4.**

**Solution:** If we let $X$ be the amount of gain, then $V(X) = 350^2 \times 1/5 + (-50)^2 \times 4/5 - 30^2 = 25,600$.

***Example 8-8:*** Find the variance for the profits in the two portfolios in **Example 8-5.**

**Solution:** Let $X$ represent the profit for portfolio A, and let $Y$ represent the profit for portfolio B. Then

$$V(X) = (-1,500)^2 \times 0.2 + (-100)^2 \times 0.1 + (500)^2 \times 0.4 + (1,500)^2$$
$$\times 0.2 + (3,500)^2 \times 0.1 - 540^2 = 1,934,400$$

$$V(Y) = (-2,500)^2 \times 0.2 + (-500)^2 \times 0.1 + (1,500)^2 \times 0.3 + (2,500)^2$$
$$\times 0.3 + (3,500)^2 \times 0.1 - 1,000^2 = 4,050,000$$

Now, if you select a portfolio based on the variance, then you should select the one with the smaller variability, since this would involve less risk. Thus, in this case, you should select portfolio A, since it has the smaller variability.

## Standard Deviation

It is easier to deal with a quantity that has the same units as the variable itself. If we take the square root of a square unit, we will get the unit itself. Thus, if we take the square root of the variance, called the *standard deviation*, we will get a quantity that has the same unit as the random variable.

**Explanation of the term—standard deviation for a random variable:** The **standard deviation for a random variable** $X$ is defined to be the positive square root of the variance. It is computed from

$$SD(X) = \sqrt{V(X)}$$

**Quick Tip**

The standard deviation is a rough estimate of the average distance of the values of the random variable from the expected value (mean).

*Example 8-9:* Find the standard deviation for the profits in the two portfolios in **Example 8-5.**

**Solution:** Let $X$ represent the profit for portfolio A, and let $Y$ represent the profit for portfolio B. Then

$$SD(X) = \sqrt{1,934,400} = 1,390.83$$

$$SD(Y) = \sqrt{4,050,000} = 2,012.46$$

A larger standard deviation indicates that there is greater variability in profits and hence an inherent larger risk for your investment, and vice versa. You would need to consider these issues and weigh the risks against the gains in order to make an informed decision as to the portfolio in which you should invest.

## 8-5  Bernoulli Trials and the Binomial Probability Distribution

When a coin is tossed once, the outcome can be classified in one of two possible mutually exclusive ways: a head (H) or a tail (T). Similarly, in selecting an item from a manufacturing process, we can have a defective (D) item or a nondefective (N) item. Such experiments are called **Bernoulli experiments.** In the case of tossing a coin, we may be interested in the outcome of a head. In such a case, we will classify a head as a successful outcome. In the case of selecting the item, a success may be the selection of a defective item.

**Explanation of the term—Bernoulli experiment:** A Bernoulli experiment is a random experiment, the outcome of which can be classified as one of two simple events.

Of course, we can repeat a Bernoulli experiment several times. When we do this under certain conditions, we say that we have a sequence of **Bernoulli trials.**

**Explanation of the term—Bernoulli trials:** Bernoulli trials occur when a Bernoulli experiment is repeated several *independent* times, so that the probability of success, say *p,* remains the same from trial to trial.

In a sequence of Bernoulli trials, we are often interested in the total number of successes. For example, if we examine 18 items off the production line, we may be interested in the number of defectives we observe in the sample of size 18. If we let $X$ be the random variable that represents the number of defective items in the sample, then $X$ is called a **binomial random variable** and the associated experiment, a **binomial experiment.**

**Explanation of the term—binomial experiment:** A binomial experiment is a random experiment which satisfies the following conditions:

- There are two possible outcomes (success or failure) on each trial.
- There is a fixed number of trials, say $n$.
- The probability of success, say $p$, is fixed from trial to trial.
- The trials are independent.
- The binomial random variable is the number of successes in the $n$ trials.

Binomial experiments occur quite frequently in the real world, and a model has been developed to help compute the probabilities associated with such experiments. Before we discuss the binomial distribution, however, we need to introduce the concept of factorials.

## Factorials

Suppose there are five job positions to be filled by five different applicants. There will be 5 choices for the first position. Once this is filled, there are only four applicants remaining; thus there will be 4 choices for the second position. We can continue in this manner until all of the positions are filled. Note that there will only be one choice for the last position. By a counting principle, there will be $5 \times 4 \times 3 \times 2 \times 1 = 120$ ways of filling the five positions. We say that there are 5 *factorial ways* of filling these positions.

**Notation:** We will use the symbol ! to represent factorial.

For example,

$$5! = 5 \times 4 \times 3 \times 2 \times 1 = 120$$
$$10! = 10 \times 9 \times 8 \times \cdots \times 1 = 3,628,800$$
$$1! = 1$$

We define $0! = 1$.

## Binomial Distribution

The function that generates binomial probabilities is given below. It represents the probability of exactly $x$ successes in $n$ trials in a binomial experiment.

$$P(x) = \frac{n!}{x!(n-x)!}\, p^x(1-p)^{n-x} \qquad \text{for } x = 0, 1, 2, \ldots, n$$

***Example 8-10:*** Five items are selected at random from a production line. What is the probability of *exactly* 2 defectives if it is known that the probability of a defective item is 0.05?

**Solution:** Let $X$ represent the number of defectives. Then $X$ is a binomial random variable with $n = 5$, $x = 2$, and $p = 0.05$. Substituting into the formula gives

$$P(2) = \frac{5!}{2!(5-2)!} \, 0.05^2(1-0.05)^{5-2} = 0.0214$$

That is, the probability of observing 2 defectives in this binomial experiment is 0.0214, correct to four decimal places.

***Example 8-11:*** A student randomly guesses at 10 multiple-choice questions. Find the probability that the student guesses *exactly* 3 correctly. Each question has four possible answers with only one correct answer, and each question is independent of every other question.

**Solution:** Observe that this can be considered as a binomial experiment. We have a fixed number of trials (10 questions) with a probability of success (probability of guessing correctly) of 0.25. Also, the trials (questions) are independent, and there are two possible outcomes on each question (correct guess or incorrect guess). Let $X$ be the number of correct guesses. Then $X$ is a binomial random variable with $n = 10$, $x = 3$, and $p = 0.25$. Substituting into the formula gives

$$P(3) = \frac{10!}{3!(10-3)!} \, 0.25^3 \, (1-0.25)^{10-3} = 0.2503$$

That is, the probability of guessing exactly 3 of the questions correctly is 0.2503, correct to four decimal places.

***Example 8-12:*** For the information in **Example 8-11,** what is the probability of guessing less than 3 correctly?

**Solution:** The probability of guessing less than 3 correctly is equivalent to finding $P(X < 3)$. That is, we need to find $P(X < 3) = P(X \leq 2) = P(X = 0 \text{ or } X = 1 \text{ or } X = 2)$. Since $X = 0$, $X = 1$, and $X = 2$ are mutually exclusive events, $P(X = 0 \text{ or } X = 1 \text{ or } X = 2) = P(X = 0) + P(X = 1) + P(X = 2)$. Thus,

$$P(X < 3) = P(X \leq 2) = 0.0563 + 0.1877 + 0.2816 = 0.5256$$

***Example 8-13:*** For the information in **Example 8-11,** what is the probability of guessing more than 8 correctly?

**Solution:** The probability of guessing more than 8 correctly is equivalent to finding $P(X > 8)$. That is, we need to find $P(X > 8) = P(X = 9) + P(X = 10)$. Thus,

$$P(X > 8) = P(X \geq 9) = 0.0000 + 0.0000 = 0.000$$

correct to four decimal places.

---

**Quick Tip**

Extensive tables can be generated and used to find probabilities of binomial random variables. The drawback, however, is that there is an infinite number of values between 0 and 1 for the probability of success, and therefore one would not have an exhaustive table for reference.

***Example 8-14:*** From the information in **Example 8-11,** a table of probabilities and cumulative probabilities was generated. This information is given in **Table 8-5.**

**Table 8-5**

| X | $P(X = x)$ | $P(X \leq x)$ |
|---|---|---|
| 0 | 0.0563 | 0.0563 |
| 1 | 0.1877 | 0.2440 |
| 2 | 0.2816 | 0.5256 |
| 3 | 0.2503 | 0.7759 |
| 4 | 0.1460 | 0.9219 |
| 5 | 0.0584 | 0.9803 |
| 6 | 0.0162 | 0.9965 |
| 7 | 0.0031 | 0.9996 |
| 8 | 0.0004 | 1.0000 |
| 9 | 0.0000 | 1.0000 |
| 10 | 0.0000 | 1.0000 |

(a) Find the probability that a student correctly guesses between 4 and 6 questions, inclusive. Use the *exact* probability values, $P(X = x)$, to solve the problem.

**Solution:** The probability of guessing between 4 and 6 questions (inclusive) correctly is equivalent to finding $P(4 \leq X \leq 6)$. That is, we need to find $P(4 \leq X \leq 6) = P(X = 4) + P(X = 5) + P(X = 6) = 0.1460 + 0.0584 + 0.0162 = 0.2206$.

(b) Find the probability that a student correctly guesses between 4 and 6 questions, inclusive. Use the *cumulative* probability values, $P(X \leq x)$, to solve the problem.

**Solution:** Note that $P(4 \leq X \leq 6) = P(X \leq 6) - P(X \leq 3)$, as shown in **Fig. 8-12.**

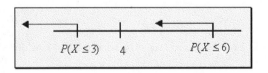

$P(X \leq 3)$    4      $P(X \leq 6)$      **Fig. 8-12**

Thus, $P(4 \leq X \leq 6) = P(X = 4) + P(X = 5) + P(X = 6) = P(X \leq 6) - P(X \leq 3) = 0.9965 - 0.7759 = 0.2206$. Note that by subtracting out $P(X \leq 3)$, we are making sure that the probability of 4 is included in the required probability computations. If we had subtracted out $P(X \leq 4)$, then the probability of 4 would not be included in the required probability computations and the solution would have been incorrect.

### Mean (Expected Value) and Variance for a Binomial Random Variable

The mean, variance, and standard deviation of a binomial random variable can be computed using the following formulas.

$$\text{Mean } \mu = np$$
$$\text{Variance} = np(1 - p)$$
$$\text{Standard deviation} = \sqrt{np(1 - p)}$$

*Example 8-15:* What is the expected value of the number of correct guesses in **Example 8-11**?

**Solution:** Since $\mu = n \times p$, then $\mu = 10 \times 0.25 = 2.5$. That is, if the exam is taken a repeated number of times, on average, the student will guess 2.5 of the questions correctly. Observe that 2.5 is an average value, since the student cannot guess 0.5 of a question correctly.

**Technology**

## Technology Corner

All of the concepts discussed in this chapter can be computed and illustrated using most statistical software packages. However, using such software for the simpler computations encountered in this chapter will be technology overkill. All scientific and graphical calculators will aid directly in the computations. In addition, the newer calculators, like the TI-83, may have the binomial distribution, from which you can compute binomial probabilities. If you own a calculator, you should consult the manual to determine what statistical features are included.

**Illustration:** **Figure 8-13** shows the solutions for **Examples 8-11** and **8-12** computed using the MINITAB software. **Figure 8-14** shows the same solutions computed by the TI-83 calculator. Other solutions can be obtained with a little mathematical manipulation by the software and the calculator.

---

### Example 8-11

#### Probability Density Function

Binomial with n = 10 and p = 0.250000

| x | P( X = x) |
|------|-----------|
| 3.00 | 0.2503 |

### Example 8-12

#### Cumulative Distribution Function

Binomial with n = 10 and p = 0.250000

| x | P( X <= x) |
|------|-----------|
| 2.00 | 0.5256 |

**Fig. 8-13:** MINITAB output for Examples 8-11 and 8-12

---

### Example 8-11

```
binompdf(10,.25,
3)
       .2502822876
■
```

### Example 8-12

```
binomcdf(10,.25,
2)
       .525592804
■
```

**Fig. 8-14:** TI-83 output for Examples 8-11 and 8-12

---

**It's a Wrap**

Discrete random variable concepts can be investigated through

✔ Discrete probability distributions
✔ Expectation and variance
✔ Special probability distributions such as the binomial distribution

Again, care should always be taken when computing probabilities. Also, care should be taken when using the expected value and the standard deviation to aid in decision making.

**Test Yourself**

## True/False Questions

1. A random variable can assume only one value with a given probability.
2. A discrete random variable can assume any set of values that can be counted or listed.
3. The values of a continuous random variable cannot be counted or listed.
4. The probability associated with a regular six-sided die falls between 0 and 6, inclusive.
5. The sum of all associated probabilities for an experiment is sometimes less than 1.
6. The expected value for any discrete random variable $X$ is always $\sum x \times P(x)$, where $P(x)$ is the probability of $x$.
7. The variance for any discrete random variable $X$ is $\sqrt{\sum (x - \mu)^2 \, P(x)}$.
8. In a binomial experiment, the number of trials is infinite.
9. In a binomial experiment, the trials can be dependent on each other.
10. In a binomial experiment, there are exactly two possible outcomes for each trial.
11. The amount of time you study for an exam is a discrete random variable.
12. The formula $\mu = np$ can be used to find the expected value for any discrete random variable.
13. Discrete random variables may assume only positive values.
14. The length of any page in any of your textbooks is a continuous random variable.
15. The height of a basketball player is a continuous random variable.
16. Sometimes a continuous random variable may be discrete.
17. The expected value of a binomial random variable consisting of 9 trials and probability of failure of 0.3 is 2.7, if a failure is considered to be a successful outcome.
18. The mean of a discrete random variable is also called the expected value of the random variable.
19. The standard deviation for a binomial distribution is equal to $np(1 - p)$, where $n$ is the number of trials, and $p$ is the probability of success.
20. A random variable is a rule that assigns one and only one numerical value to each point in the sample space for a random experiment.

## Completion Questions

1. A discrete random variable takes on different values, each with an associated _____ .
2. A random variable that can assume any of a set of possible values that can be counted or listed is a (discrete, continuous) _____ random variable.
3. A random variable that can assume any value in an interval on the real number line is a (discrete, continuous) _____ random variable.
4. For any discrete probability distribution, the sum of all associated probabilities must be equal to (0, 1) _____ .
5. The mean of a discrete random variable $X$ is given by the equation $\mu =$ _____ .
6. In a binomial experiment, the individual trials are (independent, dependent) _____ of each other.
7. The variance for a binomial distribution with $n$ trials and probability of success $p$ is obtained by computing $\{np, np(1 - p), p(1 - p)\}$ _____ .
8. The average value of an infinite number of observations for a discrete random variable is called the _____ of the random variable.
9. Name a discrete probability distribution. _____

10. The number of pages in your statistics book is a (discrete, continuous) _____ random variable, and the length of any page in the text is a (discrete, continuous) _____ random variable.

11. Every probability value associated with a discrete random variable in a probability distribution must lie between _____ and _____ , inclusive.

12. In a binomial experiment, if the probability of success is $p$, then the probability of failure is _____ .

13. A (discrete, continuous) _____ random variable is one which can assume any of an infinite number of different values in a(n) _____ that cannot be counted or listed.

14. A random variable is a rule that assigns one and only one numerical value to each point in the _____ for a random experiment.

15. The standard deviation is used instead of the variance because it has the same (value, units) _____ as the associated variable.

## Multiple-Choice Questions

1. Given the following probability distribution for a random variable $X$,

| $x$    | 1    | 2   | 3   | 4   | 5   | 6    | 7   |
|--------|------|-----|-----|-----|-----|------|-----|
| $P(x)$ | 0.15 | 0.2 | 0.1 | 0.2 | 0.1 | 0.15 | 0.1 |

the probability that $X$ is an odd number is
(a) 0.65.
(b) 0.35.
(c) 0.55.
(d) 0.45.

2. Given the following probability distribution for a random variable $X$,

| $x$    | 1    | 2   | 3   | 4   | 5   | 6    | 7   |
|--------|------|-----|-----|-----|-----|------|-----|
| $P(x)$ | 0.15 | 0.2 | 0.1 | 0.2 | 0.1 | 0.15 | 0.1 |

the mean of $X$ is
(a) 3.75.
(b) 4.
(c) 1.65.
(d) 2.1.

3. Given the following probability distribution for a random variable $X$,

| $x$    | 1    | 2   | 3   | 4   | 5   | 6    | 7   |
|--------|------|-----|-----|-----|-----|------|-----|
| $P(x)$ | 0.15 | 0.2 | 0.1 | 0.2 | 0.1 | 0.15 | 0.1 |

the standard deviation of $X$ is

(a) 1.9378.

(b) 1.9462.

(c) 3.7875.

(d) 3.7550.

4. Which of the following does *not* represent a probability distribution

| (a) | | | (b) | | | (c) | | | (d) | |
|---|---|---|---|---|---|---|---|---|---|---|
| $x$ | $P(x)$ | | $x$ | $P(x)$ | | $x$ | $P(x)$ | | $x$ | $P(x)$ |
| 3 | 0.4 | | −2 | 0.60 | | 0 | 0.2 | | 0.25 | 0.2 |
| 5 | 0.3 | | −1 | 0.30 | | 1 | 0.6 | | 0.50 | 0.3 |
| 7 | 0.3 | | 0 | 0.10 | | 2 | 0.3 | | 0.75 | 0.5 |

5. In an experiment where the probability of a success is 0.4, if you are interested in the probability of 2 successes out of 7 trials, the correct probability is

(a) 0.0774.

(b) 0.1600.

(c) 0.2613.

(d) 0.0016.

6. A statistics instructor (with at least 20 years' experience teaching the same course) has established that 10 percent of all the students who take his course receive a failing grade. If 10 students have enrolled for his course next semester, the probability that *at most one* of these students will fail is

(a) 0.387.

(b) 0.1.

(c) 0.651.

(d) 0.736.

7. A statistics instructor (with at least 20 years' experience teaching the same course) has established that 10 percent of all the students who take his course receive a failing grade. If 10 students have enrolled for his course next semester, the probability that *more than one* of these students will fail is

(a) 0.349.

(b) 0.264.

(c) 1.0.

(d) 0.613.

8. A statistics instructor (with at least 20 years' experience teaching the same course) has established that 10 percent of all the students who take his course receive a failing grade. If 10 students have enrolled for his course next semester, the mean number of students who will fail is

(a) 1.0.

(b) 2.0.

(c) 0.5.

(d) 20.

9. A statistics instructor (with at least 20 years' experience teaching the same course) has established that 10 percent of all the students who take his course receive a failing grade. If 10 students have enrolled for his course next semester, the variance for the number of students who fail is
   (a) 1.0000.
   (b) 9.0000.
   (c) 0.9000.
   (d) 0.9487.

10. A multiple-choice examination has 15 questions. Each question has four possible answers, of which only one is correct. The probability that by just guessing, a student will get *exactly* 7 correct is
    (a) 0.039.
    (b) 0.727.
    (c) 0.273.
    (d) 0.561.

11. The probability distribution for a random variable $X$ is given below.

    | $x$ | −3 | −2 | 0 | 2 | 3 |
    |------|-----|-----|-----|-----|-----|
    | $P(x)$ | 0.1 | 0.3 | 0.2 | 0.3 | 0.1 |

    The mean of the distribution is
    (a) 0.
    (b) −2.5.
    (c) +2.5.
    (d) 2.

12. The probability distribution for a random variable $X$ is given below.

    | $x$ | −3 | −2 | 0 | 2 | 3 |
    |------|-----|-----|-----|-----|-----|
    | $P(x)$ | 0.1 | 0.3 | 0.2 | 0.3 | 0.1 |

    The variance of the distribution is
    (a) 0.0000.
    (b) 4.2000.
    (c) 2.0494.
    (d) $2\sqrt{13}$.

13. Which of the following is *not* a property of a binomial experiment?
    (a) The number of trials is fixed.
    (b) There are exactly two possible outcomes for each trial.
    (c) The individual trials are dependent on each other.
    (d) The probability of success is the same for each trial.

14. If $n$ is the number of trials and $p$ is the probability of success for a binomial experiment, the standard deviation for the resulting binomial distribution is

(a) $\sqrt{n(1-p)}$.

(b) $\sqrt{p(1-p)}$.

(c) $\sqrt{np}$.

(d) $\sqrt{np(1-p)}$.

15. $n!$ ($n$ factorial) could be defined as
    (a) $n(n-1)(n-2)\cdots(3)(2)(1), n > 0$.
    (b) $n(n-1)(n-2)\cdots(3)(2)(1), n \geq 0$.
    (c) $n(n-1)(n-2)\cdots(3)(2)(1), n > 1$.
    (d) $n(n-1)(n-2)\cdots(3)(2)(1), n \geq 1$.

16. Given the following distribution function,

| $x$ | 0 | 1 | 2 | 3 | 4 |
|-----|-----|------|-----|------|-----|
| $P(x)$ | 0.3 | 0.05 | 0.1 | 0.35 | 0.2 |

the computed mean is
(a) 2.0000.
(b) 2.1000.
(c) 0.2000.
(d) 2.4000.

17. If a fair coin is tossed 5 times and the number of tails is observed, the probability that exactly 2 tails are observed is
(a) 2/5.
(b) 5/16.
(c) 1/2.
(d) 15/64.

18. A loan officer has indicated that 80 percent of all loan application forms have zero errors. If 6 forms are selected at random, the probability that *exactly* 2 of them will have at least one error is
(a) 0.150.
(b) 0.040.
(c) 0.850.
(d) 0.246.

19. A loan officer has indicated that 80 percent of all loan application forms have zero errors. If 6 forms are selected at random, the probability that *at most* 1 of them will have at least one error is
(a) 0.655.
(b) 0.002.
(c) 0.393.
(d) 0.607.

20. A loan officer has indicated that 80 percent of all loan application forms have zero errors. If 6 forms are selected at random, the mean number of forms that will have at least one error is

(a) 4.8.

(b) 1.2.

(c) 1/6.

(d) 0.

21. A loan officer has indicated that 80 percent of all loan application forms have zero errors. If 6 forms are selected at random, the standard deviation for the number of forms with at least one error is

(a) 4.8000.

(b) 1.2000.

(c) 0.9600.

(d) 0.9798.

22. Which of the following does *not* represent a probability distribution

| (a) | | | (b) | | | (c) | | | (d) | |
|---|---|---|---|---|---|---|---|---|---|---|
| *x* | *P(x)* | | *x* | *P(x)* | | *x* | *P(x)* | | *x* | *P(x)* |
| 3 | 0.5 | | −2 | 0.60 | | 0 | 0.04 | | 0.25 | 0.32 |
| 5 | 0.5 | | −1 | −0.30 | | 1 | 0.63 | | 0.50 | 0.30 |
| 7 | 0.0 | | 0 | 0.70 | | 2 | 0.33 | | −0.75 | 0.38 |

23. The value of *p* that makes the following table a probability distribution is

| *x* | −2 | −1 | 0 | 1 | 2 |
|---|---|---|---|---|---|
| *P(x)* | 0.13 | 0.17 | *p* | 0.19 | 0.22 |

(a) 0.3000.

(b) 0.1800.

(c) 0.4100.

(d) 0.2900.

24. If *X* is a random variable with the following probability distribution

| *x* | 1 | 2 | 3 | 4 |
|---|---|---|---|---|
| *P(x)* | *p* | 0.4 | 0.25 | 0.3 |

then the mean for *X*

(a) cannot be determined.

(b) is 2.75.

(c) is 2.75 + *p*.

(d) is 2.8.

25. The owner of a pawnshop knows that 80 percent of the customers who enter her store will pawn an item. If there are 13 customers in the shop, the probability that *at most* 6 will pawn an item is

(a) 0.006.

(b) 0.001.

(c) 0.930.

(d) 0.007.

## Further Exercises

If possible, you can use any technology help available to solve the following problems.

1.  Suppose that on a very large campus, 2.5 percent of the students are foreign students. If 30 students are selected at random:
    (a) Find the probability that less than 7 from this group are foreign students.
    (b) Find the probability that the number of foreign students in this group will be between 2 and 8, inclusive.
    (c) Find the mean number of foreign students if groups of 30 are selected at random.

2.  For a binomial distribution with $n = 20$ and $p = 0.4$, which of the following is (are) *not* true?
    (a) The largest probability for this distribution occurs at a random variable value of 8.
    (b) The distribution is symmetrical.
    (c) The mean for this distribution is 8.
    (d) The standard deviation for this distribution is 2.1909.

3.  From past studies, it is known that 60 percent of the students at a given campus read the weekly campus newspaper. Your statistics professor recently wrote an article on the odds of winning the state lottery in this paper and would like to refer to this article during his lecture. If the class size is 20,
    (a) what is the probability that *none* of the students read the article?
    (b) what is the probability that *less than 5* students read the article?
    (c) what is the probability that *at least 5* of the students read the article?
    (d) what is the probability that *at least 75 percent* of the students in the class read the article?
    (e) what is the mean number of students that did *not* read the article?
    (f) what is the variance for the number of students that *read* the article?

4.  Because of no-shows, airlines commonly overbook their flights. An airline sells 100 tickets for a flight that can carry at most 95 passengers. If the probability that a passenger is a no-show is 0.10 and passengers arrive for the flight independent of each other, find
    (a) the probability that every passenger that shows up will be able to get on the flight.
    (b) the probability that not every passenger that shows up will be able to get on the flight.
    (c) the probability that the plane departs with empty seats.

5.  A college finds that 40 percent of all students take a course in statistics. If a group of 8 students is considered, find the probability that
    (a) *precisely 6* of them take statistics.
    (b) *at least 6* of them take statistics.

6.  (a) Find the value of the unknown probability $p$ that makes the following table into a probability distribution.

| $x$ | 0 | 1 | 2 | 3 | 4 |
|------|-------|-----|-------|-------|-------|
| $P(x)$ | 0.226 | $p$ | 0.328 | 0.215 | 0.106 |

    (b) If $X$ represents the number of successes in an experiment, use the table to find the probability of at most three successes.

(c) Find the mean of $X$.

(d) Find the standard deviation of $X$.

7. A small town has 15 walk–do not walk traffic signals that operate independently of each other. The probability is 0.98 that at any given time, these signals will be operating properly.

(a) If $X$ is the random variable representing the number of signals that operate properly, what kind (give a name) of random variable is $X$?

(b) Find the mean of the random variable named in (a).

(c) Find the variance of the random variable named in (a).

(d) Find the probability that all 15 signals are operating properly.

(e) Find the probability that exactly 13 of the signals are operating properly.

(f) Find the probability that more than 10 of the signals are operating properly.

8. In a recent study, it was found that 1 out of every 50 Pap smears sampled was misdiagnosed by a certain lab. In a sample of 100,

(a) find the probability that exactly 3 will be misdiagnosed.

(b) find the probability that between 2 and 4 (inclusive) will be misdiagnosed.

(c) find the probability that at most 4 will be misdiagnosed.

9. In a religious survey of southerners, it was found that 82 percent believed in angels. In a sample of 20 southerners,

(a) find the probability that exactly 5 will believe in angels.

(b) find the probability that between 2 and 4 (inclusive) will believe in angels.

(c) find the probability that at most 4 will not believe in angels.

## ANSWER KEY
### True/False Questions

1. F   2. T   3. T   4. F   5. F   6. T   7. F   8. F   9. F   10. T   11. F
12. F   13. F   14. T   15. T   16. F   17. T   18. T   19. F   20. T

### Completion Questions

1. probability   2. discrete   3. continuous   4. 1   5. $\sum xP(x)$   6. independent
7. $np(1-p)$   8. mean (expected value)   9. Binomial   10. discrete, continuous
11. 0, 1   12. $(1-p)$   13. continuous, interval   14. sample space   15. units

### Multiple-Choice Questions

1. (d)   2. (a)   3. (b)   4. (c)   5. (c)   6. (d)   7. (b)   8. (a)   9. (c)
10. (a)   11. (a)   12. (b)   13. (c)   14. (d)   15. (d)   16. (b)   17. (b)
18. (d)   19. (a)   20. (b)   21. (d)   22. (b)   23. (d)   24. (d)   25. (d)

# CHAPTER 9

# The Normal Probability Distribution

**Do I Need to Read This Chapter?**

Y ou should read this chapter if you need to review or to learn about

➡ The normal probability distribution
➡ The standard normal distribution
➡ $z$ scores
➡ Probability associated with any normal distribution

**Get Started**

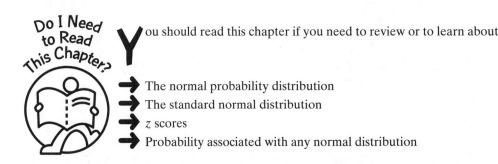

Here we will focus on a special continuous random variable that can assume any value in the interval $(-\infty, +\infty)$. That is, the random variable can assume any value on the real line. The distribution for this random variable is called the normal distribution and was originally called the gaussian distribution in honor of Karl Gauss, who in 1833 published a work describing it. This distribution is considered the most important probability distribution in all of statistics. A picture of a normal distribution is shown in Fig. 9-1.

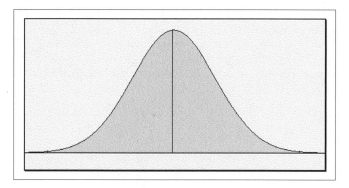

**Fig. 9-1:** The normal distribution

## 9-1 The Normal Distribution

The normal distribution can be viewed as the limiting distribution of a binomial random variable. That is, in a binomial experiment, if we use a fixed probability of success $p$, we can analyze what happens as the number of trials $n$ increases. To visualize what happens, we can construct histograms for a fixed $p$ and increasingly large $n$. **Figures 9-2** through **9-7** show histograms for $n = 5$, 10, 25, and 50 with $p = 0.1$ when the simulation was done 1,000 times. Superimposed on the histograms are smooth curves which show the shapes of the distributions. Observe that as $n$ increases for a fixed $p = 0.1$, the shape of the smooth curve becomes increasingly bell-shaped. This bell-shaped curve is associated with the normal distribution. Similar results can be observed for different $p$ values.

**Note:** If you have access to statistical software, simulate values for a binomial random variable with different $p$ and $n$ values and graph the results to observe other patterns.

**Figure 9-2** shows the histogram for 1,000 simulations of a binomial distribution with the number of trials $n = 5$ and the probability of success $p = 0.1$. The number of simulations with 0, 1, 2, 3, 4, and 5 successes was recorded; a frequency histogram with these values is shown in **Fig. 9.2.** Observe that the smooth curve that is used to approximate the distribution is skewed to the right.

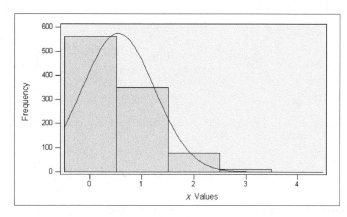

**Fig. 9-2:** Histogram with curve for $n = 5$ and $p = 0.1$

**Figure 9-3** shows the histogram for 1,000 simulations of a binomial distribution with the number of trials $n = 10$ and the probability of success $p = 0.1$. The number of simulations with 0, 1, 2, 3, 4, . . ., 10 successes was recorded; a frequency histogram with these values is shown in **Fig. 9-3.** Observe that the smooth curve that is used to approximate the distribution is still skewed to the right, but not as acutely as in **Fig. 9-2.**

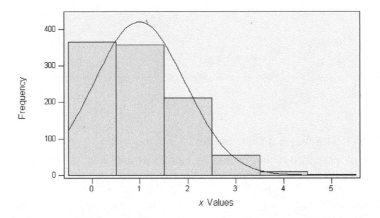

**Fig. 9-3:** Histogram with curve for $n = 10$ and $p = 0.1$

**Figure 9-4** shows the histogram for 1,000 simulations of a binomial distribution with the number of trials $n = 25$ and the probability of success $p = 0.1$. The number of simulations with 0, 1, 2, 3, 4, . . ., 25 successes was recorded; a frequency histogram with these values is shown in **Fig. 9-4.** Observe that the smooth curve that is used to approximate the distribution is slightly skewed to the right, but not as acutely as in **Fig. 9-3.**

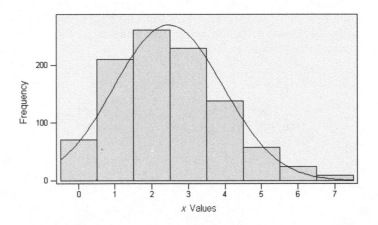

**Fig. 9-4:** Histogram with curve for $n = 25$ and $p = 0.1$

**Note:** The spaces in the histogram in **Fig. 9-5** indicate that there were no generated $x$ values in the simulation.

**Figure 9-5** shows the histogram for 1,000 simulations of a binomial distribution with the number of trials $n = 50$ and the probability of success $p = 0.1$. The number of simulations with 0, 1, 2, 3, 4, . . ., 50 successes was recorded; a frequency histogram with these values is shown in **Fig. 9-5.** Observe that the smooth curve that is used to approximate the distribution is almost symmetrical.

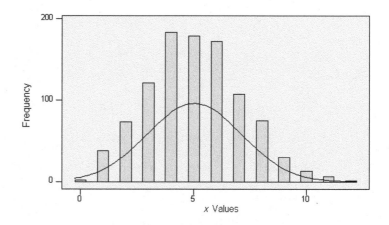

**Fig. 9-5:** Histogram with curve for $n = 50$ and $p = 0.1$

**Figures 9-2** through **9-5** show that as the number of trials increases and the probability of success is held fixed, the approximation to the frequency distribution for the number of successes becomes more and more bell-shaped.

The following two diagrams, **Figs. 9-6** and **9-7,** show distributions for $n = 10$ and $p = 0.5$ and for $n = 100$ and $p = 0.9$, respectively.

**Figures 9-6** and **9-7** show that as the probability increases and the number of trials is held fixed, the approximation to the frequency distribution for the number of successes becomes more and more bell-shaped.

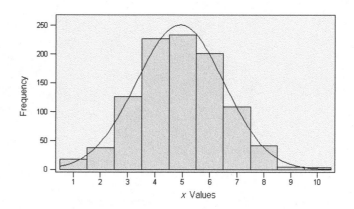

**Fig. 9-6:** Histogram with curve for $n = 10$ and $p = 0.5$

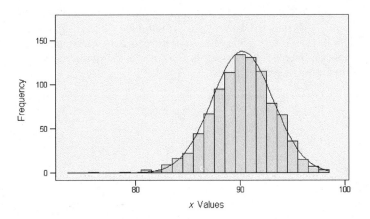

**Fig. 9-7:** Histogram with curve for $n = 100$ and $p = 0.9$

As mentioned earlier, the normal distribution is considered the most important probability distribution in all of statistics. It is used to describe the distribution of many natural phenomena, such as the height of a person, IQ score, weight, blood pressure, etc.

### 9-2 Properties of the Normal Distribution

The mathematical equation for the normal distribution is

$$y = \frac{e^{-(x-\mu)^2/2\sigma^2}}{\sigma\sqrt{2\pi}}$$

where $e \approx 2.718$, $\pi \approx 3.14$, $\mu =$ population mean, and $\sigma =$ population standard deviation. When this equation is graphed for a given $\mu$ and $\sigma$, a continuous, bell-shaped, symmetrical graph will result. Thus, we can display an infinite number of graphs for this equation, depending on the values of $\mu$ and $\sigma$. In such a case, we say we have a family of normal curves. Some representations of normal distribution curves are shown in **Figs. 9-8** through **9-10.**

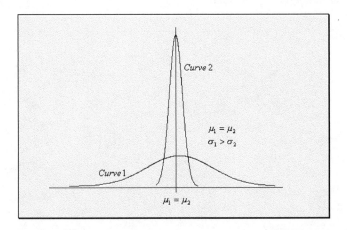

**Fig. 9-8:** Normal distributions with the same mean but with different standard deviations

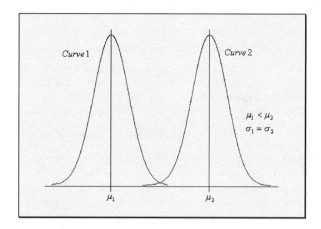

**Fig. 9-9:** Normal distributions with different means but with the same standard deviation

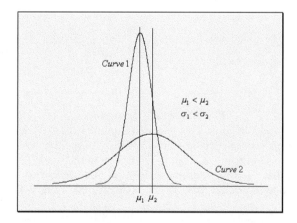

**Fig. 9-10:** Normal distributions with different means and different standard deviations

Note that these normal curves have similar shapes but are located at different points along the $x$ axis. Also, the larger the standard deviation, the more spread out is the distribution, and the curves are symmetrical about the mean value.

**Explanation of the term—normal distribution:** A **normal distribution** is a continuous, symmetrical, bell-shaped distribution of a normal random variable.

## Summary of the Properties of the Normal Distribution

- The curve is continuous.
- The curve is bell-shaped.
- The curve is symmetrical about the mean.
- The mean, median, and mode are located at the center of the distribution and are equal to each other.

- The curve is unimodal (single mode).
- The curve never touches the *x* axis.
- The total area under the normal curve is equal to 1.

A very important property of any normal distribution is that within a fixed number of standard deviations from the mean, all normal distributions have the same fraction of their probabilities. **Figures 9-11** through **9-13** illustrate this for $\pm 1\sigma$, $\pm 2\sigma$, and $\pm 3\sigma$ from the mean. Recall that this was discussed in **Chap. 3** as the *Empirical Rule.*

### Empirical Rule Revisited

**One-sigma rule:** Approximately 68 percent of the data values should lie within one standard deviation of the mean. That is, regardless of the shape of the normal distribution, the probability that a normal random variable will be within one standard deviation of the mean is approximately equal to 0.68. This is illustrated in Fig. 9-11.

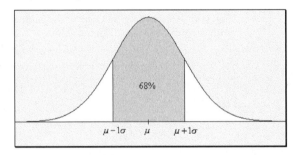

**Fig. 9-11:** One-sigma rule

**Two-sigma rule:** Approximately 95 percent of the data values should lie within two standard deviations of the mean. That is, regardless of the shape of the normal distribution, the probability that a normal random variable will be within two standard deviations of the mean is approximately equal to 0.95. This is illustrated in **Fig. 9-12.**

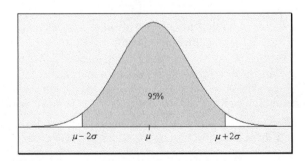

**Fig. 9-12:** Two-sigma rule

**Three-sigma rule:** Approximately 99.7 percent of the data values should lie within three standard deviations of the mean. That is, regardless of the shape of the normal distribution, the probability that a normal random variable will be within three standard deviations of the mean is approximately equal to 0.997. This is illustrated in **Fig. 9-13.**

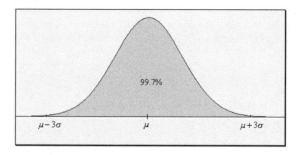

**Fig. 9-13:** Three-sigma rule

## Quick Tips

1. The total area under the normal curve is equal to 1.
2. The probability that the normal random variable is equal to a given discrete value is always zero, since the normal random variable is continuous.
3. The probability that a normal random variable is between two values is given by the area under the normal curve between the two given values and the horizontal axis.

### 9-3 The Standard Normal Distribution

Since each normally distributed random variable has its own mean and standard deviation, the shape and location of normal curves will vary. Thus, one would have to have information on the areas for all normal distributions. This, of course, is impractical. Therefore, we use the information for a special normal distribution called the **standard normal distribution** to simplify this situation.

**Explanation of the term—standard normal distribution:** The standard normal distribution is a normal distribution with a mean of 0 and a standard deviation of 1.

Any normal random variable can be converted to a standard normal random variable by computing the corresponding $z$ score. The $z$ score is computed from the following formula:

$$z = \frac{\text{value} - \text{mean}}{\text{standard deviation}} = \frac{x - \mu}{\sigma}$$

In the equation, $x$ is the value of a normal random variable $X$ with mean $\mu$ and standard deviation $\sigma$.

## Quick Tip

The $z$ score is normally distributed with a mean of 0 and a standard deviation of 1.

Recall that a $z$ score gives the number of standard deviations a specific value is above or below the mean.

Extensive tables can be constructed for the standard normal random variable to aid in finding areas under the standard normal curve. Usually, standard normal tables give the area between the mean of 0 and a value $z$, as shown in **Fig. 9-14.**

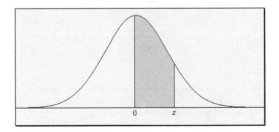

**Fig. 9-14:** Area under the standard normal curve between 0 and $z$

A sample portion of a table using four decimal places is shown in **Table 9-1.**

**Table 9-1:** Sample of the Standard Normal Table

| $z$ | 0.00 | 0.01 | 0.02 | 0.03 |
|-----|------|------|------|------|
| 0.0 | 0.0000 | 0.0040 | 0.0080 | 0.0120 |
| 0.1 | 0.0398 | 0.0438 | 0.0478 | 0.0517 |
| 0.2 | 0.0793 | 0.0832 | 0.0871 | 0.0910 |
| 0.3 | 0.1179 | 0.1217 | 0.1255 | 0.1293 |
| 0.4 | 0.1554 | 0.1591 | 0.1628 | 0.1664 |
| 0.5 | 0.1915 | 0.1950 | 0.1985 | 0.2019 |
| 0.6 | 0.2257 | 0.2291 | 0.2324 | 0.2357 |
| 0.7 | 0.2580 | 0.2611 | 0.2642 | 0.2673 |
| 0.8 | 0.2881 | 0.2910 | 0.2939 | 0.2967 |
| 0.9 | 0.3159 | 0.3186 | 0.3212 | 0.3238 |
| 1.0 | 0.3413 | 0.3438 | 0.3461 | 0.3485 |

The first column in **Table 9-1** gives the $z$ values correct to one decimal place, and the first row gives the second decimal place for a $z$ score. For example, if we want to find the area between $z = 0$ and $z = 0.92$, we will find $z = 0.9$ in the first column, then look for $z = 0.02$ along the first row. Where the corresponding row and column intersect gives the value 0.3212. This is the area between $z = 0$ and $z = 0.92$. This is equivalent to finding $P(0 \leq z \leq 0.92)$. The area is depicted in **Fig. 9-15.**

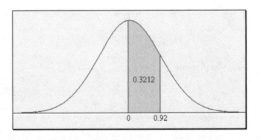

**Fig. 9-15:** Area representing $P(0 \leq z \leq 0.92)$

More extensive tables are given in the Appendix. Refer to them to solve the problems in this chapter.

***Example 9-1:*** Find the area under the standard normal curve between $z = 0$ and $z = 2.0$.

**Solution:** This is equivalent to finding $P(0 \leq z \leq 2.0)$. From the standard normal tables in the Appendix, for $z = 2.0$, the corresponding value is 0.4772. Thus, $P(0 \leq z \leq 2) = 0.4772$. This is shown in **Fig. 9-16.**

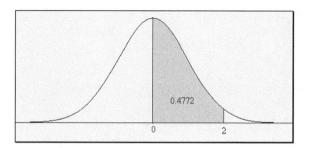

**Fig. 9-16:** Area for $P(0 \leq z \leq 2)$

***Example 9-2:*** Find the area under the standard normal curve between $z = 0$ and $z = -1.8$.

**Solution:** This is equivalent to finding $P(-1.8 \leq z \leq 0)$. Now, from the symmetry of the distribution, $P(-1.8 \leq z \leq 0) = P(0 \leq z \leq 1.8)$. From the standard normal tables at the end of the text, for $z = 1.8$, the corresponding value is 0.4641. Thus, $P(-1.8 \leq z \leq 0) = 0.4641$. This is shown in **Fig. 9-17.**

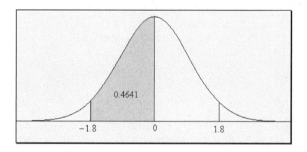

**Fig. 9-17:** Area for $P(-1.8 \leq z \leq 0)$

***Example 9-3:*** Find the area under the standard normal curve to the right of $z = 1.5$.

**Solution:** This is equivalent to finding $P(z \geq 1.5)$. From the standard normal tables at the end of the text, for $z = 1.5$, the corresponding value is 0.4332. But this is the area between 0 and 1.5. Since the total area under the curve is 1, and because of the symmetry of the distribution, the total area to the right of 0 will be equal to 0.5. Thus, the required area for $P(z \geq 1.5) = 0.5 - 0.4332 = 0.0668$. This is shown in **Fig. 9-18.**

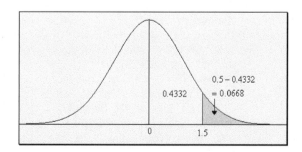

**Fig. 9-18:** Area for $P(z \geq 1.5)$

***Example 9-4:*** Find the area under the standard normal curve to the left of $z = -1.75$.

**Solution:** This is equivalent to finding $P(z \leq -1.75)$. Now, from the symmetry of the distribution, $P(z \leq -1.75) = P(z \geq 1.75)$. So now this is a similar problem to **Example 9-3.** From the standard normal tables at the end of the text, for $z = 1.75$, the corresponding value is 0.4599. But this is the area between 0 and 1.75. Since the total area under the curve is 1, and because of the symmetry of the distribution, the total area to the right of 0 will be equal to 0.5. Thus, the required area for $P(z \leq -1.75) = P(z \geq 1.75) = 0.5 - 0.4599 = 0.0401$. This is shown in **Fig. 9-19.**

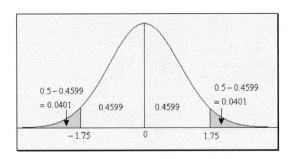

**Fig. 9-19:** Area for $P(z \leq -1.75)$

***Example 9-5:*** Find the area under the standard normal curve between $z = 1.5$ and $z = 2.5$.

**Solution:** That is, we need to find $P(1.5 \leq z \leq 2.5)$. This is equivalent to finding $P(0 \leq z \leq 2.5) - P(0 \leq z \leq 1.5)$. From the standard normal tables at the end of the text, for $z = 2.5$, the corresponding value is 0.4938, and for $z = 1.5$, the corresponding value is 0.4332. Thus, the required area for $P(1.5 \leq z \leq 2.5) = P(0 \leq z \leq 2.5) - P(0 \leq z \leq 1.5) = 0.4938 - 0.4332 = 0.0606$. This is shown in **Fig. 9-20.**

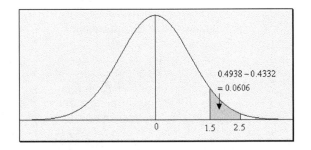

**Fig. 9-20:** Area for $P(1.5 \leq z \leq 2.5)$

***Example 9-6:*** Find the area under the standard normal curve between $z = -2.78$ and $z = -1.66$.

**Solution:** This is equivalent to finding $P(-2.78 \leq z \leq -1.66)$. Because of the symmetry of the distribution, this is equivalent to finding $P(1.66 \leq z \leq 2.78) = P(0 \leq z \leq 2.78) - P(0 \leq z \leq 1.66)$. From the standard normal tables at the end of the text, for $z = 2.78$, the corresponding value is 0.4973, and for $z = 1.66$, the corresponding value is 0.4515. Thus, the required area for $P(1.66 \leq z \leq 2.78) = P(0 \leq z \leq 2.78) - P(0 \leq z \leq 1.66) = 0.4973 - 0.4515 = 0.0458$. This is shown in **Fig. 9-21.**

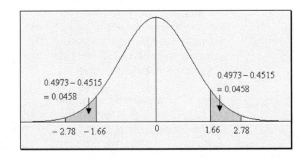

**Fig. 9-21:** Area for $P(-2.78 \leq z \leq -1.66)$

***Example 9-7:*** Find the area under the standard normal curve between $z = -2.79$ and $z = 1.71$.

**Solution:** That is, we need to find $P(-2.79 \leq z \leq 1.71)$. This is equivalent to finding $P(-2.79 \leq z \leq 0) + P(0 \leq z \leq 1.71) = P(0 \leq z \leq 2.79) + P(0 \leq z \leq 1.71) = 0.4974 + 0.4564 = 0.9538$. This is shown in **Fig. 9-22.**

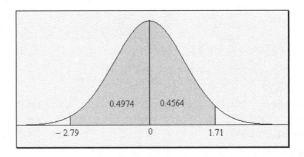

**Fig. 9-22:** Area for $(-2.79 \leq z \leq 1.71)$

*Example 9-8:* Find the area under the standard normal curve to the right of $z = -2.79$.

**Solution:** That is, we need to find $P(z \geq -2.79)$. Now, $P(z \geq -2.79) = 0.4974 + 0.5 = 0.9974$. This is shown in **Fig. 9-23.**

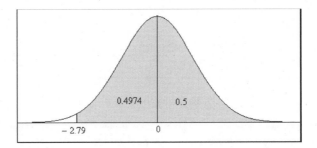

**Fig. 9-23:** Area for $P(z \geq -2.79)$

**Note:** Other problems can be illustrated which may involve any combination of the above situations.

## Quick Tip

In solving problems relating to the standard normal distribution, it may be helpful if you use the following procedure:

- Write out the equivalent probability statement.
- Draw a normal curve.
- Shade in the desired area.
- Use the standard normal distribution table to find the shaded area.

### 9-4    Applications of the Normal Distribution

We can use the standard normal distribution curve to solve problems involving variables that are normally or approximately normally distributed. To solve problems involving normal random variables, we need to transform the original normal variable into a standard normal random variable by using

$$z = \frac{\text{value} - \text{mean}}{\text{standard deviation}} = \frac{x - \mu}{\sigma}$$

In the equation, $x$ is the value of a normal random variable $X$ with mean $\mu$ and standard deviation $\sigma$. Once the transformation has been made, the problem may be reduced to one similar to those presented in the previous section.

***Example 9-9:*** If IQ scores are normally distributed with a mean of 100 and a standard deviation of 5, what is the probability that a person chosen at random will have an IQ score greater than 110?

**Solution:** Let $X$ = IQ score. Then we need to find $P(X > 110)$. The equivalent $z$ score is $z = (110 - 100)/5 = 2$. Thus, $P(X > 110) = P(z > 2) = 0.5 - 0.4772 = 0.0228$. This is shown in **Fig. 9-24.**

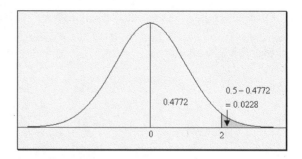

**Fig. 9-24:** Area for $P(X > 110) = P(z > 2)$

## Quick Tip

All application problems involving the normal distribution can be reduced to simple problems using $z$ scores. These reduced problems can be solved in the same way as the previous examples or combinations of them.

***Example 9-10:*** Suppose family incomes in a town are normally distributed with a mean of $1,200 and a standard deviation of $600 per month. What is the probability that a family has an income between $1,400 and $2,250?

**Solution:** Let $X$ = income. So we need to find $P(1,400 \leq X \leq 2,250)$. First we need to transform the values of 1,400 and 2,250 to $z$ scores. Let $z_1 = (1,400 - 1,200)/600 = 0.33$, and $z_2 = (2,250 - 1,200)/600 = 1.75$. Thus, $P(1,400 \leq X \leq 2,250) = P(0.33 \leq z \leq 1.75)$. Now $P(0.33 \leq z \leq 1.75) = P(0 \leq z \leq 1.75) - P(0 \leq z \leq 0.33) = 0.4599 - 0.1293 = 0.3306$. This is shown in **Fig. 9-25.**

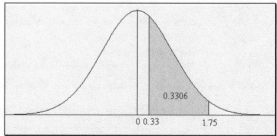

**Fig. 9-25:** Area for $P(1,400 \leq X \leq 2,250) = P(0.33 \leq z \leq 1.75)$

---

**Quick Tip**

In solving application problems relating to the normal distribution, it may be helpful if you use the following procedure:

- Define the appropriate normal variable with appropriate parameters (mean and standard deviation).
- Write out an appropriate probability statement.
- Write out an equivalent transformed probability statement using the *z* score.
- Draw a normal curve.
- Shade in the desired area.
- Use the normal distribution table to find the shaded area.

---

***Example 9-11:*** A four-year college will accept any student ranked in the top 60 percent on a national examination. If the test score is normally distributed with a mean of 500 and a standard deviation of 100, what is the cutoff score for acceptance?

**Solution:** Let $X$ = test score and let $x_0$ be the cutoff score. We need to find $x_0$ such that $P(X > x_0) = 0.6$. What we need to do is to find the corresponding $z$ score, say $z_0$, such that an area of 0.6 is to the right of $z_0$. The picture in **Fig. 9-26** illustrates the desired area.

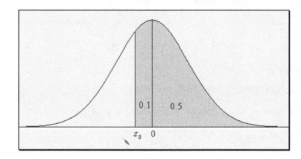

**Fig. 9-26:** Area associated with $z_0$ such that $P(z \ge z_0) = 0.6$

Using the area of 0.1, we have from the body of the table a corresponding $z$ score of 0.25. Since, from **Fig. 9-25**, $z_0$ is to the left of 0, $z_0 = -0.25$. We can use this information to solve for $x_0$ by using the equation $z_0 = (x_0 - \mu)/\sigma$ or $-0.25 = (x_0 - 500)/100$. Solving gives $x_0 = 475$. That is, the minimum score the college will accept is 475.

**Technology Corner**

All of the concepts discussed in this chapter can be computed and illustrated using some statistical software packages. All scientific and graphical calculators will aid directly in the computations. In addition, the newer calculators may have the normal distribution, from which you can compute the normal probabilities, such as the TI-83. Some calculators, like the TI-83, will allow you to shade under the normal curve as well. If you own a calculator, you should consult the manual to determine what statistical features are included.

**Illustration:** **Figure 9-27a** and **b** shows the solution for **Example 9-10** computed by the MINITAB software. **Figure 9-28** shows the same solution computed by the TI-83 calcula-

**Example 9-10--Using actual X values**

**Cumulative Distribution Function**

```
Normal with mean = 1200.00 and standard deviation = 600.000

        x       P( X <= x)
  1.40E+03        0.6306
```

**Cumulative Distribution Function**

```
Normal with mean = 1200.00 and standard deviation = 600.000

        x       P( X <= x)
  2.25E+03        0.9599
```

**So P(1400 <= X <= 2250) =P(X<=2250) - P(X<=1400) =  0.9599 - 0.6306 = 0.3293.**

**Fig. 9-27a:** MINITAB output for Example 9-10 using exact $X$ values

**Example 9-10--Using z values**

**Cumulative Distribution Function**

```
Normal with mean = 0 and standard deviation = 1.00000

        x       P( X <= x)
   0.3300         0.6293
```

**Cumulative Distribution Function**

```
Normal with mean = 0 and standard deviation = 1.00000

        x       P( X <= x)
   1.7500         0.9599
```

**Thus P(0.33 < = z < = 1.75) =P(Z <= 1.75) - P(z <=0.33) =  0.9599 - 0.6293 = 0.3306.**

**NOTE: We are using the cumulative probabilities to help solve the problem.**

**Fig. 9-27b:** MINITAB output for Example 9-10 using $z$ values

```
normalcdf(.33,1.
75,0,1)
          .3306409314
normalcdf(1400,2
250,1200,600)
          .3293822902
█
```

**Fig. 9-28:** TI-83 output for Example 9-10 using both the $X$ values and the $z$ values

tor. Other alternative approaches to the solution can be obtained with a little mathematical manipulation by the software and the calculator.

**Note:** When the actual values for the random variable are used, the computed probability is slightly different from that when the standard normal distribution is used. Using the actual values of the random variable, we get 0.3293, and using the $z$ values, we get 0.3306.

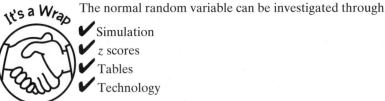

The normal random variable can be investigated through

✔ Simulation

✔ z scores

✔ Tables

✔ Technology

Again, care should always be taken when computing probabilities. Also, care should be taken when using the normal probability tables.

**Test Yourself**

## True/False Questions

1. Continuous random variables typically arise from counting some type of quantity.
2. The probability that a continuous random variable assumes any single value will always be nonzero.
3. The normal distribution is centered at its mean.
4. The area under the normal curve to the left of its mean is –0.5.
5. Approximately 68 percent of a normal population will lie within one standard deviation of its mean.
6. The total area under the normal curve is approximately 1.
7. The probability that a normal random variable $X$ is *at least* some number $a$ can be denoted as $P(X > a)$.
8. The $z$ value or $z$ score is computed from the equation $z = \dfrac{\mu - x}{\sigma}$, where $x$ is the value of a normal random variable $X$ with mean $\mu$ and standard deviation $\sigma$.
9. A positive $z$ score gives the number of standard deviations a specified value of a normal random variable is above its mean.
10. A random variable that can assume any value in the set of real numbers must be normally distributed.
11. For any continuous random variable $X$, $P(X > 2) = P(X \geq 2)$.
12. The standard normal distribution is symmetrical about 0.
13. The area under any normal curve between $a$ and $b$ gives the probability that the normal random variable lies between $a$ and $b$.
14. The standard normal distribution has a mean of 0 and a variance of 1.
15. The mean, the median, and the mode for a normal distribution are all equal.
16. If $X$ is a normal random variable, then $P(X > 0) = 0.5$ is always true.
17. Because of the right and left tails of the normal curve, we can say that the normal distribution is skewed to both the left and the right.
18. Not all normal distributions can be transformed to a standard normal distribution.
19. If the computed $z$ value is negative, you have definitely made a computational error.
20. A $z$ value of 1 indicates that 1 percent of the area under the standard normal curve is to the right of the $z$ value of 0.

## Completion Questions

1. Any normal distribution is (symmetrical, asymmetrical) _____ about its mean.
2. The total area under any normal curve is always (0.5, 1) _____.
3. The standard normal distribution is specified by a mean of (0, 1) _____ and a standard deviation of (0, 1) _____.

4. If a computed $z$ value is zero, then the value of the normal random variable must be (greater than, equal to, less than) _____ the mean of this normal random variable.

5. If $z$ is the standard normal random variable, then $P(z < 0) = P(z > 0) = (0, 0.5, 1)$ _____.

6. If $z$ is the standard normal random variable, then $P(-1 < z < 1) \approx (0.68, 0.95, 0.997)$ _____.

7. The second quartile of the standard normal random variable corresponds to a $z$ score of $(-3, -2, -1, 0, 1, 2, 3)$ _____.

8. The percentage of the standard normal distribution that lies within three standard deviations of the mean is approximately $(68, 95, 99.7)$ _____ percent.

9. A $z$ score of $z = (-2, 2)$ _____ corresponds to a value of the standard normal random variable that is above the mean by two standard deviations.

10. Transforming a normal distribution to a standard normal distribution will not change the (shape, mean, standard deviation) _____ of the distribution.

11. If two values of a normal random variable from the same distribution correspond to the same $z$ score, then these values must be (negative, positive, equal) _____.

12. When a normal random variable is transformed to a $z$ score, the resulting distribution of $z$ scores will have a standard deviation value of $(0, 1)$ _____.

13. A (negative, positive) _____ $z$ score always corresponds to a value of the normal random variable that is less than the mean of the random variable.

14. The $z$ scores for a standard normal random variable are the set of all (whole, real) _____ numbers.

15. The normal probability distribution is a (continuous, discrete) _____ probability distribution.

16. The mean, the median, and the mode for a normal random variable are all (equal, not equal) _____ to each other.

17. The probability that a continuous random variable assumes a discrete value is always $(0, 1)$ _____.

18. The total area under any standard normal curve is always $(0.5, 1)$ _____.

19. When normal scores are transformed into $z$ scores, the resulting $z$ scores will have a mean of $(0, 1)$ _____.

20. For a normal random variable, the probability of observing a value less than or equal to its mean is $(0, 0.5, 1)$ _____.

21. The $z$ value associated with a given value of a normal random variable measures how far (above, below, above or below) _____, in terms of standard deviations, the value is from the mean of the distribution.

22. When the value of the standard deviation increases, the value of the $z$ score will generally tend to (increase, decrease) _____.

23. Very large positive or negative $z$ scores will correspond to raw scores that are generally (more, less) _____ likely to occur.

## Multiple-Choice Questions

1. The area under the standard normal curve between $-2.0$ and $-1$ is
   (a) 0.0228.
   (b) 0.1359.

(c) 0.4772.

(d) 0.3413.

2. The probability that an observation taken from a standard normal population will be between −1.96 and 1.28 is

(a) 0.0753.

(b) −0.0753.

(c) 0.1253.

(d) 0.8747.

3. The value of $z_0$ such that $P(z \le z_0) = 0.8997$ is

(a) 1.28.

(b) 0.00.

(c) 0.1003.

(d) none of the above.

4. The two $z$ values such that the area bounded by them is equal to the middle 68.26 percent of the standard normal distribution is

(a) ±3.

(b) ±1.

(c) ±2.

(d) ±1.96.

5. The area under the standard normal curve between $z = -1.68$ and $z = 0$ is

(a) 0.4535.

(b) 0.0465.

(c) 0.9535.

(d) −0.4535.

6. If $X$ is a normal random variable with a mean of 15 and a variance of 9, then $P(X < 18)$ is

(a) 0.7486.

(b) 0.8413.

(c) 0.3413.

(d) 0.1587.

7. If $X$ is a normal random variable with a mean of 15 and a variance of 9, then $P(X = 18)$ is

(a) 0.8413.

(b) 0.0000.

(c) 0.3413.

(d) 0.1587.

8. The two $z$ values such that the area bounded by them is equal to the middle 90 percent of the standard normal distribution is

(a) ±1.640.

(b) ±1.650.

(c) ±2.000.

(d) ±1.645.

9. The time it takes for a dose of a certain drug to be effective as a sedative on lab animals is normally distributed with a mean of 1 hour and a standard deviation of 0.1 hour. If $X$ represents this time, then $P(X > 1.1)$ is
   (a) 0.0000.
   (b) 0.5000.
   (c) 0.3643.
   (d) 0.1587.

10. The area under any normal curve that is within two standard deviations of the mean is approximately
   (a) 0.950.
   (b) 0.680.
   (c) 0.997.
   (d) 0.500.

11. Which of the following does not apply to the normal distribution?
   (a) The normal curve is unimodal.
   (b) The total probability under the curve is 1.
   (c) The normal curve is symmetrical about its standard deviation.
   (d) The mean, the median, and the mode are all equal.

12. If $z$ is a standard normal random variable, then the probability that $z > 1$ or $z < -2$ is
   (a) 0.1587.
   (b) 0.0228.
   (c) 0.8185.
   (d) 0.1815.

13. A standard normal distribution is a normal distribution with
   (a) $\mu = 1$ and $\sigma = 0$.
   (b) $\mu = 0$ and $\sigma = 1$.
   (c) any mean and $\sigma = 0$.
   (d) any mean and any standard deviation.

14. If IQ scores are normally distributed with a mean of 100 and a standard deviation of 20, then the probability of a person's having an IQ score of at least 130
   (a) is 0.4332.
   (b) is 0.5000.
   (c) does not exist.
   (d) is 0.0668.

15. A bank finds that the balances in its savings accounts are normally distributed with a mean of $500 and a standard deviation of $50. The probability that a randomly selected account has a balance of more than $600 is
   (a) 0.4772.
   (b) 0.0228.
   (c) 0.9772.
   (d) 0.0000.

16. The lifetime of a certain brand of tires is normally distributed; they last an average of 50,000 miles with a standard deviation of 8,400. The probability that a randomly selected tire will last beyond 55,000 miles is
   (a) 0.2257.
   (b) 0.7257.

(c) 0.0000.

(d) 0.2743.

17. The waiters in a restaurant receive an average tip of $20 per table with a standard deviation of $5. The amounts of tips are normally distributed, and a waiter feels that he has provided excellent service if the tip is more than $25. The probability that a waiter has provided excellent service to a table is

   (a) 0.1587.

   (b) 0.8413.

   (c) 0.8000.

   (d) 0.6587.

18. The waiters in a bar receive an average tip of $20 per table with a standard deviation of $5. The amounts of tips are normally distributed, and a waiter feels that he has provided excellent service if the tip is more than $25. The probability that a waiter has *not* provided excellent service (according to the waiter's theory) to a table is

   (a) 0.1587.

   (b) 0.8413.

   (c) 0.8000.

   (d) 0.6587.

19. Suppose family incomes in a town are normally distributed with a mean of $1,200 and a standard deviation of $600 per month. The probability that a given family has an income over $2,000 per month is

   (a) 0.0918.

   (b) 0.9082.

   (c) 0.4082.

   (d) 0.5918.

20. Suppose family incomes in a town are normally distributed with a mean of $1,200 and a standard deviation of $600 per month. The probability that a given family has an income between $1,000 and $2,050 per month is

   (a) 0.4585.

   (b) 0.7001.

   (c) 0.4222.

   (d) 0.5515.

21. The life of a brand of battery is normally distributed with a mean of 62 hours and a standard deviation of 6 hours. The probability that a single randomly selected battery lasts more than 70 hours is

   (a) 0.0000.

   (b) 0.0918.

   (c) 0.4082.

   (d) 0.9082.

22. The life of a brand of battery is normally distributed with a mean of 62 hours and a standard deviation of 6 hours. The percentage of batteries that should last between 55 and 65 hours is

   (a) 0.4295.

   (b) 0.6875.

   (c) 0.5705.

   (d) 0.3125.

23. For any $z$ distribution, the sum of all the associated $z$ scores will always be
    (a) equal to 1.
    (b) less than 1.
    (c) greater than 1.
    (d) equal to 0.

24. The average score on one of your statistics examination was 75 with a standard deviation of 10. If your corresponding $z$ score was 2, then your corresponding raw score and percentile rank (approximate) would be
    (a) 55; 48 percent.
    (b) 65; 12 percent.
    (c) 85; 75 percent.
    (d) 95; 98 percent.

25. If the $z$ score associated with a given raw score is equal to 0, this implies that
    (a) the raw score equals 0.
    (b) the raw score does not exist.
    (c) the raw score is extremely large.
    (d) the raw score is the same as the mean.

26. Which of the following is not needed in computing the $z$ score for a normal random variable?
    (a) The raw score
    (b) The percentile rank of the raw score
    (c) The standard deviation
    (d) The mean score

27. The U.S. Bureau of the Census reports that the mean annual alimony income received by women is $3,000 with a standard deviation of $7,500. If the annual alimony income is assumed to be normally distributed, then the proportion of women who receive less than $2,000 in annual alimony income is
    (a) −0.0517.
    (b) 0.5517.
    (c) 0.8966.
    (d) 0.4483.

28. The weights of male basketball players on a certain campus are normally distributed with a mean of 180 pounds and a standard deviation of 26 pounds. If a player is selected at random, the probability that the player will weigh more than 225 pounds is
    (a) 0.0418.
    (b) 0.4582.
    (c) 0.9582.
    (d) 0.5418.

29. The weights of male basketball players on a certain campus are normally distributed with a mean of 180 pounds and a standard deviation of 26 pounds. If a player is selected at random, the probability that the player will weigh less than 225 pounds is
    (a) 0.5418.
    (b) 0.9582.
    (c) 0.4582.
    (d) 0.0418.

30. The weights of male basketball players on a certain campus are normally distributed with a mean of 180 pounds and a standard deviation of 26 pounds. If a player is selected at random, the probability that the player will weigh between 180 and 225 pounds is
    (a) 0.5000.
    (b) 0.5418.
    (c) 0.4582.
    (d) 0.9582.

31. If $X$ is a normally distributed random variable with a mean of 6 and a variance of 4, then the probability that $X$ is less than 10 is
    (a) 0.8413.
    (b) 0.9772.
    (c) 0.3413.
    (d) 0.4772.

32. If $X$ is a normally distributed random variable with a mean of 6 and a variance of 4, then the probability that $X$ is greater than 10 is
    (a) 0.1587.
    (b) 0.0228.
    (c) 0.6587.
    (d) 0.5228.

33. In a normal distribution, the distribution will be less spread out when
    (a) the mean of the raw scores is small.
    (b) the median of the raw scores is small.
    (c) the mode of the raw scores is small.
    (d) the standard deviation of the raw scores is small.

34. For the standard normal random variable $z$, $P(z = 0)$ is
    (a) 0.5.
    (b) less than 0.5.
    (c) the same as $P(-0.5 \leq z \leq 0.5)$.
    (d) 0.

## Further Exercises

If possible, you can use any technology available to help solve the following problems.

1. The Test of English as a Foreign Language (TOEFL) is required by most universities to help in their admission decisions on international students. From past records at a certain university, these scores are approximately normally distributed with a mean of 490 and a variance of 6,400. If an international student's application is selected at random,
    (a) what is the probability that the student's TOEFL score is at most 450?
    (b) what is the probability that the student's TOEFL score is at least 520?
    (c) what is the probability that the student's TOEFL score is equal to the mean score?
    (d) what is the probability that the student's TOEFL score is between 475 and 525?
    (e) what is the student's percentile rank if her score is 500?

2. The amount of hot chocolate dispensed by a hot chocolate machine is normally distributed with a mean of 16 ounces and a standard deviation of 2 ounces.
    (a) If the cups can hold a maximum of 18 ounces each, what is the probability that a selected cup will be overfilled?

(b) If the cups can hold a maximum of 18 ounces each, what is the probability that a selected cup will be underfilled?

(c) If the cups can hold a maximum of 18 ounces each, what is the probability that a selected cup will be less than half-filled?

3. The annual rainfall in a particular region of the country has a mean of 75 inches and a standard deviation of 10 inches. If the rainfall is assumed to be normally distributed, find in any given year
   (a) the probability that it rained more than 100 inches.
   (b) the probability that it rained at most 65 inches.
   (c) the probability that it rained between 60 and 95 inches.
   (d) the percentile corresponding to 75 inches of rainfall.

4. The diameters of ball bearings manufactured by a particular machine are normally distributed with a mean of 2 cm and a standard deviation of 0.2 cm. If a ball bearing is selected at random,
   (a) find the probability that the diameter is greater than 1.77 cm.
   (b) find the 70th percentile for the distribution of ball bearing diameters.
   (c) find the probability that the diameter is between 1.79 to 2.11 cm.

5. A company manufactures a certain brand of light bulb. The lifetime for these bulbs is normally distributed with a mean of 4,000 hours and a standard deviation of 200 hours.
   (a) What proportion of these bulbs will last beyond 4,200 hours?
   (b) What lifetime should the company claim for these bulbs in order for only 2 percent of the bulbs to burn out before the claimed lifetime?

## ANSWER KEY

### True/False Questions

1. F    2. F    3. T    4. F    5. T    6. F    7. T    8. F    9. T    10. F    11. T
12. T    13. T    14. T    15. T    16. F    17. F    18. F    19. F    20. F

### Completion Questions

1. symmetrical    2. 1    3. 0, 1    4. equal to    5. 0.5    6. 0.68    7. 0    8. 99.7    9. 2
10. shape    11. equal    12. 1    13. negative    14. real    15. continuous    16. equal
17. 0    18. 1    19. 0    20. 0.5    21. above or below    22. decrease    23. less

### Multiple-Choice Questions

1. (b)    2. (d)    3. (a)    4. (b)    5. (a)    6. (b)    7. (b)    8. (d)    9. (d)
10. (a)    11. (c)    12. (d)    13. (b)    14. (d)    15. (b)    16. (d)    17. (a)
18. (b)    19. (a)    20. (d)    21. (b)    22. (c)    23. (d)    24. (d)    25. (d)
26. (b)    27. (d)    28. (a)    29. (b)    30. (c)    31. (b)    32. (b)    33. (d)
34. (d)

# CHAPTER 10

# Sampling Distributions and the Central Limit Theorem

**Do I Need to Read This Chapter?**

You should read this chapter if you need to review or to learn about

➤ The sampling distribution of a sample proportion

➤ The sampling distribution of a sample mean

➤ The Central Limit Theorem

➤ The sampling distribution of a difference between two independent sample proportions

➤ The sampling distribution of a difference between two independent sample means

---

### Get Started

Here we will focus on sampling distributions and the Central Limit Theorem. Sampling distributions for the sample mean, the sample proportion, differences of sample proportions from two independent populations, and differences of sample means from two independent populations will be investigated, along with the Central Limit Theorem for these situations. We will consider only sampling populations that are infinite. The Central Limit Theorem will lay the foundation for the broad area of statistical inference.

---

## 10-1 Sampling Distribution of a Sample Proportion

Suppose we are interested in the true proportion of Americans who favor doctor-assisted suicide. If we let the population proportion be denoted by $p$, then $p$ can be defined by

$$p = \frac{\text{number of Americans who favor doctor-assisted suicide}}{\text{total number of Americans}}$$

Since the population of interest is too large for us to observe all Americans, we can estimate the true proportion by observing a random sample from the population. If we let the sample proportion be denoted by $\hat{p}$ (read as "$p$ hat"), then the point estimate $\hat{p}$ can be defined by

$$\hat{p} = \frac{\text{number of Americans who favor doctor-assisted suicide in the sample}}{\text{sample size}}$$

**Explanation of the term—point estimate:** A **point estimate** is a single number that is used to estimate a population parameter.

Suppose we assume that the true proportion of Americans who favor doctor-assisted suicide is 68 percent (*source: USA Today* Snapshot). (In general, we will not know the true population proportion.) If we select a random sample of, say, 50 Americans, we may observe that 35 of them favor doctor-assisted suicide. Thus, our sample proportion of Americans who favor doctor-assisted suicide will be $\hat{p} = 35/50 = 0.7$, or 70 percent. If we were to select another random sample of size 50, we would most likely obtain a different value for $\hat{p}$. If we selected 50 different samples of the same sample size and computed the proportion of Americans who favor doctor-assisted suicide for all 50 samples, we should not expect these values to all be the same. That is, there will be some variability in these computed proportions. Pictorially, the situation is demonstrated in **Fig. 10-1.**

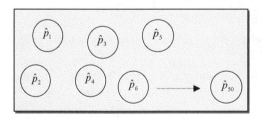

**Fig. 10-1:** Fifty samples of size 50

These 50 sample proportions constitute a **sampling distribution of a sample proportion.**

**Explanation of the term—sampling distribution of a sample proportion:** A sampling distribution of a sample proportion is a distribution obtained by using the proportions computed from random samples of a specific size obtained from a population.

In order to investigate properties of the sampling distribution of a sample proportion, simulations of the situation can be done. For example, the MINITAB statistical software can be used for the simulation. In the example, 50 samples of size 50 were generated. The distribution used in the simulation was the binomial distribution with parameters $n = 50$ and $p = 0.68$. This assumed distribution is reasonable, since we are interested in the proportion (number) of persons in the sample of size 50 who support doctor-assisted suicide. You may try your own simulation if you have access to such statistical software. The descriptive statistics for a simulation are shown in **Fig. 10-2.**

**Descriptive Statistics**

| Variable | N | Mean | Median | TrMean | StDev | SE Mean |
|---|---|---|---|---|---|---|
| Sample Proportion | 50 | 0.67600 | 0.68000 | 0.67682 | 0.06337 | 0.00896 |

| Variable | Minimum | Maximum | Q1 | Q3 |
|---|---|---|---|---|
| Sample Proportion | 0.54000 | 0.82000 | 0.64000 | 0.72500 |

**Fig. 10-2:** MINITAB descriptive statistics of simulation for sample proportion

Let $\mu_{\hat{p}}$ represent the mean of the sample proportions and $\sigma_{\hat{p}}$ represent the standard deviation of the sample proportions. **Table 10-1** shows some summary information, obtained from **Fig. 10-2,** for the 50 simulated sample proportions.

**Table 10-1:** Some Summary Information for the Simulation on Sample Proportions

| True proportion $p = 0.68$ | Mean of sample proportions $\mu_{\hat{p}} = 0.6760$ |
|---|---|
| $\sqrt{p(1-p)/n} = 0.0660$ | Standard deviation of the sample proportions $\sigma_{\hat{p}} = 0.0634$ |

Observe that $p \approx \mu_{\hat{p}}$ and $\sqrt{p(1-p)/n} \approx \sigma_{\hat{p}}$, where the symbol $\approx$ represents "approximately equal to." Of course, if we do a large number of these simulations and take averages, we should expect that these values would be closer, if not equal, to each other. The main purpose of this illustration was to help in understanding the stated properties given in **Table 10-2.**

**Table 10-2:** Properties of the Sampling Distribution for the Sample Proportion

**SAMPLING DISTRIBUTION OF THE SAMPLE PROPORTION $\hat{p}$**

❖ Suppose that random samples of size *n* are selected from a population (distribution) in which the true proportion of the attribute of interest is *p*. Then the sampling distribution of the sample proportions $\hat{p}$ has the following properties:

➢ The mean of the sample proportions is equal to the true population proportion. That is, symbolically, $\mu_{\hat{p}} = p$.

➢ The standard deviation of the sample proportions is given by $\sigma_{\hat{p}} = \sqrt{p(1-p)/n}$.

Next we can investigate the shape of the distribution for these sample proportions. **Figure 10-3** shows a histogram for the simulation.

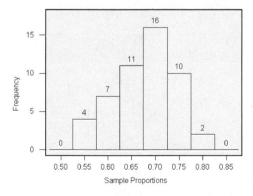

**Fig. 10-3:** Histogram for simulated sample proportions

Observe that the shape of the distribution of the simulated sample proportions is approximately bell-shaped. That is, the distribution of the sample proportions is approximately normally distributed.

We can investigate with other sample sizes and probabilities $p$. However, we will generally observe the same properties when the sample size is "large enough" ($n \geq 30$).

We can generalize the observations in a very important theorem called the Central Limit Theorem for Sample Proportions. This theorem is given in **Table 10-3.**

**Table 10-3:**  Central Limit Theorem for Sample Proportions

| CENTRAL LIMIT THEOREM FOR SAMPLE PROPORTIONS |
| --- |
| Suppose that random samples of size $n$ are selected from a population (distribution) in which the true proportion of the attribute of interest is $p$. Then, provided that $np > 5$ *and* $n(1 - p) > 5$, the sampling distribution of the sample proportion $\hat{p}$ will be approximately normally distributed with mean of $\mu_{\hat{p}} = p$, and a standard deviation of $$\sigma_{\hat{p}} = \sqrt{p(1 - p)/n}$$ **Note:** The normality assumption will improve with larger sample size. |

---

**Quick Tip**

Since the sampling distribution of the sample proportion $\hat{p}$ is approximately <u>normally</u> distributed for large enough sample sizes, with mean $\mu_{\hat{p}} = p$ and standard deviation $\sigma_{\hat{p}} = \sqrt{p(1 - p)/n}$, we can compute $z$ scores for observed $\hat{p}$ values. Also, we will be able to compute probabilities associated with these $\hat{p}$ values. The equation that we use to compute the associated $z$ score is

$$z = \frac{\hat{p} - \mu_{\hat{p}}}{\sigma_{\hat{p}}} = \frac{\hat{p} - p}{\sqrt{p(1 - p)/n}}$$

---

***Example 10-1:***  In a survey, it was reported that 33 percent of women believe in the existence of aliens. If 100 women are selected at random, what is the probability that more than 45 percent of them will say that they believe in aliens?

**Solution:**  We need to find $P(\hat{p} > 0.45)$. Now $n = 100$, $np = 33 > 5$, $n(1 - p) = 67 > 5$, $\hat{p} = 0.45$, $\mu_{\hat{p}} = p = 0.33$, and $\sigma_{\hat{p}} = \sqrt{p(1 - p)/n} = \sqrt{0.33(1 - 0.33)/100} = 0.0470$. The corresponding $z$ score is $z = (0.45 - 0.33)/0.0470 = 2.55$.

Thus, $P(\hat{p} > 0.45) = P(z > 2.55) = 0.5 - 0.4946 = 0.0054$ or 0.54 percent. That is, the probability is rather small (less than 1 percent) that more than 45 percent of the women in the sample will believe in aliens. This is depicted in **Fig. 10-4.**

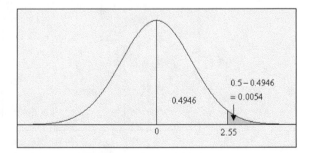

**Fig. 10-4:**  Area for $P(\hat{p} > 0.45)$

**Example 10-2:** It is estimated that approximately 53 percent of college students graduate in 5 years or less. This figure is affected by the fact that more students are attending college on a part-time basis. If 500 students on a large campus are selected at random, what is the probability that between 50 and 60 percent of them will graduate in 5 years or less?

**Solution:** We need to find $P(0.5 \leq \hat{p} \leq 0.6)$. Now $n = 500$, $np = 265 > 5$, $n(1 - p) = 235 > 5$, $\hat{p} = 0.5$ and $0.6$, $\mu_{\hat{p}} = p = 0.53$, and $\sigma_{\hat{p}} = \sqrt{p(1 - p)/n} = \sqrt{0.53(1 - 0.53)/500} = 0.0223$. The corresponding $z$ scores are $z = (0.5 - 0.53)/0.0223 = -1.35$ and $z = (0.6 - 0.53)/0.0223 = 3.14$. Thus, $P(0.5 \leq \hat{p} \leq 0.6) = P(-1.35 \leq z \leq 3.14) = 0.4115 + 0.4990 = 0.9105$. This is depicted in **Fig. 10-5.**

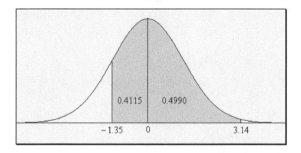

**Fig. 10-5:** Area for $P(0.5 \leq \hat{p} \leq 0.6)$

## 10-2  Sampling Distribution of a Sample Mean

Suppose we are interested in the true daily mean time men spend driving their motor vehicles in the United States. If we let the population mean be denoted by $\mu$, then $\mu$ can be defined by

$$\mu = \frac{\text{total daily amount of time spent driving by American males}}{\text{total number of American males who drive}}$$

Since the population of interest is too large for us to observe all American males who drive, we can estimate the true mean by observing a random sample from the population of American males who drive. If we let the sample mean be denoted by the point estimate $\bar{x}$ (read as "$x$ bar"), then $\bar{x}$ can be defined by

$$\bar{x} = \frac{\text{total daily amount of time spent driving by American males in the sample}}{\text{sample size}}$$

Suppose we assume that the true daily mean time American males spend driving is 81 minutes (*source:* Federal Highway Administration). (In general, we will not know the mean of the population.) If we select a random sample of 50 American males who drive, we may observe that the average daily time spent behind the wheel for this sample is 85 minutes. If we were to select another random sample of size 50, we would most likely obtain a different value for $\bar{x}$. If we selected 100 different samples, say, of the same sample size and computed the average time spent behind the wheel for each, we should not expect these 100 sample means to all be the same. That is, there will be some variability in these computed sample means. Pictorially, the situation is demonstrated in **Fig. 10-6.**

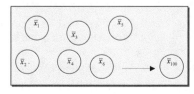

**Fig. 10-6:** One hundred random samples of size 50

These 100 sample means constitute a **sampling distribution of a sample mean.**

**Explanation of the term—sampling distribution of a sample mean:** A sampling distribution of a sample mean is a distribution obtained by using the means computed from random samples of a specific size obtained from a population.

In order to investigate the properties of the sampling distribution of a sample mean, simulations of the situation can be done. Again the MINITAB statistical software was used for the simulation, and 100 samples of size of 50 were used. Here we will assume that the time spent driving is normally distributed with a mean of 81 and a standard deviation of 1, for the sake of the simulation. You may try your own simulation if you have access to such statistical software. The descriptive statistics for the simulation are shown in **Fig. 10-7.**

```
Descriptive Statistics


Variable              N        Mean      Median     TrMean     StDev    SE Mean
Sample Means         100      81.001     81.018     81.003     0.140     0.014

Variable          Minimum    Maximum       Q1         Q3
Sample Means       80.667     81.342     80.890     81.090
```

**Fig. 10-7:** MINITAB descriptive statistics of simulation for sample mean

Let $\mu_{\bar{x}}$ represent the mean of the sample means and $\sigma_{\bar{x}}$ represent the standard deviation of the sample means. **Table 10-4** shows some summary statistics for the 100 simulated sample means.

**Table 10-4:** Some Summary Information for the Simulation on Sample Means

| True mean $\mu = 81$ | Mean of sample means $\mu_{\bar{x}} = 81.007$ |
|---|---|
| $\dfrac{\sigma}{\sqrt{n}} = \dfrac{1}{\sqrt{50}} = 0.141$ | Standard deviation of the sample means $\sigma_{\bar{x}} = 0.140$ |

Observe that $\mu \approx \mu_{\bar{x}}$ and $\dfrac{\sigma}{\sqrt{n}} \approx \sigma_{\bar{x}}$, where again the symbol $\approx$ represents "approximately equal to." Of course, if we do a large number of these simulations and take averages, we should expect that these values would be closer, if not equal, to each other. Also, if we assume different standard deviations for the drive time distribution, we will observe similar results. The main purpose of this illustration was to help in understanding the stated properties given in **Table 10-5.**

**Table 10-5:** Properties of the Sampling Distribution for the Sample Mean

**SAMPLING DISTRIBUTION OF THE SAMPLE MEAN $\bar{x}$**

❖ If all possible random samples of size $n$ are selected from a population with mean $\mu$ and standard deviation $\sigma$, then the sampling distribution of $\bar{x}$ has the following properties:

➤ The mean of the sample means is equal to the population mean. That is, symbolically, $\mu_{\bar{x}} = \mu$.

➤ The standard deviation of the sample means is equal to the standard deviation of the sampling population divided by the square root of the sample size. That is, symbolically, $\sigma_{\bar{x}} = \dfrac{\sigma}{\sqrt{n}}$.

Next we can investigate the shape of the distribution for these sample means. **Figure 10-8** shows a histogram of the simulation for this situation.

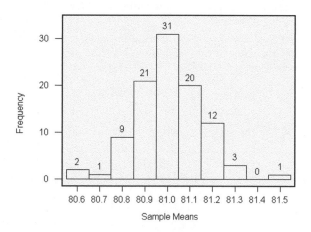

**Fig. 10-8:** Histogram for simulated sample means

Observe that the shape of the distribution of the simulated sample means is approximately bell-shaped. That is, the distribution of the sample means is approximately normally distributed.

We can investigate with other sample sizes, other population means, and other distributions. However, we will generally observe the same properties when the sample size is "large enough" ($n \geq 30$).

We can generalize the observations in a very important theorem called the Central Limit Theorem for Sample Means. This is given in **Table 10-6.**

**Table 10-6:** Central Limit Theorem for Sample Means

**THE CENTRAL LIMIT THEOREM FOR SAMPLE MEANS**

As the sample size *n* increases, the shape of the distribution of the sample means obtained from *any* population (distribution) with mean $\mu$ and standard deviation $\sigma$ will approach a normal distribution. This distribution (the distribution of the sample means) will have a mean $\mu_{\bar{x}} = \mu$ and a standard deviation

$$\sigma_{\bar{x}} = \frac{\sigma}{\sqrt{n}}.$$

**Note:** If the sampling distribution is an exact normal distribution, the distribution of the sample means will also be an exact normal distribution for *any* sample size.

**Note:** In applying the Central Limit Theorem for Sample Means, the sampling population can be *any* distribution.

**Quick Tip**

Since the sampling distribution of the sample mean $\bar{x}$ is approximately normally distributed, with mean $\mu_{\bar{x}} = \mu$ and standard deviation $\sigma_{\bar{x}} = \sigma/\sqrt{n}$, we can compute *z* scores for observed $\bar{x}$ values. Also, we will be able to compute probabilities that are associated with these $\bar{x}$ values. The equation that is used to compute the *z* score is given next.

$$z = \frac{\bar{x} - \mu_{\bar{x}}}{\sigma_{\bar{x}}} = \frac{\bar{x} - \mu}{\sigma/\sqrt{n}}$$

***Example 10-3:*** A tire manufacturer claims that its tires will last an average of 60,000 miles with a standard deviation of 3,000 miles. Sixty-four tires were placed on test and the average failure miles for these tires was recorded. What is the probability that the average failure miles will be more than 59,500 miles?

**Solution:** Observe here that we do not know the distribution of failure miles, but the sample size is large, so we can apply the Central Limit Theorem for the sample means. We need to find $P(\overline{x} > 59{,}500)$. Now $\overline{x} = 59{,}500$, $\mu = 60{,}000$, $\sigma = 3{,}000$, and $n = 64$. From this information,

$$z = \frac{\overline{x} - \mu}{\sigma/\sqrt{n}} = \frac{59{,}500 - 60{,}000}{3{,}000/\sqrt{64}} = -1.33.$$ Thus, $P(\overline{x} > 59{,}500) = P(z > -1.33) = 0.4082 + 0.5 = 0.9082.$

This area is displayed in **Fig. 10-9.**

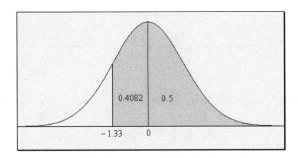

**Fig. 10-9:** Area for $P(\overline{x} > 59{,}500)$

***Example 10-4:*** A supervisor has determined that the average salary of the employees in his department is \$40,000 with a standard deviation of \$15,000. A sample of 25 of the employees' salaries was selected at random. Assuming that the distribution of the salaries is normal, what is the probability that the average for this sample is between \$36,000 and \$42,000?

**Solution:** Here we know that the salaries are normally distributed, so the size of the sample does not matter. Thus, we can proceed to apply the Central Limit Theorem for the sample means. We need to find $P(36{,}000 \leq \overline{x} \leq 42{,}000)$. Now $n = 25$, $\overline{x} = 36{,}000$ and $42{,}000$, $\mu = 40{,}000$,

and $\sigma = 15{,}000$. The corresponding $z$ scores are $z = \dfrac{(36{,}000 - 40{,}000)}{15{,}000/\sqrt{25}} = -1.33$ and

$z = \dfrac{(42{,}000 - 40{,}000)}{15{,}000/\sqrt{25}} = 0.67.$ Thus, $P(36{,}000 \leq \overline{x} \leq 42{,}000) = P(-1.33 \leq z \leq 0.67) = 0.4082 +$

$0.2486 = 0.6568.$ The probability is depicted in **Fig. 10-10.**

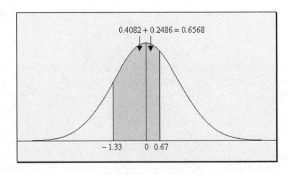

**Fig. 10-10:** Area for $P(36{,}000 \leq \overline{x} \leq 42{,}000)$

## 10-3 Sampling Distribution of a Difference between Two Independent Sample Proportions

We may be interested in comparing the proportions of two populations. For example, we may have to compare the effectiveness of two different drugs, drug 1 and drug 2, on a certain medical condition. One way of doing this is to select a homogeneous group of people with the given medical condition and randomly divide them into two groups. These groups can then be treated with the different medications over a period of time, and then the effectiveness of the medication for these two groups can be determined.

In the above illustration, we may consider the two homogeneous groups as samples from two different populations who were treated with the two drugs. Information obtained about the number of patients who were helped in the two samples can then be used to make comparisons concerning the proportion of patients who were helped by the two different medications.

To be specific, we will let the subscript 1 be associated with population 1, and let the subscript 2 be associated with population 2. We will let $n_1$ and $n_2$ denote the sample sizes for the two independent samples from the two populations. We will let $p_1$ and $p_2$ denote the population proportions of interest. Also, we will let $\hat{p}_1$ and $\hat{p}_2$ represent the sample proportions from population 1 and population 2, respectively. We can then investigate the sampling distribution of $\hat{p}_1 - \hat{p}_2$. Through simulations and theory, we can state some properties of the sampling distribution $\hat{p}_1 - \hat{p}_2$.

**Explanation of the term—sampling distribution of the difference between two independent sample proportions:** A **sampling distribution of the difference between two independent sample proportions** is a distribution obtained by using the difference of the proportions computed from random samples obtained from the two populations.

In order to investigate properties of the sampling distribution of the difference between two sample proportions, simulations of the situation can be done. Again the MINITAB statistical software was used for the simulation, and 100 samples of size of 100 were simulated from binomial distributions with $p_1 = 0.8$ and $p_2 = 0.5$. The descriptive statistics for the simulation are shown in **Fig. 10-11.**

```
Descriptive Statistics


Variable              N       Mean        Median      TrMean      StDev       SE Mean
Differences          100     0.30920      0.30000     0.30867     0.06100     0.00610
of Proportions

Variable          Minimum     Maximum         Q1          Q3
Differences       0.18000     0.45000      0.27000     0.35000
of Proportions
```

**Fig. 10-11:** MINITAB descriptive statistics of simulation for difference between two sample proportions

Let $\mu_{\hat{p}_1 - \hat{p}_2}$ represent the mean of the differences of the sample proportions and $\sigma_{\hat{p}_1 - \hat{p}_2}$ represent the standard deviation of the differences of the sample proportions. **Table 10-7** shows some summary statistics for the 100 simulated sample proportion differences.

**Table 10-7:**  Some Summary Information for the Simulation on the Difference between Two Sample Proportions

| True mean $p_1 - p_2 = 0.3$ | Mean of the differences of the sample proportions $\mu_{\hat{p}_1 - \hat{p}_2} = 0.3092$ |
|---|---|
| $\sqrt{\dfrac{p_1(1-p_1)}{n_1} + \dfrac{p_2(1-p_2)}{n_2}} = 0.0640$ | Standard deviation of the differences sample proportions $\sigma_{\hat{p}_1 - \hat{p}_2} = 0.0610$ |

Observe that $p_1 - p_2 \approx \mu_{\hat{p}_1 - \hat{p}_2}$ and $\sqrt{\dfrac{p_1(1-p_1)}{n_1} + \dfrac{p_2(1-p_2)}{n_2}} \approx \sigma_{\hat{p}_1 - \hat{p}_2}$. Again, if we do a large number of these simulations and take averages, we should expect that these values will be close, if not equal, to each other in the long run. Also, if we assume different sample sizes and different population proportions, we will observe similar results. The main purpose of this illustration was to help in understanding the stated properties given in **Table 10-8.**

**Table 10-8:**  Properties of the Sampling Distribution for the Difference between Two Independent Sample Proportions

**SAMPLING DISTRIBUTION OF $\hat{p}_1 - \hat{p}_2$**

❖ If all possible random samples of size $n_1$ and $n_2$ are selected from two populations with given proportions $p_1$ and $p_2$, then the sampling distribution of $\hat{p}_1 - \hat{p}_2$ has the following properties:

➢ The mean of the differences of the sample proportions is equal to $p_1 - p_2$, the difference of the population proportions.

➢ The standard deviation of the differences of sample proportions is approximately equal to $\sqrt{\dfrac{p_1(1-p_1)}{n_1} + \dfrac{p_2(1-p_2)}{n_2}}$.

Next, we can investigate the shape of the distribution for these differences of the sample proportions. **Figure 10-12** shows a histogram for the simulation for this situation.

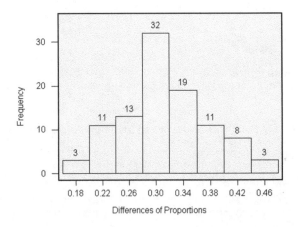

**Fig. 10-12:** Histogram for simulated differences of sample proportions

Observe that the shape of the distribution of the simulated differences of sample proportions is approximately bell-shaped, or normal.

We can investigate with other sample sizes, other population proportions, and other distributions. However, we will generally observe the same properties when $n_1 p_1 > 5$, $n_1(1 - p_1) > 5$, $n_2 p_2 > 5$, and $n_2(1 - p_2) > 5$.

We can generalize the observations in a very important theorem called the Central Limit Theorem for the Difference between Two Sample Proportions. This theorem is given in **Table 10-9.**

**Table 10-9:** Central Limit Theorem for the Difference between Two Sample Proportions

---

**CENTRAL LIMIT THEOREM FOR THE DIFFERENCE BETWEEN TWO SAMPLE PROPORTIONS**

As the sample sizes $n_1$ and $n_2$ increase, the shape of the distribution of the differences of the sample proportions obtained from *any* population (distribution) will approach a normal distribution. This distribution (the distribution of the differences of the sample proportions) will have a mean $\mu_{\hat{p}_1 - \hat{p}_2} = p_1 - p_2$ and a standard deviation

$$\sigma_{\hat{p}_1 - \hat{p}_2} = \sqrt{\frac{p_1(1 - p_1)}{n_1} + \frac{p_2(1 - p_2)}{n_2}},$$ where $p_1$ and $p_2$ are the respective population proportions of interest.

---

**Quick Tip**

Since the sampling distribution of the differences of the sample proportions $\hat{p}_1 - \hat{p}_2$ is approximately normally distributed with mean $\mu_{\hat{p}_1 - \hat{p}_2} = p_1 - p_2$ and standard deviation

$$\sigma_{\hat{p}_1 - \hat{p}_2} = \sqrt{\frac{p_1(1 - p_1)}{n_1} + \frac{p_2(1 - p_2)}{n_2}},$$

we can compute $z$ scores for observed $\hat{p}_1 - \hat{p}_2$ values. Also, we will be able to compute probabilities that are associated with these $\hat{p}_1 - \hat{p}_2$ values. The equation that is used to compute the $z$ score is given by

$$z = \frac{(\hat{p}_1 - \hat{p}_2) - \mu_{\hat{p}_1 - \hat{p}_2}}{\sigma_{\hat{p}_1 - \hat{p}_2}} = \frac{(\hat{p}_1 - \hat{p}_2) - (p_1 - p_2)}{\sqrt{\frac{p_1(1 - p_1)}{n_1} + \frac{p_2(1 - p_2)}{n_2}}}$$

---

***Example 10-5:*** A study was conducted to determine whether remediation in basic mathematics enabled students to be more successful in an elementary statistics course. Success here means that a student received a grade of C or better, and remediation was for one year. **Table 10-10** shows the summary results of the study.

**Table 10-10:** Information Related to Example 10-5

|  | REMEDIAL | NONREMEDIAL |
|---|---|---|
| Sample size | 100 | 40 |
| Number of successes | 70 | 16 |

Based on past history, it is known that 75 percent of students who enroll in remedial mathematics are successful, whereas only 50 percent of nonremedial students are successful. What is the probability that the difference in the proportion of success for the remedial and nonremedial students is at least 10 percent?

**Solution:** From the information given, we have $n_1 = 100$, $n_2 = 40$, $p_1 = 0.75$, $p_2 = 0.5$, $\hat{p}_1 = \frac{70}{100} = 0.7$, and $\hat{p}_2 = \frac{16}{40} = 0.4$. We need to determine $P(\hat{p}_1 - \hat{p}_2 \geq 0.1)$. Substituting in the above equation to find the $z$ score, we find that the $z$-score value is 0.56. Thus, $P(\hat{p}_1 - \hat{p}_2 \geq 0.1) = P(z \geq 0.56)$. Using the standard normal distribution table, $P(z \geq 0.56) = 0.5 - 0.2123 = 0.2877$. That is, the probability that the difference in the proportion of success for the remedial and nonremedial students is at least 10 percent is 0.2877.

## 10-4 Sampling Distribution of a Difference between Two Independent Sample Means

We may be interested in comparing the means of two populations. For example, we may have to compare the effectiveness of two different diets, diet 1 and diet 2, for weight loss. One way of doing this is to select a homogeneous group of people who are classified as overweight and randomly divide them into two groups. These groups can then be treated with the different diets over a period of time, and the effectiveness of the diets for these two groups can be determined.

Again in the above illustration, we may consider the two homogeneous groups as samples from two different independent populations who were treated with the two diets. Information obtained about the weight loss in the two samples can then be used to make comparisons concerning the average weight loss with the two diets.

To be specific, we will let the subscript 1 be associated with population 1, and the subscript 2 be associated with population 2. We will let $n_1$ and $n_2$ denote the sample sizes for the two independent samples from the two populations. We will let $\mu_1$ and $\mu_2$ denote the population means. Also, we will let $\bar{x}_1$ and $\bar{x}_2$ represent the sample means from population 1 and population 2, respectively. We can then investigate the sampling distribution of $\bar{x}_1 - \bar{x}_2$. Through simulations and theory, we can state some properties of the sampling distribution $\bar{x}_1 - \bar{x}_2$.

**Explanation of the term—sampling distribution of the difference between two independent sample means:** A **sampling distribution of the difference between two independent sample means** is a distribution obtained by using the difference of the means computed from random samples obtained from the two populations.

In order to investigate properties of the sampling distribution of the difference between two sample means, simulations of the situation can be done. Again the MINITAB statistical software was used for the simulation, and 100 samples of size of 25 were simulated from normal distributions with means $\mu_1 = 2$ and $\mu_2 = 5$ and standard deviations $\sigma_1 = 1$ and $\sigma_2 = 4$. The descriptive statistics for the differences of the sample means for the simulation are given in **Fig. 10-13**.

### Descriptive Statistics

| Variable | N | Mean | Median | TrMean | StDev | SE Mean |
|---|---|---|---|---|---|---|
| Differences of Means | 100 | 3.0096 | 2.9089 | 2.9856 | 0.7302 | 0.0730 |

| Variable | Minimum | Maximum | Q1 | Q3 | | |
|---|---|---|---|---|---|---|
| Differences of Means | 1.1910 | 5.1450 | 2.4402 | 3.4479 | | |

**Fig. 10-13:** MINITAB descriptive statistics of simulation for differences between two sample means

Let $\mu_{\bar{x}_1 - \bar{x}_2}$ represent the mean of the differences of the sample means, and $\sigma_{\bar{x}_1 - \bar{x}_2}$ represent the standard deviation of the differences of the sample means. **Table 10-11** shows some summary statistics for the differences of the 100 simulated sample means.

**Table 10-11:** Some Summary Information for the Simulation on the Difference between Two Sample Means

| True mean difference $\mu_1 - \mu_2 = 3$ | Mean of the differences of the sample means $\mu_{\bar{x}_1 - \bar{x}_2} = 3.0067$ |
|---|---|
| $\sqrt{\dfrac{\sigma_1^{2}}{n_1} + \dfrac{\sigma_2^{2}}{n_2}} = 0.8246$ | Standard deviation of the differences of the sample means $\sigma_{\bar{x}_1 - \bar{x}_2} = 0.7302$ |

Observe that $\mu_1 - \mu_2 \approx \mu_{\bar{x}_1 - \bar{x}_2}$ and $\sqrt{\dfrac{\sigma_1^{2}}{n_1} + \dfrac{\sigma_2^{2}}{n_2}} \approx \sigma_{\bar{x}_1 - \bar{x}_2}$. Again, if we do a large number of these simulations and take averages, we should expect that these values will be close, if not equal, to each other in the long run. Also, if we assume different sample sizes and different population means and variances, we will observe similar results. The main purpose of this illustration was to help in understanding the stated properties given in **Table 10-12.**

**Table 10-12:** Properties of the Sampling Distribution for the Difference between Two Independent Sample Means

**SAMPLING DISTRIBUTION OF $\bar{x}_1 - \bar{x}_2$**

❖ If all possible random samples of size $n_1$ and $n_2$ are selected from two populations with given means $\mu_1$ and $\mu_2$, then the sampling distribution of $\bar{x}_1 - \bar{x}_2$ has the following properties:

➤ The mean of the differences of the sample means is equal to $\mu_1 - \mu_2$, the difference of the population means.

➤ The standard deviation of the differences of sample means is approximately equal to $\sqrt{\dfrac{\sigma_1^{2}}{n_1} + \dfrac{\sigma_2^{2}}{n_2}}$.

Next we can investigate the shape of the distribution for these differences of the sample means. **Figure 10-14** shows a histogram of the simulation for this situation.

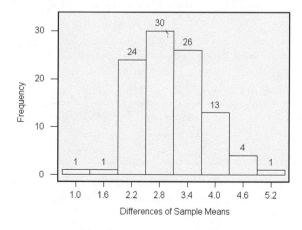

**Fig. 10-14:** Histogram for simulated differences of sample means

Observe that the shape of the distribution of the simulated differences of sample means is approximately bell-shaped, or normal.

We can investigate with other sample sizes, population means, and variances. However, we will generally observe the same properties.

We can generalize the observations in a very important theorem called the Central Limit Theorem for the Difference between Two Sample Means. This theorem is stated in **Table 10-13.**

**Table 10-13:** Central Limit Theorem for the Difference between Two Sample Means

**THE CENTRAL LIMIT THEOREM FOR THE DIFFERENCE BETWEEN TWO SAMPLE MEANS**

As the sample sizes $n_1$ and $n_2$ increase, the shape of the distribution of the differences of the sample means obtained from *any* population (distribution) will approach a normal distribution. This distribution (the distribution of the differences of the sample means) will have a mean $\mu_{\bar{x}_1 - \bar{x}_2} = \mu_1 - \mu_2$ and a standard deviation $\sigma_{\bar{x}_1 - \bar{x}_2} = \sqrt{\dfrac{\sigma_1^2}{n_1} + \dfrac{\sigma_2^2}{n_2}}$, where $\mu_1$ and $\mu_2$ are the respective population means of interest.

**Note:** If the population standard deviations are unknown but the sample sizes are large ($\geq 30$), then we can approximate the population variances by corresponding sample variances.

**Quick Tip**

Since the sampling distribution of the differences of the sample means $\bar{x}_1 - \bar{x}_2$ is approximately normally distributed, with mean $\mu_{\bar{x}_1 - \bar{x}_2} = \mu_1 - \mu_2$ and standard deviation $\sigma_{\bar{x}_1 - \bar{x}_2} = \sqrt{\dfrac{\sigma_1^2}{n_1} + \dfrac{\sigma_2^2}{n_2}}$, we can compute $z$ scores for observed $\bar{x}_1 - \bar{x}_2$ values. Also, we will be able to compute probabilities that are associated with these $\bar{x}_1 - \bar{x}_2$ values. The equation that is used to compute the $z$ score is given next.

$$z = \frac{(\bar{x}_1 - \bar{x}_2) - \mu_{\bar{x}_1 - \bar{x}_2}}{\sigma_{\bar{x}_1 - \bar{x}_2}} = \frac{(\bar{x}_1 - \bar{x}_2) - (\mu_1 - \mu_2)}{\sqrt{\dfrac{\sigma_1^2}{n_1} + \dfrac{\sigma_2^2}{n_2}}}$$

***Example 10-6:*** Based on extensive use of two methods of teaching a high school algebra course, the summary information for final scores given in **Table 10-14** was obtained.

**Table 10-14:** Information Related to Example 10-6

|                     | METHOD 1 | METHOD 2 |
| ------------------- | -------- | -------- |
| Mean                | 85       | 76       |
| Standard deviation  | 16       | 19       |

In a sample of 45 scores using method 1 and 55 scores using method 2, what will be the probability that method 1, on average, is more successful than method 2?

**Solution:** From the information given, we need to determine $P(\bar{x}_1 - \bar{x}_2 \geq 0)$. We have $n_1 = 45$, $n_2 = 55$, $\mu_1 = 85$, $\mu_2 = 76$, $\sigma_1 = 16$, $\sigma_2 = 19$, $\sqrt{\dfrac{\sigma_1^2}{n_1} + \dfrac{\sigma_2^2}{n_2}} = 3.5$, and $z = [0 - (85 - 76)]/3.5 = -2.57$. Substituting, $P(\bar{x}_1 - \bar{x}_2 \geq 0) = P(z \geq -2.57)$. Using the standard normal distribution table in the Appendix, $P(z \geq -2.57) = 0.4949 + 0.5 = 0.9949$. This is a rather large probability, and so if one is given the choice of the two methods, one should choose method 1. It seems, based on this probability value, method 1 is more effective.

---

## Quick Tip

If the sample sizes are large enough to apply the Central Limit Theorem ($n_1 \geq 30$ and $n_2 \geq 30$), then the assumption of normal populations is less crucial, since the distribution of $\bar{x}_1 - \bar{x}_2$ will be approximately normal.

---

Technology

## Technology Corner

All of the concepts discussed in this chapter can be computed and illustrated using some statistical software packages. All scientific and graphical calculators can be used for the computations. In addition, some of the newer calculators will allow you to simulate data from different distributions to aid in understanding the concepts presented in this chapter. If you own a calculator, you should consult the manual to determine what statistical features are included.

It's a Wrap

The sampling distributions of proportions and means can be investigated through

✔ Simulation

✔ Histograms

✔ Descriptive statistics

✔ Technology

Here, we consider only sampling populations that are infinitely large. Other formulas can be presented that would accommodate a finite population. Much fuller texts deal with such situations. Again, care should always be taken when computing probabilities. Also, care should be taken when using the normal probability tables.

Test
Yourself

## True/False Questions

1. A single-value prediction for a parameter is called a point estimate.
2. The distribution of the difference of sample means is obtained by observing a fixed number of sample means from two populations.
3. The expected value (the average of all possible samples of a given size) of the sample mean is equal to the mean of the population from which the random samples are taken.
4. If we take every possible random sample of a fixed size from a normal population with a given variance, then the variance of the distribution of the sample means will be larger than the variance of the given normal distribution.
5. The sampling distribution of the sample mean is approximately normal for all sample sizes.

6. One of the properties of the Central Limit Theorem (for all situations) is that if the sampling population is *not* normally distributed, then the sampling distribution will be approximately normally distributed provided that the sample size(s) is (are) large enough ($\geq 30$).

7. One property of the distribution of sample means is that if the original population is normally distributed, then the distribution of the sample means is also normally distributed, regardless of the sample size.

8. If $\sigma$ is the standard deviation of the sampling population, then the variance of the sample means is $\dfrac{\sigma}{\sqrt{n}}$, where $n$ is the sample size of the random samples selected from the sampling population.

9. The standard deviation of a set of differences of sample proportions is approximately equal to $\sqrt{\dfrac{p_1(1-p_1)}{n_1} + \dfrac{p_2(1-p_2)}{n_2}}$, where $p_1$ and $p_2$ are the population proportions of interest and $n_1$ and $n_2$ are the respective sample sizes.

10. If we sample from a normal population with mean $\mu$ and standard deviation $\sigma$, then the $z$ score associated with $\bar{x}$ is $z = \dfrac{\bar{x} - \mu}{\sigma/\sqrt{n}}$, where $n$ is the sample size.

11. If the sample sizes are large enough to apply the Central Limit Theorem ($n_1 \geq 30$ and $n_2 \geq 30$), then the assumption of normal populations is less crucial in considering the sampling distribution of the difference between two sample means $\bar{x}_1 - \bar{x}_2$, since the distribution of $\bar{x}_1 - \bar{x}_2$ will be approximately normal.

12. The Central Limit Theorem applies only to normal distributions from which samples are taken.

13. If all possible samples of the same size have the same chance of being selected, these samples are said to be random samples.

14. The distribution of the sample mean is obtained by considering a single sample.

15. The Central Limit Theorem cannot be applied to sampling distributions when the samples are obtained from discrete distributions.

16. The smaller the variance for the distribution of the sample mean, the closer the sample mean is to the population mean.

17. The probability distribution of the sample means is referred to as the sampling distribution of the mean.

18. The Central Limit Theorem applies only to continuous distributions.

19. The sampling distribution of the sample proportion is approximately normal for large enough sample sizes.

20. When we consider the sampling distribution for the sample proportions from a single population, the mean of the sample means, for large enough sample sizes, will equal the population proportion of interest.

## Completion Questions

1. A single-value estimate for $\mu$ is known as a (point, interval) _____ estimate.

2. The population of sample means of every possible sample size $n$ is known as the _____ distribution of the mean.

3. In the Central Limit Theorem for the sample mean, if the original population is not normally distributed, then the sampling distribution will be (exactly, approximately) _____ normally distributed provided that the sample size is large enough.

4. The distribution of sample proportions will be approximately normal if the sample size is (greater, smaller) _____ than 30.

5. If the sampling population is normal, then for any sample size, the distribution of the sample mean will be (approximately, exactly) _____ normally distributed.

6. The mean of the distribution of sample means is always equal to the _____ of the population from which the samples are obtained.

7. As the sample size increases, the standard deviation for the sampling distribution of sample means will (increase, decrease) _____.

8. If we sample from a normal population with mean $\mu$ and standard deviation $\sigma$, then the $z$ score associated with $\bar{x}$ for sample size $n$ is $z =$ _____.

9. The Central Limit Theorem for the sample mean applies to (some, only discrete, only continuous, all) _____ distributions from which samples are obtained.

10. For a sample size of 36 and a standard deviation for the distribution of sample means of 5, the standard deviation for the population will be _____.

11. List two statistics that we can find the sampling distributions of: _____ and _____ .

12. If $\sigma$ is the standard deviation of the sampling population, then for a fixed sample size $n$, the standard deviation of the sample means can be computed from (give a formula) _____ .

13. When we consider the sampling distribution of the difference between two proportions, the mean for the sampling distribution will be the difference between the two (sample, population) _____ proportions.

14. When we consider the sampling distribution of the difference between two sample proportions, as the sample sizes $n_1$ and $n_2$ increase, the shape of the sampling distribution obtained from (any, some, discrete, continuous) _____ populations will approach a normal distribution.

15. When we consider the sampling distribution of the difference between two sample means, if the sample sizes are large enough to apply the Central Limit Theorem ($n_1 \geq 30$ and $n_2 \geq 30$), then the assumption of normal populations is (less, very) _____ crucial, since the distribution of $\bar{x}_1 - \bar{x}_2$ will be approximately normal.

16. If we sample from any population with a proportion of interest $p$, the $z$ score associated with $\hat{p}$ for sample size $n$ is $z =$ _____.

## Multiple-Choice Questions

1. As the sample size increases,
   (a) the population mean decreases.
   (b) the population standard deviation decreases.
   (c) the standard deviation for the distribution of the sample means increases.
   (d) the standard deviation for the distribution of the sample means decreases.

2. The concept of sampling distribution applies to
   (a) only discrete probability distributions from which random samples are obtained.
   (b) only continuous probability distributions from which random samples are obtained.
   (c) only the normal probability distribution.
   (d) any probability distribution from which random samples are obtained.

3. When we consider sampling distributions, if the sampling population is normally distributed, then the distribution of the sample means
   (a) will be exactly normally distributed.
   (b) will be approximately normally distributed.
   (c) will have a discrete distribution.
   (d) will be none of the above.

4. The expected value of the sampling distribution of the sample mean is equal to
   (a) the standard deviation of the sampling population.
   (b) the mean of the sampling population.
   (c) the mean of the sample.
   (d) the population size.

5. The sample statistic $\bar{x}$ is the point estimate of
   (a) the population standard deviation $\sigma$.
   (b) the population median.
   (c) the population mean $\mu$.
   (d) the population mode.

6. If repeated random samples of size 40 are taken from an infinite population, the distribution of sample means
   (a) will always be normal because we do not know the distribution of the population.
   (b) will always be normal because the sample mean is always normal.
   (c) will always be normal because the population is infinite.
   (d) will be approximately normal because of the Central Limit Theorem.

7. The mean TOEFL score of international students at a certain university is normally distributed with a mean of 490 and a standard deviation of 80. Suppose groups of 30 students are studied. The mean and the standard deviation for the distribution of sample means will respectively be
   (a) 490, 8/3.
   (b) 16.33, 80.
   (c) 490, 14.61.
   (d) 490, 213.33.

8. A certain brand of light bulb has a mean lifetime of 1500 hours with a standard deviation of 100 hours. If the bulbs are sold in boxes of 25, the parameters of the distribution of sample means are
   (a) 1,500, 100.
   (b) 1,500, 4.
   (c) 1,500, 2.
   (d) 1,500, 20.

9. Samples of size 49 are drawn from a population with a mean of 36 and a standard deviation of 15. Then $P(\bar{x} < 33)$ is
   (a) 0.5808.
   (b) 0.4192.
   (c) 0.1608.
   (d) 0.0808.

10. A tire manufacturer claims that its tires will last an average of 40,000 miles with a standard deviation of 3,000 miles. Forty-nine tires were placed on test and the average failure miles was recorded. The probability that the average failure miles was less than 39,500 is
    (a) 0.3790.
    (b) 0.8790.
    (c) 0.1210.
    (d) 0.6210.

11. A tire manufacturer claims that its tires will last an average of 40,000 miles with a standard deviation of 3,000 miles. Forty-nine tires were placed on test and the average failure miles was recorded. The probability that the average failure miles was equal to 39,500 is
    (a) 0.4525.
    (b) 0.9525.
    (c) 0.0475.
    (d) 0.0000.

12. A tire manufacturer claims that its tires will last an average of 40,000 miles with a standard deviation of 3,000 miles. Forty-nine tires were placed on test and the average failure miles was recorded. The probability that the average failure miles was more than 39,500 is
    (a) 0.3790.
    (b) 0.8790.
    (c) 0.1210.
    (d) 0.6210.

13. A tire manufacturer claims that its tires will last an average of 40,000 miles with a standard deviation of 3,000 miles. Forty-nine tires were placed on test and the average failure miles was recorded. The probability that the average failure miles was between 39,500 and 40,000 is
    (a) 0.3790.
    (b) 0.8790.
    (c) 0.1210.
    (d) 0.6210.

14. Lloyd's Cereal Company packages cereal in 1-pound boxes (1 pound = 16 ounces). It is assumed that the amount of cereal per box varies according to a normal distribution with a standard deviation of 0.05 pound. One box is selected at random from the production line every hour, and if the weight is less than 15 ounces, the machine is adjusted to increase the amount of cereal dispensed. The probability that the amount dispensed per box will have to be increased during a 1-hour period is
    (a) 0.3944.
    (b) 0.8944.
    (c) 0.1056.
    (d) 0.6056.

15. Lloyd's Cereal Company packages cereal in 1-pound boxes (1 pound = 16 ounces). It is assumed that the amount of cereal per box varies according to a normal distribution. A sample of 16 boxes is selected at random from the production line every hour, and if the

average weight is less than 15 ounces, the machine is adjusted to increase the amount of cereal dispensed. If the mean for an hour is 1 pound and the standard deviation is 0.1 pound, the probability that the amount dispensed per box will have to be increased is

(a) 0.5062.

(b) 0.0062.

(c) 0.4938.

(d) 0.9938.

16. Lloyd's Cereal Company packages cereal in 1-pound boxes (1 pound = 16 ounces). It is assumed that the amount of cereal per box varies according to a normal distribution. A sample of 16 boxes is selected at random from the production line every hour, and if the average weight is less than 15 ounces, the machine is adjusted to increase the amount of cereal dispensed. If the mean for an hour is 1 pound and the standard deviation is 0.1 pound, the probability that the amount dispensed per box will not have to be increased is

(a) 0.5062.

(b) 0.0062.

(c) 0.4938.

(d) 0.9938.

17. Suppose that a very large number of random samples of size 25 are selected from a population with mean $\mu$ and standard deviation $\sigma$. If the mean of all the $\bar{x}$'s found is 300 and the standard deviation of these $\bar{x}$'s is 20, the estimates of the true mean $\mu$ and the true standard deviation $\sigma$ of the distribution from which the samples were drawn are, respectively,

(a) 300 and 100.

(b) 300 and 4.

(c) 300 and 16.

(d) 300 and 80.

18. Suppose that a very large number of random samples of size 25 are selected from a population with mean $\mu$ and standard deviation $\sigma$. If the mean of all the $\bar{x}$'s found is 300 and the standard deviation of these $\bar{x}$'s is 20, the estimates of the true mean and the true standard deviation of the distribution of sample means are, respectively,

(a) 300 and 100.

(b) 300 and 4.

(c) 300 and 20.

(d) 300 and 16.

19. A waiter estimates that his average tip per table is $20 with a standard deviation of $4. If his tables seat 9 customers, the probability that the average tip for one table is less than $21 when the tip per table is normally distributed is

(a) 0.2734.

(b) 0.2266.

(c) 0.7734.

(d) 0.7266.

20. A waiter estimates that his average tip per table is $20 with a standard deviation of $4. If his tables seat 9 customers, the probability that the average tip for one table is more than $21 when the tip per table is normally distributed is

(a) 0.2734.

(b) 0.2266.

(c) 0.7734.

(d) 0.7266.

21. A waiter estimates that his average tip per table is $20 with a standard deviation of $4. If his tables seat 9 customers, the probability that the average tip for one table is equal to $21 when the tip per table is normally distributed is

(a) 0.2734.

(b) 0.2266.

(c) 0.7734.

(d) 0.0000.

22. A waiter estimates that his average tip per table is $20 with a standard deviation of $4. If his tables seat 9 customers, the probability that the average tip for one table is between $19 and $21 when the tip per table is normally distributed is

(a) 0.2734.

(b) 0.2266.

(c) 0.7734.

(d) 0.5468.

23. Samples of size 49 are selected at random from an infinite population whose mean and variance are both 25. It is assumed that the distribution of the population is unknown. The mean and the standard deviation of the distribution of sample means are, respectively,

(a) 49 and 3.5714.

(b) 25 and 5.

(c) 25 and 0.7143.

(d) 25 and 0.5102.

24. Two machines are used to fill 50-pound bags of dog food. Sample information for these two machines is given below.

|  | MACHINE 1 | MACHINE 2 |
|---|---|---|
| Sample size | 81 | 64 |
| Sample mean (pounds) | 51 | 48 |
| Sample variance | 16 | 12 |

The point estimate for the difference between the two population means $(\mu_1 - \mu_2)$ is

(a) 17.

(b) 3.

(c) 4.

(d) −4.

25. Two machines are used to fill 50-pound bags of dog food. Sample information for these two machines is given below.

|                       | MACHINE 1 | MACHINE 2 |
|-----------------------|-----------|-----------|
| Sample size           | 81        | 64        |
| Sample mean (pounds)  | 51        | 48        |
| Sample variance       | 16        | 12        |

The standard deviation for the distribution of differences of sample means $(\bar{x}_1 - \bar{x}_2)$ is

(a) 0.6205.

(b) 0.1931.

(c) 0.3850.

(d) 0.3217.

26. Two machines are used to fill 50-pound bags of dog food. Sample information for these two machines is given below.

|                       | MACHINE 1 | MACHINE 2 |
|-----------------------|-----------|-----------|
| Sample size           | 81        | 64        |
| Sample mean (pounds)  | 51        | 48        |
| Sample variance       | 16        | 12        |

Find $P(\bar{x}_1 - \bar{x}_2 \geq 2)$.

(a) 0.4463

(b) 0.0537

(c) 0.9463

(d) 0.5537

27. A study was conducted to determine whether remediation in mathematics enabled students to be more successful in college algebra. Success here means that a student received a grade of C or better, and remediation was for one year (students took an equivalent of one year of high school algebra). The following table shows the results of this study.

|                     | REMEDIAL (1) | NONREMEDIAL (2) |
|---------------------|--------------|-----------------|
| Sample size         | $n_1 = 34$   | $n_2 = 150$     |
| Number of successes | $x_1 = 28$   | $x_2 = 104$     |

The point estimate for $p_1 - p_2$ is

(a) −0.1302

(b) 0.1302

(c) 0.2280

(d) −0.2280

28. A study was conducted to determine whether remediation in mathematics enabled students to be more successful in college algebra. Success here means that a student received a grade of C or better, and remediation was for one year (students took an equivalent of one year of high school algebra). The following table shows the results of this study.

| | REMEDIAL (1) | NONREMEDIAL (2) |
|---|---|---|
| Sample size | $n_1 = 34$ | $n_2 = 150$ |
| Number of successes | $x_1 = 28$ | $x_2 = 104$ |

If it is known from past history that the success rates for students in the remedial and nonremedial groups are 90 percent and 75 percent, respectively, then the standard deviation for the sampling distribution of $\hat{p}_1 - \hat{p}_2$ is

(a) 0.0057.

(b) 0.0624.

(c) 0.0755.

(d) 0.0039.

29. A study was conducted to determine whether remediation in mathematics enabled students to be more successful in college algebra. Success here means that a student received a grade of C or better, and remediation was for one year (students took an equivalent of one year of high school algebra). The following table shows the results of this study.

| | REMEDIAL (1) | NONREMEDIAL (2) |
|---|---|---|
| Sample size | $n_1 = 34$ | $n_2 = 150$ |
| Number of successes | $x_1 = 28$ | $x_2 = 104$ |

If it is known from past history that the success rates for students in the remedial and nonremedial groups are 90 percent and 75 percent, respectively, find $P(\hat{p}_1 - \hat{p}_2 \geq 0.2)$.

(a) 0.2881

(b) 0.2119

(c) 0.7881

(d) 0.7119

## Further Exercises

If possible, you can use any technology help available to solve the following problems.

1. Consider a population that consists of the elements {1, 3, 5}. Write down all possible samples of size 2 (chosen with replacement) from this population and give the sample mean $\bar{x}$ in each case. Show that the mean of all possible samples of size 2 equals the mean of the population.

2. A population has a distribution with $\mu = 50$ and $\sigma = 12$. A random sample of size 100 is selected.
   (a) Calculate the standard deviation of the sample mean.
   (b) Find $P(51 < \bar{x} < 53)$

3. Suppose that the high daily temperatures in a small town in the eastern United States are normally distributed with a mean of 58.6°F and a standard deviation of 9.8°F.
   (a) Find the probability that the average high daily temperature is between 45 and 55°F.
   (b) Find the probability that the average high daily temperature is less than 60°F.
   (c) If a random sample of size 4 of average high daily temperatures is selected, find the probability that the mean of this sample of average high daily temperatures is less than 57°F.

   (d) If a random sample of size 4 of average high daily temperatures is selected, find the probability that the mean of this sample of average high daily temperatures is between 57 and 61°F.

   (e) Suppose that a random sample of 16 high temperatures was chosen and the sample mean was recorded. Give the values of the mean and the standard deviation of the sample mean.

4. Suppose that the heights of female adults in the United States are normally distributed with a mean of 65.4 inches and a standard deviation of 2.8 inches. Let $X$ denote the height of a randomly chosen adult female.

   (a) Find the probability that $X$ is between 66 and 70 inches.

   (b) Suppose that a random sample of 10 adult females was chosen and the sample mean was recorded. Give the values of the mean and standard deviation of the sample mean, and describe the shape of the distribution.

   (c) Using the information in part (b), find the probability that the sample mean is greater than 68 inches.

5. A manufacturing company produces steel bolts to be used on a certain truck. The lengths of the bolts are normally distributed with a mean length of 6 inches and a standard deviation of 0.1 inch. Samples of size 20 are examined at random, and if the average length is outside the interval 5.98 to 6.02 inches, the entire production for the day is examined. Find the probability that a day's production will have to be examined.

6. Two instructors offer extra help designed to improve students' scores on the MCAT exam. Suppose that 85 percent of the students under tutelage from instructor 1 improve their scores, while 76 percent of the students under tutelage from instructor 2 improve their scores. For a random sample of 55 students of instructor 1 and 60 students of instructor 2, compute the probability that the difference between the percentages of students improving their scores is more than 25 percent.

7. In a highly publicized murder trial, it was estimated that 25 percent of the TV viewers watched (at least three hours) the trial on TV. A statistics student who felt that the estimate was too small for his community selected a random sample of 100 residents from the community and found that 35 of them actually watched the trial on TV. Find the probability that more than 33 percent watched the proceedings on TV.

## ANSWER KEY

### True/False Questions

   1. T   2. F   3. T   4. F   5. F   6. T   7. T   8. F   9. T   10. T   11. T
   12. F   13. T   14. F   15. F   16. T   17. T   18. F   19. T   20. T

### Completion Questions

   1. point   2. sampling   3. approximately   4. greater   5. exactly   6. mean

   7. decrease   8. $\dfrac{\bar{x} - \mu}{\sigma/\sqrt{n}}$   9. all   10. 5/6   11. sample mean, sample proportion

   12. $\sigma/\sqrt{n}$   13. population   14. any   15. less   16. $\dfrac{\hat{p} - p}{\sqrt{p(1 - p)/n}}$

### Multiple-Choice Questions

   1. (d)   2. (d)   3. (a)   4. (b)   5. (c)   6. (d)   7. (c)   8. (d)   9. (d)
   10. (c)   11. (d)   12. (b)   13. (a)   14. (c)   15. (b)   16. (d)   17. (a)   18. (c)
   19. (c)   20. (b)   21. (d)   22. (d)   23. (c)   24. (b)   25. (a)   26. (b)
   27. (b)   28. (b)   29. (b)

## PART III

Statistical
Inference

# CHAPTER 11

◆◆◆◆◆◆◆◆◆◆◆◆◆◆◆◆◆◆◆◆◆◆◆◆◆◆◆◆◆◆◆◆◆◆◆

# Confidence Intervals— Large Samples

◆◆◆◆◆◆◆◆◆◆◆◆◆◆◆◆◆◆◆◆◆◆◆◆◆◆◆◆◆◆◆◆◆◆◆

**Y**ou should read this chapter if you need to review or to learn about

→ Large-sample confidence intervals for proportions

→ Large-sample confidence intervals for means

→ Large-sample confidence interval for the difference between two proportions

→ Large-sample confidence interval for the difference between two means

---

**Get Started**

Here we will focus on confidence intervals. When a point estimate is used to estimate the parameter of interest, it is unlikely that the value of the point estimate will be equal to the value of the parameter. Therefore, we will use the value of the point estimate to help construct an interval estimate for the parameter. We will be able to state, with some confidence, that the parameter lies within the interval, and because of this, we refer to these intervals as confidence intervals. Typically, we consider 90 percent, 95 percent, and 99 percent confidence interval estimates for parameters, but any other percentage can be considered. Here, we will consider large-sample confidence intervals for a population proportion and mean and for the difference between two population proportions and means.

---

### 11-1 Large-Sample Confidence Interval for a Proportion

From **Chapter 10,** we can summarize the properties of the Central Limit Theorem for Sample Proportions with the following statements:

- Random samples of size $n$ are selected from a population in which the true proportion of the attribute of interest is $p$.

- Provided that $np > 5$ and $n(1 - p) > 5$, the sampling distribution of the sample proportion $\hat{p}$ will be approximately normally distributed with a mean of $\mu_{\hat{p}} = p$ and a standard deviation of $\sigma_{\hat{p}} = \sqrt{p(1 - p)/n}$.

Now, to find a confidence interval estimate for the unknown parameter $p$, we would need to compute $\sigma_{\hat{p}} = \sqrt{p(1 - p)/n}$, the standard deviation for the sampling distribution of the sample proportion $\hat{p}$. The question then is, How do we compute $\sigma_{\hat{p}}$, since we are estimating $p$, and $p$ is unknown? A reasonable approach would be to replace $p$ with $\hat{p}$, the point estimate for $p$, in the formulas. Thus, we will use $\sigma_{\hat{p}} \approx \sqrt{\hat{p}(1 - \hat{p})/n}$.

Before we state the formula relating to confidence intervals for a population proportion, let us consider the following example.

***Example 11-1:*** In a random sample of 100 men in the United States, 55 of them were married. Determine an approximate 95 percent confidence interval estimate for the true proportion of men who are married.

**Solution:** Since $n = 100$, $x$(number of successes) $= 55$, then $\hat{p} = 0.55$. Also, $\sigma_{\hat{p}} \approx \sqrt{\hat{p}(1 - \hat{p})/n} = 0.0497$. The sampling distribution for $\hat{p}$ is approximately normally distributed, and so from the Empirical Rule, approximately 95 percent of the observed $\hat{p}$ values will lie within 2 standard deviations of the mean, $p \approx \hat{p}$. This situation can be described by **Fig. 11-1.**

**Fig. 11-1:** Display of two standard deviations from the mean of $\hat{p}$

Thus, the approximate 95 percent confidence interval estimate for the proportion of men in the United States who are married is $0.55 \pm 2 \times 0.0497 = 0.55 \pm 0.0994$. That is, we are (approximately) 95 percent confident that the proportion of males in the United States who are married will lie between 0.4506 and 0.5997, or between 45.06 and 59.97 percent.

**Note:** We say the margin of error is $\pm 0.0994$ or $\pm 9.94\%$.

In order to state a general formula that can be used to compute the confidence interval for a population proportion, we need to be familiar with the notation $z_{\alpha}$, read as "z sub alpha." This will enable us to use any number of standard deviations from the mean in constructing the confidence interval.

**Explanation of the notation—$z_{\alpha}$:** $z_{\alpha}$ is a $z$ score such that $\alpha$ area is to the right of the $z$-score value, where $0 \leq \alpha \leq 1$.

**Figure 11-2** shows a diagram that explains the notation.

**Table 11-1** lists values for $z_{\alpha}$ and $z_{\alpha/2}$ when $\alpha = 0.1$, 0.05, and 0.01. These values can be obtained from the standard normal tables in the appendix.

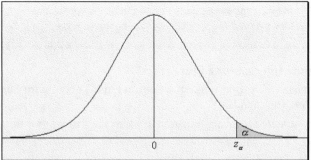

**Fig. 11-2:** Area associated with $z_{\alpha}$

**Table 11-1:** Selected $z$ Values for Selected $\alpha$ Values

| $\alpha$ | $z_\alpha$ | $z_{\alpha/2}$ |
|------|-------|--------|
| 0.1  | 1.28  | 1.645  |
| 0.05 | 1.645 | 1.96   |
| 0.01 | 2.33  | 2.576  |

**Note:** The notation $z_{\alpha/2} \neq \dfrac{z_\alpha}{2}$. First you have to divide $\alpha$ by 2, then you find the corresponding $z$-score value. For example, if we are given that $\alpha = 0.1$, and we need to find $z_{\alpha/2}$, first we find $\alpha/2$. This is equivalent to $0.1/2 = 0.05$. We then would find the value of $z_{0.05}$. From **Table 11-1,** this value will be 1.645.

The relationship between $\alpha$ and the confidence level is that the stated confidence level is the percentage equivalent to the decimal value $1 - \alpha$. For example, if we are constructing a 98 percent confidence interval, then $0.98 = 1 - \alpha$, from which $\alpha = 0.02$.

The general equation used in constructing a $(1 - \alpha) \times 100$ percent confidence interval for the population proportion for large samples is given next.

$$\hat{p} \pm z_{\alpha/2} \sqrt{\frac{\hat{p}(1 - \hat{p})}{n}}$$

**Note:** The margin of error is given by $E = z_{\alpha/2} \sqrt{\dfrac{\hat{p}(1 - \hat{p})}{n}}$.

***Example 11-2:*** In a random sample of 100 persons, 77 percent of them said that when they pray, they pray for world peace. Determine a 90 percent confidence interval estimate for the true proportion of people who pray for world peace.

**Solution:** Since $\hat{p} = 0.77$, then $\sigma_{\hat{p}} \approx \sqrt{\hat{p}(1 - \hat{p})/n} = 0.0421$. Since we need to find the 90 percent confidence interval, $\alpha = 10$ percent $= 0.1$. From **Table 11-1,** $z_{\alpha/2} = 1.645$. Thus, the 90 percent confidence interval estimate for the proportion of people who pray for world peace, using the formula, is $0.77 \pm 1.645 \times 0.0421 = 0.77 \pm 0.0692$. That is, we are 90 percent confident that the true proportion of people who pray for world peace will lie between 0.7008 and 0.8392, or between 70.08 and 83.92 percent.

### Repeated Sample Interpretation of a Confidence Interval

In **Example 11-2,** 77 percent of the sample said that they prayed for world peace. If another sample of size 100 is taken, it is unlikely that the sample proportion will again be 77 percent. Thus, in repeated sampling, we should not expect the sample proportions to all be the same. If we use these sample proportions to construct confidence intervals, we should expect most, if not all, to be different. However, we should expect 90 percent of them to contain the true proportion of people who pray for world peace. This is the reason why we say that we are 90 percent confident that the population proportion will lie between 0.7008 and 0.8392, in the example.

**Note:** We *do not* say that the probability is 0.90 that the population proportion of people who pray for world peace is between 0.7008 and 0.8392. Once the sample is obtained and the confidence interval is constructed, the population proportion of people who pray for world peace will lie in the interval or it will not lie in the interval.

**Quick Tip**

The repeated sampling interpretation of a confidence interval can be generally applied to any confidence interval for any population parameter.

**Sample size:** Sample size determination is closely related to estimation. You may need to know how large a sample is necessary in order to make an accurate estimate for the population proportion $p$. The answer depends on

- The margin of error
- The point estimate for the population proportion
- The degree of confidence

For example, you may need to know how far away from the population proportion you would like the estimate to be, and how confident you are of this. Since $E = z_{\alpha/2} \sqrt{\dfrac{\hat{p}(1 - \hat{p})}{n}}$, we can solve to find $n$, the sample size. Solving, we have

$$n = \hat{p}(1 - \hat{p})\left(\frac{z_{\alpha/2}}{E}\right)^2$$

**Note:** This will be the minimum sample size that is required.

***Example 11-3:*** A statistician wishes to estimate, with 99 percent confidence, the proportion of people who trust DNA testing. A previous study showed that 91 percent of those who were surveyed trusted DNA testing. The statistician wishes to be accurate to within 3 percent of the true proportion. What is the minimum sample size necessary for the statistician to carry out the analysis?

**Solution:** We are given that $\alpha = 0.01$, $z_{\alpha/2} = 2.576$, $\hat{p} = 0.91$, and $E = 0.03$. Substituting into the formula, we get that the sample size

$$n = \hat{p}(1 - \hat{p})\left(\frac{z_{\alpha/2}}{E}\right)^2 = (0.91)(1 - 0.91)\left(\frac{2.576}{0.03}\right)^2 = 603.86 \approx 604$$

That is, in order for the statistician to be 99 percent certain that the estimate is within 3 percent of the true proportion of people who trust DNA testing, a sample of at least 604 is needed.

**Quick Tip**

When computing the sample size to make an accurate estimate for the population proportion $p$, if $\hat{p}$ is unknown, use a value of 0.5 in the formula for $\hat{p}$.

## 11-2 Large-Sample Confidence Interval for a Mean

From **Chap. 10,** we can summarize the properties of the Central Limit Theorem for Sample Means with the following statements:

- Sampling is from any distribution with mean $\mu$ and standard deviation $\sigma$.
- Provided that $n$ is large ($n \geq 30$, as a rule of thumb), the sampling distribution of the sample mean $\bar{x}$ will be approximately normally distributed with a mean $\mu_{\bar{x}} = \mu$ and a standard deviation $\sigma_{\bar{x}} = \dfrac{\sigma}{\sqrt{n}}$.
- If the sampling distribution is normal, the sampling distribution of the sample means will be an exact normal distribution for any sample size.

Now, in finding confidence interval estimates for the unknown parameter $\mu$, we would need to compute $\sigma_{\bar{x}} = \dfrac{\sigma}{\sqrt{n}}$. The general equation used in constructing a $(1 - \alpha) \times 100$ percent confidence interval for the population mean is given below.

$$\boxed{\bar{x} \pm z_{\alpha/2}\,\frac{\sigma}{\sqrt{n}}}$$

**Note:** The margin of error is given by $E = z_{\alpha/2}\,\dfrac{\sigma}{\sqrt{n}}$.

***Example 11-4:*** A random sample of 100 public school teachers in a particular state has a mean salary of \$31,578. It is known from past history that the standard deviation of the salaries for the teachers in the state is \$4,415. Construct a 99 percent confidence interval estimate for the true mean salary for public school teachers for the given state.

**Solution:** Given $\alpha = 0.01$, $z_{\alpha/2} = 2.576$, $\bar{x} = 31{,}578$, $n = 100$, $\sigma = 4{,}415$, and $\sigma_{\bar{x}} = \dfrac{\sigma}{\sqrt{n}} = 441.5$.

Thus, the 99 percent confidence interval estimate for the mean salary, using the formula, is $31{,}578 \pm 2.576 \times 441.5 = 31{,}578 \pm 1{,}137.3$. That is, we are 99 percent confident that the average salary for public school teachers for the given state will lie between \$30,440.9 and \$32,715.3.

---

## Quick Tip

In computing a confidence interval for the population mean, when the population standard deviation $\sigma$ is unknown, it can be replaced with the sample standard deviation $s$ if the sample size is large ($n \geq 30$).

---

***Example 11-5:*** The president of a large community college wishes to estimate the average distance commuting students travel to the campus. A sample of 64 students was randomly selected and yielded a mean of 35 mi and a standard deviation of 5 mi. Construct a 95 percent confidence interval estimate for the true mean distance commuting students travel to the campus.

**Solution:** Given $\alpha = 0.05$, $z_{\alpha/2} = 1.96$, $\bar{x} = 35$, $s = 5$, $n = 64$, and $\sigma_{\bar{x}} \approx \dfrac{s}{\sqrt{n}} = 0.625$. The 95 percent confidence interval estimate for the mean distance, using the formula, is $35 \pm 1.96 \times 0.625 = 35 \pm 1.225$. That is, we are 95 percent confident that the average distance commuting students travel to the campus will lie between 33.775 and 36.225 mi.

**Sample size:** In considering large-sample confidence intervals for the mean, since the error of estimate is given by $E = z_{\alpha/2}\,\dfrac{\sigma}{\sqrt{n}}$, we can solve to find $n$, the sample size. Solving, we have

$$\boxed{n = \left(\frac{z_{\alpha/2} \times \sigma}{E}\right)^2}$$

*Example 11-6:* What sample size should be selected to estimate the mean age of workers in a large factory to within $\pm 1$ year at a 95 percent confidence level if the standard deviation for the ages is 3.5 years?

**Solution:** We are given that $\alpha = 0.05$, $z_{\alpha/2} = 1.96$, $\sigma = 3.5$, and $E = 1$. Substituting into the formula, we get that the sample size

$$n = \left(\frac{z_{\alpha/2} \times \sigma}{E}\right)^2 = \left(\frac{1.96 \times 3.5}{1}\right)^2 = 47.0596 \approx 48$$

That is, in order to be 95 percent certain that the estimate is within 1 year of the true mean age, a sample of at least 48 is needed.

**Note:** For a large enough sample size ($n \geq 30$), when the population standard deviation $\sigma$ is unknown, we can replace $\sigma$ with $s$ in the above equations.

## 11-3 Large-Sample Confidence Interval for the Difference between Two Population Proportions

From **Chap. 10,** we can summarize the properties of the sampling distribution for the difference between two independent sample proportions with the following statements:

- As the sample sizes $n_1$ and $n_2$ increase, the shape of the distribution of the differences of the sample proportions obtained from any population (distribution) will approach a normal distribution.
- The distribution of the differences of the sample proportions will have a mean $\mu_{\hat{p}_1 - \hat{p}_2} = p_1 - p_2$.
- The distribution of the differences of the sample proportions will have a standard deviation $\sigma_{\hat{p}_1 - \hat{p}_2} = \sqrt{\dfrac{p_1(1 - p_1)}{n_1} + \dfrac{p_2(1 - p_2)}{n_2}}$, where $p_1$ and $p_2$ are the respective population proportions of interest.

These properties can aid us in the construction of a $(1 - \alpha) \times 100$ percent confidence interval for the difference of two population proportions. Again, since we do not know the values of the true proportions, we will use the corresponding estimates for these true proportions. The general equation used in constructing a $(1 - \alpha) \times 100$ percent confidence interval for the difference between two population proportions for large samples, is given next.

$$(\hat{p}_1 - \hat{p}_2) \pm z_{\alpha/2} \sqrt{\frac{\hat{p}_1(1 - \hat{p}_1)}{n_1} + \frac{\hat{p}_2(1 - \hat{p}_2)}{n_2}}$$

*Example 11-7:* A study was conducted to determine whether remediation in basic mathematics enabled students to be more successful in an elementary statistics course. Success here means that a student received a grade of C or better, and remediation was for one year. **Table 11-2** shows the results of the study.

**Table 11-2:** Information Related to Example 11-7

|  | REMEDIAL | NONREMEDIAL |
| --- | --- | --- |
| Sample size | 100 | 40 |
| Number of successes | 70 | 16 |

Construct a 95 percent confidence interval for the difference between the proportions for the remedial and nonremedial groups.

**Solution:** From the information given, we have $n_1 = 100$, $n_2 = 40$, $\hat{p}_1 = \dfrac{70}{100} = 0.7$, $\hat{p}_2 = \dfrac{16}{40} =$

$0.4$, $\sqrt{\dfrac{\hat{p}_1(1-\hat{p}_1)}{n_1} + \dfrac{\hat{p}_2(1-\hat{p}_2)}{n_2}} = \sqrt{\dfrac{0.7(1-0.7)}{100} + \dfrac{0.4(1-0.4)}{40}} = 0.09$, $\alpha = 0.05$, $z_{\alpha/2} = 1.96$. Thus,

the 95 percent confidence interval estimate for the difference of the proportions is $(0.7 - 0.4) \pm$ $1.96 \times 0.09 = 0.3 \pm 0.1764$. That is, we are 95 percent confident that the difference between the proportions for the remedial and nonremedial groups will lie between 0.1236 and 0.4764, or between 12.36 and 47.64 percent. Since both limits for the interval are positive, one may conclude that the proportion for the remedial group is larger than the proportion for the nonremedial group. That is, for the elementary statistics course, we can conclude that remediation seems to help the students do better than those students who do not obtain remediation.

**Sample sizes:** In considering a large-sample confidence interval for the difference between two population proportions, that is, when $n_1 p_1 > 5$, $n_1(1 - p_1) \geq 5$, $n_2 p_2 > 5$, and $n_2(1 - p_2) > 5$, we may need to estimate the sample sizes needed in order to collect data. The formula given here is for when the sample sizes are the same. Also, this will be the equation for the minimum sample size. The formula is given below.

$$n = 0.5 \times \left(\frac{z_{\alpha/2}}{E}\right)^2$$

***Example 11-8:*** A researcher wants to determine the difference between the proportions of males and females who believe in aliens. If a margin of error of $\pm 0.02$ is acceptable at the 95 percent confidence level, what is the minimum sample size that should be taken? Assume that equal sample sizes are selected for the two sample proportions.

**Solution:** We are given $\alpha = 0.05$ and $E = 0.02$. Since $\alpha = 0.05$, then $z_{\alpha/2} = 1.96$. Thus,

$$n = 0.5 \times \left(\frac{z_{\alpha/2}}{E}\right)^2 = 0.5 \times \left(\frac{1.96}{0.02}\right)^2 = 4802$$

That is, the researcher should sample at least 4,802 males and 4,802 females in the research study.

## 11-4 Large-Sample Confidence Interval for the Difference between Two Population Means

From **Chap. 10**, we can summarize the properties of the sampling distribution for the difference between two independent sample means with the following statements:

- As the sample sizes $n_1$ and $n_2$ increase, the shape of the distribution of the differences of the sample means obtained from any population (distribution) will approach a normal distribution.
- The distribution of the differences of the sample means will have a mean $\mu_{\bar{x}_1 - \bar{x}_2} = \mu_1 - \mu_2$, where $\mu_1$ and $\mu_2$ are the respective population means of interest.
- The distribution of the differences of the sample means will have a standard deviation

  $\sigma_{\bar{x}_1 - \bar{x}_2} = \sqrt{\dfrac{\sigma_1^2}{n_1} + \dfrac{\sigma_2^2}{n_2}}$, where $\sigma_1$ and $\sigma_2$ are the respective population standard deviations of interest.

These properties can aid us in the construction of a $(1 - \alpha) \times 100$ percent confidence interval for the difference of two population means. The general equation used in constructing a $(1 - \alpha) \times 100\%$ confidence interval for the difference between two population means for large samples, is given next.

$$(\bar{x}_1 - \bar{x}_2) \pm z_{\alpha/2} \sqrt{\frac{\sigma_1^2}{n_1} + \frac{\sigma_2^2}{n_2}}$$

## Quick Tip

For large samples ($n_1 \geq 30$, and $n_2 \geq 30$), the population variances can be replaced by the sample variances if the population variances are unknown.

***Example 11-9:*** A random sample of size $n_1 = 36$ selected from a normal distribution with standard deviation $\sigma_1 = 4$ has a mean $\bar{x}_1 = 75$. A second random sample of size $n_2 = 25$ selected from a different normal distribution with a standard deviation $\sigma_2 = 6$ has a mean $\bar{x}_2 = 85$. Find a 95 percent confidence interval for $\mu_1 - \mu_2$.

**Solution:** From the information given, we have $n_1 = 36$, $n_2 = 25$, $\bar{x}_1 = 75$, $\bar{x}_2 = 85$, $\sigma_1 = 4$, $\sigma_2 = 6$,

$\sqrt{\frac{\sigma_1^2}{n_1} + \frac{\sigma_2^2}{n_2}} = 1.3728$, $\alpha = 0.05$, $z_{\alpha/2} = 1.96$. Thus, the 95 percent confidence interval estimate

for $\mu_1 - \mu_2$ is $(75 - 85) \pm 1.96 \times 1.3728 = -10 \pm 2.6907$. That is, we are 95 percent confident that the difference between the means will lie between $-12.6907$ and $-7.3093$. Since both limits are negative, one may conclude that the mean from population 2 is larger than the mean from population 1.

***Example 11-10:*** Two methods were used to teach a high school algebra course. A sample of 75 scores was selected for method 1, and a sample of 60 scores was selected for method 2, with the summary results given in **Table 11-3.**

**Table 11-3:** Information Related to Example 11-10

|                            | METHOD 1 | METHOD 2 |
|----------------------------|----------|----------|
| Sample mean                | 85       | 76       |
| Sample standard deviation  | 3        | 2        |

Construct a 99 percent confidence interval for the difference in the mean scores for the two methods. Assume that the scores are normally distributed.

**Solution:** From the information given, we have $n_1 = 75$, $n_2 = 60$, $\bar{x}_1 = 85$, $\bar{x}_2 = 76$, $s_1 = 3$,

$s_2 = 2$, $\sqrt{\frac{s_1^2}{n_1} + \frac{s_2^2}{n_2}} = 0.4320$, $\alpha = 0.01$, $z_{\alpha/2} = 2.576$. Thus, the 99 percent confidence interval esti-

mate for the difference of the means for the two methods (method 1 − method 2) is $(85 - 76) \pm 2.576 \times 0.3981 = 9 \pm 1.1130$. That is, we are 99 percent confident that the difference between the mean scores for the two teaching methods will lie between 10.1130 and 7.8870. Since both limits are positive, one may conclude that method 1 seems to be the better of the two methods.

**Sample sizes:** In considering large-sample confidence intervals for the difference between two population means, we may need to estimate the sample sizes needed in order to collect data. The formula given here is for the minimum sample sizes when the sample sizes are the same.

$$n = \left(\frac{z_{\alpha/2}}{E}\right)^2 \times (\sigma_1^2 + \sigma_2^2)$$

***Example 11-11:*** A researcher wishes to study the difference between the average score on a standardized test for students who major in marketing and art. The standard deviation for the scores is 5 for both groups of students. How large a sample (equal in this case) must the researcher use if she wishes to be 99 percent certain of knowing the difference of the average scores for the two populations to within ±3 points?

**Solution:** We are given $\alpha = 0.01$, $E = 3$, $\sigma_1 = \sigma_2 = 5$. Since $\alpha = 0.01$, then $z_{\alpha/2} = 2.576$. Thus, $n = \left(\frac{z_{\alpha/2}}{E}\right)^2 \times (\sigma_1^2 + \sigma_2^2) = \left(\frac{2.576}{3}\right)^2 \times (5^2 + 5^2) = 36.8654 \approx 37$. That is, the researcher should sample at least 37 students from each group.

**Note:** The formulas for nonequal sample sizes are much more complex when considering confidence intervals (inferences) for the differences of parameters.

**Technology**

## Technology Corner

All of the concepts discussed in this chapter can be computed and illustrated using most statistical software packages. All scientific and graphical calculators can be used for the computations. In addition, some of the newer calculators, like the TI-83, will allow you to compute confidence intervals directly. If you own a calculator, you should consult the manual to determine what statistical features are included.

**Illustration: Figure 11-3** shows the outputs computed by the MINITAB software for **Examples 11-2** and **11-7.** The MINITAB software does not allow you to directly construct confidence intervals for a population mean or for the difference of two means when you have summary data, as in **Examples 11-4** and **11-9.** **Figure 11-4** shows the outputs computed by the TI-83 calculator for **Examples 11-2, 11-4, 11-7,** and **11-9.** The TI-83 calculator allows you to use both summary data and raw data to compute confidence intervals for both

```
Worksheet size: 100000 cells

Test and Confidence Interval for One Proportion--Example 11-2

Test of p = 0.5 vs p not = 0.5

Sample      X     N  Sample p        90.0 % CI       Z-Value  P-Value
1          77   100  0.770000  (0.700779, 0.839221)    5.40    0.000


Test and Confidence Interval for Two Proportions--Example 11-7

Sample      X     N  Sample p
1          70   100  0.700000
2          16    40  0.400000

Estimate for p(1) - p(2):  0.3
95% CI for p(1) - p(2):  (0.123603, 0.476397)
Test for p(1) - p(2) = 0 (vs not = 0):  Z = 3.33  P-Value = 0.001
```

**Fig. 11-3:** MINITAB output for Examples 11-2 and 11-7

**Example 11-2**

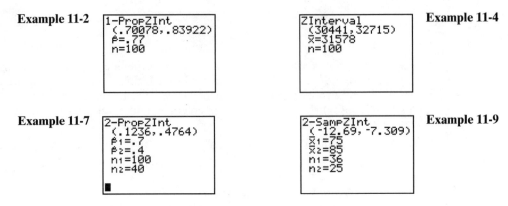

**Example 11-4**

**Example 11-7**

**Example 11-9**

**Fig. 11-4:**  TI-83 outputs for Examples 11-2, 11-4, 11-7, and 11-9

means and proportions. Care should always be taken when using the formulas in computing confidence intervals. One can use other features of the technologies to illustrate other concepts discussed in the chapter.

Here, the focus was on confidence intervals and sample sizes. We considered large-sample confidence intervals for a population proportion and mean and for the difference between two population proportions and means. We also considered formulas for finding sample sizes under different conditions. We computed 90, 95, and 99 percent confidence interval estimates for these parameters, but any other percentage can also be considered. These concepts were presented through

✔ Formulas

✔ Examples

## True/False Questions

1. The best point estimate for the population mean $\mu$ is the sample mean $\bar{x}$.
2. As the length of a confidence interval increases, the degree of confidence in its actually containing the population parameter being estimated also increases.
3. If the length of a confidence interval is very large, then the corresponding prediction is very meaningful.
4. The $z$ score corresponding to a 98 percent confidence level is 1.96.
5. The confidence interval for the population mean $\mu$ can be computed from $\bar{x} \pm z_{\alpha/2}\dfrac{\sigma}{n}$, where $\sigma$ is the population standard deviation and $n$ is the sample size.
6. For a fixed confidence level, when the sample size increases, the length of the confidence interval for a population mean decreases.
7. For a fixed confidence level, when the sample size decreases, the length of the confidence interval for a population mean decreases.
8. The distribution of sample proportions is approximately normal provided that the sample size $n \geq 30$.
9. The confidence interval for the population proportion $p$ can be computed from $\hat{p} \pm z_{\alpha/2}\hat{p}(1-\hat{p})/n$, where $\sigma$ is the population standard deviation.

10. A 90 percent confidence interval for a population mean implies that there is a 0.90 probability that the population mean will be contained in the confidence interval.
11. A 90 percent confidence interval for a population parameter means that if a large number of confidence intervals were constructed from repeated samples, then on average, 90 percent of these intervals would contain the true parameter.
12. The point estimate of a population parameter is always at the center of the confidence interval for the parameter.
13. When repeated samples are selected from a population, the point estimate for a given parameter will always be the same value.
14. The larger the level of confidence, the shorter the confidence interval.
15. The best point estimate for a population parameter is its corresponding sample statistic.
16. In order to determine the sample size when considering confidence intervals for the population mean $\mu$, it is necessary to know the level of confidence, the margin of error, and either an estimate of the population standard deviation or the standard deviation itself.
17. In order to determine the sample size when determining the population proportion, it is necessary to know the level of confidence, the margin of error, and an estimate of the population mean.
18. The maximum error of estimate gives a measure of accuracy when computing the sample size required to make inferences.
19. Based on the Central Limit Theorem for the difference of two population proportions, we can assume, for large enough sample sizes, that the sampling distribution for the difference between two sample proportions is exactly normally distributed.
20. In computing equal sample sizes when considering confidence intervals for the difference between two population proportions, the sample size will increase when the margin of error is decreased and the significance level is held fixed.
21. When computing large-sample confidence intervals for the difference between two population means, it is necessary to know the variances for the two populations.

## Completion Questions

1. As the length of the confidence interval for the population mean increases, the degree of confidence in the interval's actually containing the population mean (increases, decreases) _____.
2. The $z$ score associated with the 99 percent confidence level is (1.28, 1.645, 2.33, 2.576) _____.
3. The confidence interval for the population mean $\mu$, when the population standard deviation $\sigma$ is known, will be given by the relation _____.
4. For a fixed level of confidence, when the sample size increases, the length of the confidence interval for a population mean will (increase, decrease) _____.
5. For a fixed level of confidence, when the sample size decreases, the length of the confidence interval for a population proportion will (increase, decrease) _____.
6. When constructing confidence intervals for the population mean, if the population standard deviation is unknown, but the sample size is large enough and the sampling population is normally distributed, then the distribution that is used to compute the maximum error of estimate is the _____ distribution.
7. Provided that the sample size is large enough, the distribution of the sample proportions is approximately normal with standard deviation $\sigma_{\hat{p}} = \{\hat{p}(1 - \hat{p})/n, \sqrt{\hat{p}(1 - \hat{p})/n}, p(1 - p)/n, \sqrt{p(1 - p)/n}\}$ _____.

8. Provided that the sample size is large enough, the distribution of the sample proportions is approximately normal with mean $\mu_{\hat{p}} = $ _____.

9. When constructing confidence intervals for the population mean $\mu$ based on large samples ($n \geq 30$), when the population standard deviation is unknown, the maximum error of estimate is given by $E = $ _____.

10. The (point, interval) _____ estimate of a population parameter is always at the center of the confidence interval for that parameter.

11. A point estimate for a population parameter is the value of the corresponding sample (parameter, statistic) _____.

12. The $z$ value for a 97.8 percent confidence interval estimation is _____.

13. If we change a 90 percent confidence interval estimate to a 95 percent confidence interval estimate, the width of the confidence interval will (increase, decrease) _____.

14. A confidence interval is a range of values used to estimate a population (parameter, statistic) _____.

15. The maximum error of estimate $E$ will (decrease, increase) _____ when larger sample sizes are used in constructing confidence intervals for the difference between two population means.

16. The maximum error of estimate $E$ will (decrease, increase) _____ when a larger confidence level is used.

17. When constructing confidence intervals for the population proportion, for a given confidence level and maximum error of estimate, the maximum sample size is obtained when $p = (0.25, 0.50, 0.75)$ _____.

18. The best point estimate for the population standard deviation is the _____ standard deviation.

19. The best point estimate for the population mean is the _____ mean.

20. As the sample sizes increase, the shape of the distribution of the differences of the sample means obtained from any population will approach a(n) _____ distribution.

21. In estimating equal sample sizes when inferences are made about the difference of two population proportions, the sample size will (increase, decrease) _____ when the error of estimate decreases and the level of confidence remains fixed.

22. When confidence intervals for the difference between two population means are constructed, if the population variances are unknown, they can be estimated by their respective _____ variances for large enough sample sizes.

## Multiple-Choice Questions

1. If we are constructing a 98 percent confidence interval for the population mean, the confidence level will be
   (a) 2 percent.
   (b) 2.29.
   (c) 98 percent.
   (d) 2.39.

2. The $z$ value corresponding to a 97 percent confidence interval is
   (a) 1.88.
   (b) 2.17.

(c) 1.96.

(d) 3 percent.

3. As the sample size increases, the confidence interval for the population mean will
   (a) decrease.
   (b) increase.
   (c) stay the same.
   (d) decrease and then increase.

4. If we change the confidence level from 98 percent to 95 percent when constructing a confidence interval for the population mean, we can expect the size of the interval to
   (a) increase.
   (b) decrease.
   (c) stay the same.
   (d) do none of the above.

5. Generally, lower confidence levels will yield
   (a) smaller standard deviations for the sampling distribution.
   (b) larger margins of error.
   (c) broader confidence intervals.
   (d) narrower confidence intervals.

6. If the 98 percent confidence limits for the population mean $\mu$ are 73 and 80, which of the following could be the 95 percent confidence limits?
   (a) 73 and 81
   (b) 72 and 79
   (c) 72 and 81
   (d) 74 and 79

7. A 90 percent confidence interval for a population mean indicates that
   (a) we are 90 percent confident that the interval will contain all possible sample means with the same sample size taken from the given population.
   (b) we are 90 percent confident that the population mean will be the same as the sample mean used in constructing the interval.
   (c) we are 90 percent confident that the population mean will fall within the interval.
   (d) none of the above is true.

8. Interval estimates of a parameter provide information on
   (a) how close an estimate of the parameter is to the parameter.
   (b) what proportion of the estimates of the parameter are contained in the interval.
   (c) exactly what values the parameter can assume.
   (d) the $z$ score.

9. Which of the following confidence intervals will be the widest?
   (a) 90 percent
   (b) 95 percent
   (c) 80 percent
   (d) 98 percent

10. The best point estimate for the population variance is
   (a) a statistic.
   (b) the sample standard deviation.
   (c) the sample mean.
   (d) the sample variance.

11. When determining the sample size in constructing confidence intervals for the population mean $\mu$, for a fixed maximum error of estimate and level of confidence, the sample size will
   (a) increase when the population standard deviation is decreased.
   (b) increase when the population standard deviation is increased.
   (c) decrease when the population standard deviation is increased.
   (d) decrease and then increase when the population standard deviation is increased.

12. When computing the sample size to help construct confidence intervals for the population proportion, for a fixed margin of error of estimate and level of confidence, the sample size will be maximum when
   (a) $p = 0.25$.
   (b) $(1 - p) = 0.25$.
   (c) $p(1 - p) = 0.5$.
   (d) $p = 0.5$.

13. What value of the population proportion $p$ will maximize $p(1 - p)$?
   (a) 0.50
   (b) 0.25
   (c) 0.75
   (d) 0.05

14. Suppose that a sample of size 100 is selected from a population with unknown variance. If this information is used in constructing a confidence interval for the population mean, which of the following statements is true?
   (a) The sample must have a normal distribution.
   (b) The population is assumed to have a normal distribution.
   (c) Only 95 percent confidence intervals may be computed.
   (d) The sample standard deviation cannot be used to estimate the population standard deviation because the sample size is large.

15. A 95 percent confidence interval for the mean of a population is to be constructed and must be accurate to within 0.3 unit. A preliminary sample standard deviation is 2.9. The smallest sample size $n$ that provides the desired accuracy with 95 percent confidence is
   (a) 253.
   (b) 359.
   (c) 400.
   (d) 380.

16. A 95 percent confidence interval for the population proportion is to be constructed and must be accurate to within 0.1 unit. The largest sample size $n$ that provides the desired accuracy with 95 percent confidence
   (a) cannot be determined.
   (b) is 73.

(c) is 97.

(d) is 100.

17. In a survey about a murder case that was widely reported by the TV networks, 201 out of 300 persons surveyed said that they believed that the accused was guilty. The 95 percent confidence interval for the proportion of people who did *not* believe that the accused was guilty is

(a) 0.617 to 0.723.

(b) 0.277 to 0.383.

(c) 0.285 to 0.375.

(d) 0.625 to 0.715.

18. In constructing a confidence interval for the population mean $\mu$, if the level of confidence is changed from 98 percent to 90 percent, the standard deviation of the mean will

(a) be equal to 90 percent of the original standard deviation of the mean.

(b) increase.

(c) decrease.

(d) remain the same.

19. Suppose the heights of the population of basketball players at a certain college are normally distributed with a standard deviation of 2 ft. If a sample of heights of size 16 is randomly selected from this population with a mean of 6.2 ft, the 90 percent confidence interval for the mean height of these basketball players is

(a) 4.555 to 7.845 ft.

(b) 5.378 to 7.022 ft.

(c) 4.447 to 7.953 ft.

(d) 5.324 to 7.077 ft.

20. A 99 percent confidence interval is to be constructed for a population mean from a random sample of size 22. If the population standard deviation is known, the table value to be used in the computation is

(a) 2.518.

(b) 2.330.

(c) 2.831.

(d) 2.580.

21. The most common confidence levels and the corresponding $z$ values are listed below. Which corresponding $z$ value is incorrect?

(a) 99 percent, $z$ value = 1.280

(b) 95 percent, $z$ value = 1.960

(c) 98 percent, $z$ value = 2.330

(d) 90 percent, $z$ value = 1.645

22. The heights (in inches) of the students on a campus are assumed to have a normal distribution with a standard deviation of 4 in. A random sample of 49 students was taken and yielded a mean of 68 in. The 95 percent percent confidence interval for the population mean $\mu$ is

(a) 67.06 to 68.94 in.

(b) 66.88 to 69.12 in.

(c) 63.42 to 72.48 in.

(d) 64.24 to 71.76 in.

23. The length of time it takes a car salesperson to close a deal on a car sale is assumed to be normally distributed. A random sample of 100 such times was selected and yielded a mean of 3 h and variance of 30 min. The 98 percent confidence interval for the mean length of time it takes a car salesperson to sell a car is
   (a) 2.8835 to 3.1165 h.
   (b) 2.8176 to 3.1824 h.
   (c) 2.8352 to 3.1648 h.
   (d) 2.8710 to 3.1290 h.

24. In a recent study, it was found that 11 out of every 100 Pap smears sampled were misdiagnosed by a certain lab. If a sample of 100 is taken, the 99 percent confidence interval for the proportion of misdiagnosed Pap smears is
   (a) 0.1075 to 0.1125.
   (b) 0.0371 to 0.1829.
   (c) 0.1077 to 0.1123.
   (d) 0.0293 to 0.1908.

25. In a religious survey of southerners, it was found that 164 out of 200 believed in angels. The 90 percent confidence interval for the true proportion of southerners who believe in angels
   (a) is 0.7753 to 0.8647.
   (b) is 0.7499 to 0.8901.
   (c) is 0.8188 to 0.8212.
   (d) cannot be computed since the assumptions that are required to compute this confidence interval are violated.

26. In a random sample of 150 drunk drivers, 91 percent were males. The 99 percent confidence interval for the proportion of drunk drivers who are male is
   (a) 0.8716 to 0.9484.
   (b) 0.8498 to 0.9702.
   (c) 0.8641 to 0.9559.
   (d) 0.8555 to 0.9645.

27. The heights (in inches) of the students on a campus are assumed to have a normal distribution with a variance of 25 in. Suppose that we want to construct a 95 percent confidence interval for the population mean $\mu$ and have it accurate to within 0.5 in. The minimum sample size required is
   (a) 9,604.
   (b) 269.
   (c) 98.
   (d) 385.

28. The 95 percent confidence interval for the proportion of drunk drivers who are female is to be constructed and must be accurate to within 0.08. A preliminary sample provides an initial estimate of $\hat{p} = 0.09$. The smallest sample size that will provide the desired accuracy with 95 percent confidence is
   (a) 26.
   (b) 77.
   (c) 50.
   (d) 151.

29. If the population proportion is being estimated, the sample size needed in order to be 90 percent confident that the estimate is within 0.05 of the true proportion is
    (a) 20.
    (b) 271.
    (c) 196.
    (d) 400.

30. In a random sample of 100 observations, $\hat{p} = 0.1$. The 84 percent confidence interval for $p$ is
    (a) $0.1 \pm 0.578$.
    (b) $0.1 \pm 0.282$.
    (c) $0.1 \pm 0.0423$.
    (d) $0.1 \pm 0.001$.

31. When computing the sample size needed to estimate the population proportion $p$, which of the following is not necessary?
    (a) The required confidence level
    (b) The margin of error
    (c) An estimate of $p$
    (d) An estimate of the population variance

32. In a religious survey of southerners, it was found that 82 out of 100 believed in angels. If we wanted to construct a 99 percent confidence interval for the true proportion of southerners who believe in angels, what would be the margin of error?
    (a) 0.0991
    (b) 0.9191
    (c) 0.0381
    (d) 0.7209

33. A statistician wishes to investigate the difference between the proportions of males and the proportion of females who believe in aliens. How large a sample should be taken (equal sample size for each group) to be 95 percent certain of knowing the difference to within $\pm 0.02$?
    (a) 3,383
    (b) 4,802
    (c) 2,048
    (d) 6,787

34. A researcher wishes to investigate the difference between the mean scores on a standardized test for students who were exposed to two different methods of teaching. How large a sample should the researcher take (equal sample size for each method) to be 99 percent certain of knowing the difference of the average scores to be within $\pm 3$ points if the standard deviations for the populations are 5 and 8?
    (a) 66
    (b) 38
    (c) 27
    (d) 54

35. In 1973, the Graduate Division at the University of California, Berkeley, did an observational study on sex bias in admissions to the graduate school. It was found that in a particular major, out of 800 male applicants, 65 percent were admitted, and out of 120

female applicants, 85 percent were admitted. Establish a 95 percent confidence interval estimate of the difference in the proportions of females and males for this particular major.

(a) $0.2 \pm 0.09$

(b) $0.2 \pm 0.07$

(c) $0.2 \pm 0.11$

(d) $0.2 \pm 0.12$

36. Two brands of similar tires were tested, and their lifetimes, in miles, were compared. The data are given below. Find the 95 percent confidence interval for the true difference in the means. Assume that the lifetimes are normally distributed.

| BRAND A | BRAND B |
| --- | --- |
| $\bar{x}_1 = 41,000$ | $\bar{x}_2 = 39,600$ |
| $s_1 = 3,000$ | $s_2 = 2,600$ |
| $n_1 = 100$ | $n_2 = 100$ |

(a) $1,400 \pm 1,022.5$

(b) $1,400 \pm 653$

(c) $1,400 \pm 508.1$

(e) $1,400 \pm 778.1$

37. When will it be reasonable to construct a confidence interval for a parameter if the values for the entire population are known?

(a) Never

(b) When the population size is greater than 30

(c) When the population size is less than 30

(d) When only lower confidence levels are used

38. A researcher wants to determine the difference between the proportions of males and females who do volunteer work. If a margin of error of $\pm 0.02$ is acceptable at the 90 percent confidence level, what is the maximum sample size that should be taken? Assume equal sample sizes are selected for the two sample proportions.

(a) 3,383

(b) 2,048

(c) 6,787

(d) 8,295

# Further Exercises

If possible, you could use any technology available to help solve the following questions.

1. State in your own words what the phrase "90 percent confidence for $\mu$" tells you about the relationship between the population mean $\mu$ and the confidence interval.

2. A study is to be conducted to investigate the proportion of college professors who wear eyeglasses.

(a) In a random sample of 250 professors, 139 of them wore eyeglasses. From this information, generate a 90 percent confidence interval for the proportion of professors who wear eyeglasses.

(b) Suppose that a 90 percent confidence interval for this proportion of professors who wear eyeglasses is to have a margin of error of ±2 percent. A preliminary random sample of 100 professors resulted in 52 of them wearing eyeglasses. How many additional professors must be surveyed?

3. The average annual salary of public school teachers who graduated from a certain college is to be studied.

   (a) A random sample of 100 teachers has a mean of $31,578 with a standard deviation of $4,415. Construct a 99 percent confidence interval for the average salary of the population of public school teachers who graduated from this specific college.

   (b) Interpret in your own words what the confidence interval in (a) means.

   (c) If 100 different 99 percent confidence intervals were generated, how many of them would you expect to contain the average salary of the teachers of this population?

4. A study was conducted to determine the type of car owned. A sample of 150 females revealed that 50 of them owned a foreign car.

   (a) Construct a 90 percent confidence interval for the proportion of females who own foreign cars.

   (b) Interpret in your own words what the confidence interval in (a) means.

   (c) If the information given in this problem can be considered as a preliminary study, compute the sample size necessary to construct a 95 percent confidence interval for the true proportion of females who own a foreign car with a maximum error of ±2 percentage points.

5. (a) In a survey of 800 adults, it was found that 72 ate the recommended amount of fruits and vegetables each day. Construct a 99 percent confidence interval for the proportion of this population that follows these recommendations.

   (b) Express in your own words what the interval in (a) means.

   (c) Suppose no preliminary study was done. Compute the sample size necessary to construct a 95 percent confidence interval for the population proportion with a margin of error of ±3 percentage points.

6. The TOEFL (Test of English as a Foreign Language) scores for international students from two different countries were studied. The information is given below. Construct a 90 percent confidence interval for the difference in the average scores for the two countries.

| COUNTRY 1 | COUNTRY 2 |
|---|---|
| $\bar{x}_1 = 490$ | $\bar{x}_2 = 462$ |
| $s_1 = 80$ | $s_2 = 85$ |
| $n_1 = 110$ | $n_2 = 120$ |

7. Two different brands of light bulbs were tested to establish the manufacturers' claim that the bulbs lasted more than 2,000 hours. The number of successes and sample sizes are given below. Construct a 99 percent confidence interval for the difference in the average scores for the two countries.

| BRAND A | BRAND B |
|---|---|
| $x_1 = 500$ | $x_2 = 462$ |
| $n_1 = 1,200$ | $n_2 = 1,000$ |

## ANSWER KEY
### True/False Questions

1. T   2. T   3. F   4. F   5. F   6. T   7. F   8. T   9. F   10. F   11. T
12. T   13. F   14. F   15. T   16. T   17. F   18. T   19. F   20. T   21. F

### Completion Questions

1. increases   2. 2.576   3. $\bar{x} \pm z_{\alpha/2}\sigma/\sqrt{n}$   4. decrease   5. increase   6. standard
normal or $z$   7. $\sqrt{p(1-p)/n}$   8. $p$   9. $z_{\alpha/2} \times \left(s/\sqrt{n}\right)$   10. point   11. statistic
12. 2.29   13. increase   14. parameter   15. decrease   16. increase   17. 0.50
18. sample   19. sample   20. normal   21. increase   22. sample

### Multiple-Choice Questions

1. (c)   2. (b)   3. (a)   4. (b)   5. (d)   6. (d)   7. (c)   8. (a)   9. (d)
10. (d)   11. (b)   12. (d)   13. (a)   14. (b)   15. (b)   16. (c)   17. (b)
18. (d)   19. (b)   20. (d)   21. (a)   22. (b)   23. (c)   24. (d)   25. (a)
26. (b)   27. (d)   28. (c)   29. (b)   30. (c)   31. (d)   32. (a)   33. (b)
34. (a)   35. (b)   36. (d)   37. (a)   38. (a)

# CHAPTER 12

# Hypothesis Tests— Large Samples

**Y**ou should read this chapter if you need to review or to learn about

➡ Large-sample hypothesis tests for a population proportion

➡ Large-sample hypothesis tests for a population mean

➡ Large-sample hypothesis tests for the difference between two population proportions

➡ Large-sample hypothesis tests for the difference between two population means

---

**Get Started**

Here we will focus on large-sample hypothesis tests for population proportions and population means. We will, through these tests, make inferences about these parameters.

---

## 12-1 Some Terms Associated with Hypothesis Testing

Every situation that requires a hypothesis test starts with a statement of a hypothesis.

**Explanation of the term—statistical hypothesis:** A **statistical hypothesis** is an opinion about a population parameter. The opinion may or may not be true.

There are two types of statistical hypotheses: (a) the null hypothesis, and (b) the alternative hypothesis.

---

**Explanation of the term—null hypothesis:** The **null hypothesis** states that there is no difference between a parameter (or parameters) and a specific value.

**Notation:** The null hypothesis is denoted by $H_0$.

**Explanation of the term—alternative hypothesis:** The **alternative hypothesis** states that there is a precise difference between a parameter (or parameters) and a specific value.

**Notation:** The alternative hypothesis is denoted by $H_1$.

After the hypotheses are stated, the next step is to design the study. An appropriate statistical test will be selected, the level of significance will be chosen, and a plan to conduct the study will be formulated. To make an inference for the study, the statistical test and level of significance are used.

**Explanation of the term—statistical test:** A **statistical test** uses the data collected from the study to make a decision about the null hypothesis. This decision will be to reject or not to reject the null hypothesis.

We will need to compute a *test value* or a *test statistic* in order to make the decision. The formula that is used to compute this value will vary depending on the statistical test.

## Possible Outcomes for a Hypothesis Test

When a test is done, there are four possible outcomes. These outcomes are summarized in **Table 12-1.**

**Table 12-1**

|                     | $H_0$ IS TRUE      | $H_0$ IS FALSE     |
| ------------------- | ------------------ | ------------------ |
| Reject $H_0$        | **Error** <br> Type I | Correct <br> decision |
| Do not reject $H_0$ | Correct <br> decision | **Error** <br> Type II |

Observe that there are two possibilities for a correct decision and two possibilities for an incorrect decision.

## Types of Error for a Hypothesis Test

From **Table 12-1,** we can observe that there are two ways of making a mistake when doing a hypothesis test. These two errors are called **Type I error** and **Type II error.**

**Explanation of the term—type I error:** A **type I error** occurs if the null hypothesis is rejected when it is true.

**Explanation of the term—type II error:** A **type II error** occurs if the null hypothesis is not rejected when it is false.

When we reject or do not reject the null hypothesis, how confident are we that we are making the correct decision? This question can be answered by the specified **level of significance.**

**Explanation of the term—level of significance:** The level of significance, denoted by $\alpha$, is the probability of a Type I error.

Typical values for $\alpha$ are 0.1, 0.05, and 0.01. For example, if $\alpha = 0.1$ for a test, and the null hypothesis is rejected, then one will be 90% certain that this is the correct decision.

**Note:** We will not address the probability of a Type II error in the text because of its complexity.

Once the level of significance is selected, a critical value for the appropriate test is selected from a table. If a $z$ test is used, the $z$ table will be used to obtain the appropriate critical value.

**Explanation of the term—critical value:** A **critical value** separates the critical region from the noncritical region.

**Explanation of the term—critical or rejection region:** A **critical or rejection region** is a range of test statistic values for which the null hypothesis should be rejected. This range of values will indicate that there is a significant or large enough difference between the postulated parameter value and the corresponding point estimate for the parameter.

**Explanation of the term—noncritical or nonrejection region:** A **noncritical or nonrejection region** is a range of test statistic values that indicates that the difference between the postulated value for the parameter and the corresponding point estimate value is probably due to chance, and so the null hypothesis should not be rejected.

**Figure 12-1** illustrates the idea of critical and noncritical regions. Here we assume that a $z$ test is used.

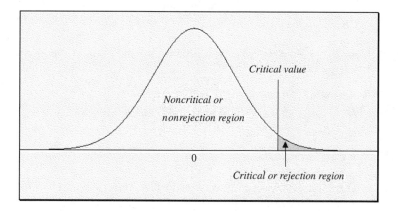

**Fig. 12-1:** Diagram depicting the critical and noncritical regions

There are two broad classifications of hypothesis tests: (a) one-tailed tests, and (b) two-tailed tests.

A **one-tailed test** points out that the null hypothesis should be rejected when the test statistic value is in the critical region on one side of the parameter value being tested.

A **two-tailed test** points out that the null hypothesis should be rejected when the test statistic value is in either of the two critical regions.

## 12-2 Large-Sample Test for a Proportion

Tests presented in this and subsequent chapters will follow a five-step procedure.

***Step 1:*** State the null hypothesis $H_0$.

***Step 2:*** State the alternative hypothesis $H_1$.

***Step 3:*** State the formula for the test statistic (T.S.) and compute its value.

***Step 4:*** State the decision rule (D.R.) for rejecting the null hypothesis for a given level of significance.

***Step 5:*** State a conclusion in the context of the information given in the problem.

Here we will present tests of hypotheses that will enable us to determine, based on sample data, whether the true value of a proportion equals a given constant. Samples of size $n$ will be obtained, and the number or proportion of successes observed. It will be assumed that the trials in the experiment are independent and that the probability of success is the same for each trial. That is, we are assuming that we have a binomial experiment, and we are testing hypotheses about the parameter $p$ of a binomial population.

## Summary of Hypothesis Tests

### (a) One-tailed (right-tailed)

$H_0$: $p \leq p_0$ (where $p_0$ is a specified proportion value)

$H_1$: $p > p_0$

T.S.: $z = \dfrac{x - np_0}{\sqrt{np_0(1 - p_0)}}$

D.R.: For a specified significance level $\alpha$, reject the null hypothesis if the computed test statistic value $z$ is greater than $+z_\alpha$.

Conclusion: . . .

**Note:** This is a right-tailed test because the direction of the inequality sign in the alternative hypothesis is to the right.

### (b) One-tailed (left-tailed)

$H_0$: $p \geq p_0$ (where $p_0$ is a specified proportion value)

$H_1$: $p < p_0$

T.S.: $z = \dfrac{x - np_0}{\sqrt{np_0(1 - p_0)}}$

D.R.: For a specified significance level $\alpha$, reject the null hypothesis if the computed test statistic value $z$ is less than $-z_\alpha$.

Conclusion: . . .

**Note:** This is a left-tailed test because the direction of the inequality sign in the alternative hypothesis is to the left.

### (c) Two-tailed

$H_0$: $p = p_0$ (where $p_0$ is a specified proportion value)

$H_1$: $p \neq p_0$

T.S.: $z = \dfrac{x - np_0}{\sqrt{np_0(1 - p_0)}}$

D.R.: For a specified significance level $\alpha$, reject the null hypothesis if the computed test statistic value $z$ is less than $-z_{\alpha/2}$ or if it is greater than $+z_{\alpha/2}$.

Conclusion: . . .

**Note:** This is a two-tailed test because of the not-equal-to symbol in the alternative hypothesis. Also, note that the level of significance is shared equally when finding the critical $z$ value ($z_{\alpha/2}$).

**Note:** The test statistic in the above tests, $z = \dfrac{x - np_0}{\sqrt{np_0(1 - p_0)}}$, is equivalent to $z = \dfrac{x/n - p_0}{\sqrt{p_0(1 - p_0)/n}} =$

$\dfrac{\hat{p} - p_0}{\sqrt{p_0(1 - p_0)/n}}$. Observe that this is similar to the $z$ score discussed in **Chap. 10** in connection with the sampling distribution for a population proportion.

***Example 12-1:*** Your teacher claims that 60 percent of American males are married. You feel that the proportion is higher. In a random sample of 100 American males, 65 of them were married. Test your teacher's claim at the 5 percent level of significance.

**Solution:** We are given $n = 100$, $x$ (number of successes) $= 65$, $\alpha = 0.05$, $z_\alpha = 1.645$, and $p_0 = 0.6$. Also, $\sqrt{np_0(1 - p_0)} = 4.8990$. Since you would like to establish that the proportion is higher, the alternative hypothesis should reflect your counterbelief. Thus,

$H_0$: $p \le 0.6$

$H_1$: $p > 0.6$

T.S.: $z = \dfrac{x - np_0}{\sqrt{np_0(1 - p_0)}} = \dfrac{65 - 100 \times 0.6}{4.8990} = 1.0206$

D.R.: For a significance level of $\alpha = 0.05$, reject the null hypothesis if the computed test statistic value $z = 1.0206 > z_\alpha = 1.645$.

Conclusion: Since $1.0206 < 1.645$, do not reject $H_0$. There is insufficient sample evidence to refute your teacher's claim. That is, there is insufficient sample evidence to claim that more than 60 percent of American males are married at the 5 percent level of significance. Any difference between the sample proportion and the postulated proportion of 0.6 may be due to chance.

**Figure 12-2** depicts the rejection region.

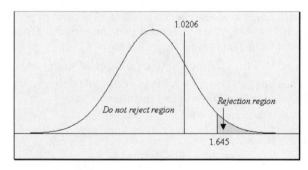

**Fig. 12-2:** Diagram depicting the rejection region for Example 12-1

***Example 12-2:*** A preacher would like to establish that of people who pray, less than 80 percent pray for world peace. In a random sample of 110 persons, 77 of them said that when they pray, they pray for world peace. Test at the 10 percent level of significance.

**Solution:** We are given $n = 110$, $x$ (number of successes) $= 77$, $\alpha = 0.1$, $z_\alpha = 1.28$, and $p_0 = 0.8$. Also, $\sqrt{np_0(1 - p_0)} = 4.1952$. Since the preacher would like to establish that less than 80 percent of people pray for world peace, the alternative hypothesis should reflect this. Thus,

$H_0$: $p \geq 0.8$

$H_1$: $p < 0.8$

T.S.: $z = \dfrac{x - np_0}{\sqrt{np_0(1 - p_0)}} = \dfrac{77 - 110 \times 0.8}{4.1952} = -2.6220$

D.R.: For a significance level of $\alpha = 0.10$, reject the null hypothesis if the computed test statistic value $z = -2.6220 < -z_\alpha = -1.28$.

Conclusion: Since $-2.6220 < -1.28$, reject $H_0$. There is sufficient sample evidence to support the notion that less than 80 percent of people pray for world peace at the 10 percent level of significance. That is, there is a significant difference between the sample proportion and the postulated proportion of 0.8.

**Figure 12-3** depicts the rejection region.

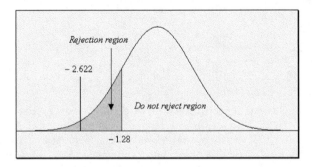

**Fig. 12-3:** Diagram depicting the rejection region for Example 12-2

***Example 12-3:*** A researcher claims that 90 percent of people trust DNA testing. In a survey of 100 people, 91 of them said that they trusted DNA testing. Test the researcher's claim at the 1 percent level of significance.

**Solution:** We are given $n = 100$, $x$ (number of successes) = 91, $\alpha = 0.01$, $z_{\alpha/2} = 2.576$, and $p_0 = 0.9$. Also, $\sqrt{np_0(1 - p_0)} = 3$. Since we are asked to test the researcher's claim, the alternative hypothesis should contradict this claim. Thus,

$H_0$: $p = 0.9$

$H_1$: $p \neq 0.9$

T.S.: $z = \dfrac{x - np_0}{\sqrt{np_0(1 - p_0)}} = \dfrac{91 - 100 \times 0.9}{3} = 0.3333$

D.R.: For a significance level of $\alpha = 0.01$, reject the null hypothesis if the computed test statistic value, $z = 0.3333 < -z_{\alpha/2} = -2.576$ or if $z = 0.3333 > z_{\alpha/2} = 2.576$.

Conclusion: Since neither of the conditions is satisfied in the decision rule, do not reject $H_0$. There is insufficient sample evidence to refute the researcher's claim that the proportion of people who believe in DNA testing is equal to 90 percent at the 1 percent level of significance. That is, there is not a significant difference between the sample proportion and the postulated proportion of 0.9.

**Figure 12-4** depicts the rejection region.

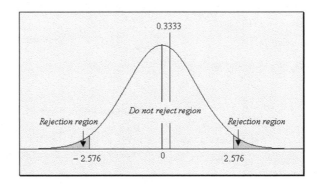

**Fig. 12-4:** Diagram depicting the rejection region for Example 12-3

## 12-3 Large-Sample Test for a Mean

We refer to tests based on the statistic $z = \dfrac{\overline{x} - \mu_0}{\sigma/\sqrt{n}}$ or $z = \dfrac{\overline{x} - \mu_0}{s/\sqrt{n}}$ as large-sample tests because we are assuming that the sampling distribution for the sample means is approximately normally distributed. The test requires that the sample size $n \geq 30$ when $\sigma$ is unknown, unless the sampling population is exactly normally distributed. If the sampling distribution is normal, the test is appropriate for any sample size.

Below is a summary of the tests for a population mean.

### Summary of Hypothesis Tests

**(a) One-tailed (right-tailed)**

$H_0$: $\mu \leq \mu_0$ (where $\mu_0$ is a specified value of the population mean)

$H_1$: $\mu > \mu_0$

T.S.: $z = \dfrac{\overline{x} - \mu_0}{\sigma/\sqrt{n}}$ or $z = \dfrac{\overline{x} - \mu_0}{s/\sqrt{n}}$, for $\sigma$ unknown and $n \geq 30$

D.R.: For a specified significance level $\alpha$, reject the null hypothesis if the computed test statistic value $z$ is greater than $+z_\alpha$.

Conclusion: . . .

**Note:** This is a right-tailed test because the direction of the inequality sign in the alternative hypothesis is to the right.

**(b) One-tailed (left-tailed)**

$H_0$: $\mu \geq \mu_0$ (where $\mu_0$ is a specified value of the population mean)

$H_1$: $\mu < \mu_0$

T.S.: $z = \dfrac{\overline{x} - \mu_0}{\sigma/\sqrt{n}}$ or $z = \dfrac{\overline{x} - \mu_0}{s/\sqrt{n}}$, for $\sigma$ unknown and $n \geq 30$

D.R.: For a specified significance level $\alpha$, reject the null hypothesis if the computed test statistic value $z$ is less than $-z_\alpha$.

Conclusion: . . .

**Note:** This is a left-tailed test because the direction of the inequality sign in the alternative hypothesis is to the left.

### (c) Two-tailed

$H_0$: $\mu = \mu_0$ (where $\mu_0$ is a specified value of the population mean)

$H_1$: $\mu \neq \mu_0$

T.S.: $z = \dfrac{\bar{x} - \mu_0}{\sigma/\sqrt{n}}$ or $z = \dfrac{\bar{x} - \mu_0}{s/\sqrt{n}}$, for $\sigma$ unknown and $n \geq 30$

D.R.: For a specified significance level $\alpha$, reject the null hypothesis if the computed test statistic value $z$ is less than $-z_{\alpha/2}$ or if it is greater than $+z_{\alpha/2}$.

Conclusion: . . .

**Note:** This is a two-tailed test because of the not-equal-to symbol in the alternative hypothesis. Also, note that the level of significance is shared equally when finding the critical $z$ value ($z_{\alpha/2}$).

***Example 12-4:*** A teachers' union would like to establish that the average salary for high school teachers in a particular state is less than $32,500. A random sample of 100 public high school teachers in the particular state has a mean salary of $31,578. It is known from past history that the standard deviation of the salaries for the teachers in the state is $4,415. Test the union's claim at the 5 percent level of significance.

**Solution:** Given $\alpha = 0.05$, $z_\alpha = 1.645$, $\bar{x} = 31,578$, $n = 100$, $\sigma = 4,415$, $\mu_0 = 32,500$, and $\dfrac{\sigma}{\sqrt{n}} = \dfrac{4,415}{\sqrt{100}} = 441.5$. Since the union would like to establish that the average salary is less than $32,500, this will be a left-tailed test. Thus,

$H_0$: $\mu \geq 32,500$

$H_1$: $\mu < 32,500$

T.S.: $z = \dfrac{\bar{x} - \mu_0}{\sigma/\sqrt{n}} = \dfrac{31,578 - 32,500}{441.5} = -2.0883$

D.R.: For a significance level of $\alpha = 0.05$, reject the null hypothesis if the computed test statistic value $z = -2.0883 < -z_\alpha = -1.645$.

Conclusion: Since $-2.0883 < -1.645$, reject $H_0$. There is sufficient sample evidence to support the claim that the average salary for high school teachers in the state is less than $32,500 at the 5 percent level of significance. That is, there is a significant difference between the sample mean and the postulated value of the population mean.

**Figure 12-5** depicts the rejection region.

***Example 12-5:*** The dean of students of a large community college claims that the average distance that commuting students travel to the campus is 32 mi. The commuting students feel otherwise.

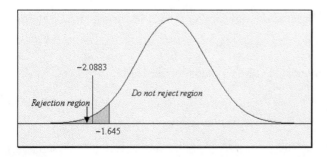

**Fig. 12-5:** Diagram depicting the rejection region for Example 12-4

A sample of 64 students was randomly selected and yielded a mean of 35 mi and a standard deviation of 5 mi. Test the dean's claim at the 5 percent level of significance.

**Solution:** Given $\alpha = 0.05$, $z_{\alpha/2} = 1.96$, $\bar{x} = 35$, $s = 5$, $n = 64$, $\mu_0 = 32$, and $\dfrac{s}{\sqrt{n}} = 0.625$. This will

be a two-tailed test, since the students feel that the dean's claim is not correct, but whether they feel that the average distance is less than 32 mi or more than 32 mi is not specified. Thus,

$H_0$: $\mu = 32$

$H_1$: $\mu \neq 32$

T.S.: $z = \dfrac{\bar{x} - \mu_0}{s/\sqrt{n}} = \dfrac{35 - 32}{0.625} = 4.8$

D.R.: For a significance level $\alpha = 0.05$, reject the null hypothesis if the computed test statistic value $z = 4.8 < -z_{\alpha/2} = -1.96$ or if $z = 4.8 > z_{\alpha/2} = 1.96$.

Conclusion: Since $4.8 > 1.96$, reject $H_0$. There is sufficient sample evidence to refute the dean's claim. The sample evidence supports the students' claim that the average distance commuting students travel to the campus is not equal to 32 mi at the 5 percent level of significance. That is, there is a significant difference between the sample mean and the postulated value of the population mean of 32 mi.

**Figure 12-6** depicts the rejection region.

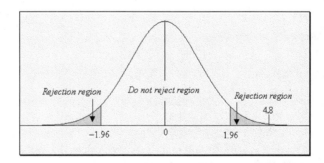

**Fig. 12-6:** Diagram depicting the rejection region for Example 12-5

## 12-4 Large-Sample Test for the Difference between Two Population Proportions

There may be problems in which one must decide whether the observed difference between two sample proportions is due to chance or whether the difference is due to the fact that the corresponding population proportions are not the same or that the proportions are from different populations. Here we will discuss such problems.

Recall from **Chap. 11** that the differences of the sample proportions will have a mean of $\mu_{\hat{p}_1 - \hat{p}_2} = p_1 - p_2$ and a standard deviation (or standard error)

$$\sigma_{\hat{p}_1 - \hat{p}_2} = \sqrt{\frac{p_1(1 - p_1)}{n_1} + \frac{p_2(1 - p_2)}{n_2}},$$

where $p_1$ and $p_2$ are the respective population proportions of interest. When we test the null hypothesis $p_1 = p_2(= p)$ against an appropriate alternative hypothesis, the mean $\mu_{\hat{p}_1 - \hat{p}_2} = p_1 - p_2$ will be zero and the standard error can be written as $\sqrt{p(1-p)\left(\dfrac{1}{n_1} + \dfrac{1}{n_2}\right)}$. Since the value of $p$ is unknown, we can estimate the value by $\hat{p} = \dfrac{x_1 + x_2}{n_1 + n_2}$.

Below is a summary of the tests for the difference between two population proportions using the above information.

## Summary of Hypothesis Tests

### (a) One-tailed (right-tailed)

$H_0$: $p_1 \le p_2$

$H_1$: $p_1 > p_2$

T.S.: $z = \dfrac{\hat{p}_1 - \hat{p}_2}{\sqrt{\hat{p}(1-\hat{p})\left(\dfrac{1}{n_1} + \dfrac{1}{n_2}\right)}}$

D.R.: For a specified significance level $\alpha$, reject the null hypothesis if the computed test statistic value $z$ is greater than $+z_\alpha$.

Conclusion: . . .

**Note:** This is a right-tailed test because the direction of the inequality sign in the alternative hypothesis is to the right.

### (b) One-tailed (left-tailed)

$H_0$: $p_1 \ge p_2$

$H_1$: $p_1 < p_2$

T.S.: $z = \dfrac{\hat{p}_1 - \hat{p}_2}{\sqrt{\hat{p}(1-\hat{p})\left(\dfrac{1}{n_1} + \dfrac{1}{n_2}\right)}}$

D.R.: For a specified significance level $\alpha$, reject the null hypothesis if the computed test statistic value $z$ is less than $-z_\alpha$.

Conclusion: . . .

**Note:** This is a left-tailed test because the direction of the inequality sign in the alternative hypothesis is to the left.

### (c) Two-tailed

$H_0$: $p_1 = p_2$

$H_1$: $p_1 \ne p_2$

T.S.: $z = \dfrac{\hat{p}_1 - \hat{p}_2}{\sqrt{\hat{p}(1-\hat{p})\left(\dfrac{1}{n_1} + \dfrac{1}{n_2}\right)}}$

D.R.: For a specified significance level $\alpha$, reject the null hypothesis if the computed test statistic value $z$ is less than $-z_{\alpha/2}$ or if it is greater than $+z_{\alpha/2}$.

Conclusion: . . .

**Note:** This is a two-tailed test because of the not-equal-to symbol in the alternative hypothesis. Also, note that the level of significance is shared equally when finding the critical $z$ value ($z_{\alpha/2}$).

***Example 12-6:*** A study was conducted to determine whether remediation in basic mathematics enabled students to be more successful in an elementary statistics course. Success here means that a student received a grade of C or better, and remediation was for one year. **Table 12-2** shows the results of the study.

**Table 12-2**

|                    | REMEDIAL | NONREMEDIAL |
|--------------------|----------|-------------|
| Sample size        | 100      | 40          |
| Number of successes| 70       | 16          |

Test, at the 5 percent level of significance, whether remediation helped the students to be more successful.

**Solution:** From the information given, we have $n_1 = 100$, $n_2 = 40$, $\hat{p}_1 = \dfrac{70}{100} = 0.7$, $\hat{p}_2 = \dfrac{16}{40} = 0.4$, $\hat{p} = \dfrac{x_1 + x_2}{n_1 + n_2} = \dfrac{70 + 16}{100 + 40} = 0.6143$, $\alpha = 0.05$, and $z_\alpha = 1.645$. Since we would like to establish whether remediation helped the students to be more successful, this is equivalent to establishing whether the proportion of students under remediation who were successful is greater than the proportion of those who were not under remediation. Thus we have a right-tailed test in favor of remediation.

$H_0$: $p_1 \leq p_2$

$H_1$: $p_1 > p_2$

T.S.: $z = \dfrac{\hat{p}_1 - \hat{p}_2}{\sqrt{\hat{p}(1 - \hat{p})\left(\dfrac{1}{n_1} + \dfrac{1}{n_2}\right)}} = \dfrac{0.7 - 0.4}{\sqrt{0.6143 \times (1 - 0.6143)\left(\dfrac{1}{100} + \dfrac{1}{40}\right)}} = 3.2944$

D.R.: For a significance level $\alpha = 0.05$, reject the null hypothesis if the computed test statistic value $z = 3.2944 > z_\alpha = 1.645$.

Conclusion: Since $3.2944 > 1.645$, reject the null hypothesis. That is, at the 5 percent level of significance, we can conclude that remediation is helping the students to do better in elementary statistics.

***Example 12-7:*** A researcher wants to determine whether there is a difference between the proportions of males and females who believe in aliens. The sample information is given in **Table 12-3.**

**Table 12-3**

|                    | MALES | FEMALES |
|--------------------|-------|---------|
| Sample size        | 75    | 100     |
| Number who said yes| 50    | 45      |

Test at the 1 percent level of significance.

**Solution:** From the information given, we have $n_1 = 75, n_2 = 100, \hat{p}_1 = \dfrac{50}{75} = 0.6667, \hat{p}_2 = \dfrac{45}{100} =$

$0.45, \hat{p} = \dfrac{x_1 + x_2}{n_1 + n_2} = \dfrac{50 + 45}{75 + 100} = 0.5429, \alpha = 0.01$, and $z_{\alpha/2} = 2.576$. This is a two-tailed test,

since we are just testing whether or not there is a difference. Thus,

$H_0$: $p_1 = p_2$

$H_1$: $p_1 \neq p_2$

T.S.: $z = \dfrac{\hat{p}_1 - \hat{p}_2}{\sqrt{\hat{p}(1 - \hat{p})\left(\dfrac{1}{n_1} + \dfrac{1}{n_2}\right)}} = \dfrac{0.6667 - 0.45}{\sqrt{0.5429 \times (1 - 0.5429) \times \left(\dfrac{1}{75} + \dfrac{1}{100}\right)}} = 2.8478$

D.R.: For a significance level $\alpha = 0.01$, reject the null hypothesis if the computed test statistic value $z = 2.8478 < -z_{\alpha/2} = -2.576$ or if $z = 2.8478 > z_{\alpha/2} = 2.576$.

Conclusion: Since $2.8478 > 2.576$, reject the null hypothesis. That is, we can conclude, at the 1 percent level of significance, that the proportions of males and females who believe in aliens are significantly different from each other.

## 12-5 Large-Sample Test for the Difference between Two Population Means

There may be problems in which one must decide whether the observed difference between two sample means is due to chance or whether the difference is due to the fact that the population means are not the same or that the samples are not from the same population. Here we will discuss such problems.

Recall from **Chap. 11** that the differences of the sample means will have a mean of $\mu_{\bar{x}_1 - \bar{x}_2} = \mu_1 - \mu_2$ and a standard deviation (or standard error) $\sigma_{\bar{x}_1 - \bar{x}_2} = \sqrt{\dfrac{\sigma_1^2}{n_1} + \dfrac{\sigma_2^2}{n_2}}$, where $\sigma_1$ and $\sigma_2$ are the respective population standard deviations of interest. Recall that the samples are assumed to have been obtained from normal populations. Also, when $\sigma_1$ and $\sigma_2$ are unknown, if the sample sizes are large enough ($n_1 \geq 30$ and $n_2 \geq 30$), we can estimate the population variances with the sample variances $s_1$ and $s_2$.

Below is a summary of the tests for the difference between two population means using the above information.

### Summary of Hypothesis Tests

**(a) One-tailed (right-tailed)**

$H_0$: $\mu_1 \leq \mu_2$

$H_1$: $\mu_1 > \mu_2$

T.S.: $z = \dfrac{\bar{x}_1 - \bar{x}_2}{\sqrt{\dfrac{\sigma_1^2}{n_1} + \dfrac{\sigma_2^2}{n_2}}}$ or $\dfrac{\bar{x}_1 - \bar{x}_2}{\sqrt{\dfrac{s_1^2}{n_1} + \dfrac{s_2^2}{n_2}}}$ for $n_1 \geq 30$ and $n_2 \geq 30$

D.R.: For a specified significance level $\alpha$, reject the null hypothesis if the computed test statistic value $z$ is greater than $+z_\alpha$.

Conclusion: . . .

**Note:** This is a right-tailed test because the direction of the inequality sign in the alternative hypothesis is to the right.

**(b) One-tailed (left-tailed)**

$H_0$: $\mu_1 \geq \mu_2$

$H_1$: $\mu_1 < \mu_2$

T.S.: $z = \dfrac{\bar{x}_1 - \bar{x}_2}{\sqrt{\dfrac{\sigma_1^2}{n_1} + \dfrac{\sigma_2^2}{n_2}}}$ or $\dfrac{\bar{x}_1 - \bar{x}_2}{\sqrt{\dfrac{s_1^2}{n_1} + \dfrac{s_2^2}{n_2}}}$ for $n_1 \geq 30$ and $n_2 \geq 30$

D.R.: For a specified significance level $\alpha$, reject the null hypothesis if the computed test statistic value $z$ is less than $-z_\alpha$.

Conclusion: . . .

**Note:** This is a left-tailed test because the direction of the inequality sign in the alternative hypothesis is to the left.

**(c) Two-tailed**

$H_0$: $\mu_1 = \mu_2$

$H_1$: $\mu_1 \neq \mu_2$

T.S.: $z = \dfrac{\bar{x}_1 - \bar{x}_2}{\sqrt{\dfrac{\sigma_1^2}{n_1} + \dfrac{\sigma_2^2}{n_2}}}$ or $\dfrac{\bar{x}_1 - \bar{x}_2}{\sqrt{\dfrac{s_1^2}{n_1} + \dfrac{s_2^2}{n_2}}}$ for $n_1 \geq 30$ and $n_2 \geq 30$

D.R.: For a specified significance level $\alpha$, reject the null hypothesis if the computed test statistic value $z$ is less than $-z_{\alpha/2}$ or if it is greater than $+z_{\alpha/2}$.

Conclusion: . . .

**Note:** This is a two-tailed test because of the not-equal-to symbol in the alternative hypothesis. Also, note that the level of significance is shared equally when finding the critical $z$ value ($z_{\alpha/2}$).

***Example 12-8:*** A random sample of size $n_1 = 36$ selected from a normal distribution with standard deviation $\sigma_1 = 4$ has a mean $\bar{x}_1 = 75$. A second random sample of size $n_2 = 25$ selected from a different normal distribution with a standard deviation $\sigma_2 = 6$ has a mean $\bar{x}_2 = 85$. Is there a significant difference between the population means at the 5 percent level of significance?

**Solution:** From the information given, we have $n_1 = 36$, $n_2 = 25$, $\bar{x}_1 = 75$, $\bar{x}_2 = 85$, $\sigma_1 = 4$, $\sigma_2 = 6$,

$\sqrt{\dfrac{\sigma_1^2}{n_1} + \dfrac{\sigma_2^2}{n_2}} = 1.3728$, $\alpha = 0.05$, and $z_{\alpha/2} = 1.96$. Since we are just determining whether there is

a difference between the population means, this will be a two-tailed test. Hence,

$H_0$: $\mu_1 = \mu_2$

$H_1$: $\mu_1 \neq \mu_2$

T.S.: $z = \dfrac{\bar{x}_1 - \bar{x}_2}{\sqrt{\dfrac{\sigma_1^2}{n_1} + \dfrac{\sigma_2^2}{n_2}}} = \dfrac{75 - 85}{1.3728} = -7.2844$

D.R.: For a significance level $\alpha = 0.05$, reject the null hypothesis if the computed test statistic value $z = -7.2844 < -z_{\alpha/2} = -1.96$ or if $z = -7.2844 > +z_{\alpha/2} = 1.96$.

Conclusion: Since $-7.2844 < -1.96$, reject the null hypothesis. That is, at the 5 percent significance level, we can conclude that the means are significantly different from each other.

***Example 12-9:*** Two methods were used to teach a high school algebra course. A sample of 75 scores was selected for method 1, and a sample of 60 scores was selected for method 2. The summary results are given in **Table 12-4.**

**Table 12-4**

|                            | METHOD 1 | METHOD 2 |
| -------------------------- | -------- | -------- |
| Sample mean                | 85       | 83       |
| Sample standard deviation  | 3        | 2        |

Test whether method 1 was more successful than method 2 at the 1 percent significance level.

**Solution:** From the information given, we have $n_1 = 75$, $n_2 = 60$, $\bar{x}_1 = 85$, $\bar{x}_2 = 83$, $s_1 = 3$, $s_2 = 2$, $\sqrt{\dfrac{s_1^2}{n_1} + \dfrac{s_2^2}{n_2}} = 0.4321$, $\alpha = 0.01$, and $z_\alpha = 2.33$. Since we are testing whether method 1 was more successful than method 2, this is equivalent to establishing whether the mean score for method 1 is greater than that for method 2. Thus,

$H_0$: $\mu_1 \le \mu_2$

$H_1$: $\mu_1 > \mu_2$

T.S.: $z = \dfrac{\bar{x}_1 - \bar{x}_2}{\sqrt{\dfrac{s_1^2}{n_1} + \dfrac{s_2^2}{n_2}}} = \dfrac{85 - 83}{0.4321} = 4.6291$

D.R.: For a significance level $\alpha = 0.01$, reject the null hypothesis if the computed test statistic value $z = 4.6291 > z_\alpha = 2.33$.

Conclusion: Since $4.6291 > 2.33$, reject the null hypothesis. At the 1 percent significance level, we can conclude that method 1 was more successful than method 2. That is, we can conclude that the average score for the algebra course using method 1 is significantly greater than the average score using method 2.

## 12-6  *P*-Value Approach to Hypothesis Testing

With the advent of the computer, certain probabilities can be readily computed and used to help make decisions in hypothesis tests. One such probability value is called the *P* value.

**Explanation of the term—*P* value:** A ***P* value** is the smallest significance level at which a null hypothesis may be rejected.

### Determining *P* Values

For each of the above tests, we can determine a *P* value.

(a) Right-tailed test: *P* value $= P(z > z^*)$, where $z^*$ is the computed value for the test statistic.

(b) Left-tailed test: *P* value $= P(z < z^*)$, where $z^*$ is the computed value for the test statistic.

(c) Two-tailed test: *P* value $= 2 \times P(z > |z^*|)$, where $z^*$ is the computed value for the test statistic.

***Example 12-10:*** Compute the *P* value for the test in **Example 12-1.**

**Solution:** The test was a right-tailed test with $z^* = 1.0206$. Thus, the $P$ value will be $P$ value $= P(z > 1.0206) = 0.5 - 0.3461 = 0.1539$. Note we have to subtract 0.3461 from 0.5 since the standard tables are set up so that we find the area from $z = 0$ to $z = 1.02$.

***Example 12-11:*** Compute the $P$ value for the test in **Example 12-2.**

**Solution:** The test was a left-tailed test with $z^* = -2.622$. Thus, the $P$ value will be $P$ value $= P(z < -2.662) = 0.5 - 0.4961 = 0.0039$.

***Example 12-12:*** Compute the $P$ value for the test in **Example 12-3.**

**Solution:** The test was a two-tailed test with $z^* = 0.3333$. Thus, the $P$ value will be $P$ value $= 2 \times P(z > |0.3333|) = 2 \times (0.5 - 0.1293) = 0.3707 = 0.7414$.

**Note:** The $P$ values for any of the presented tests can be computed using the above procedures.

## Interpreting $P$ Values

We can use the value of the $P$ value for a statistical test to measure the strength of the evidence against the null hypothesis for that test. The smaller the $P$ value, the stronger the evidence against the null hypothesis. That is, the smaller the $P$ value, the stronger is the evidence that we should reject the null hypothesis in favor of the alternative hypothesis. Several cutoff points can be used for $P$ values. Typical cutoff points are

➡ $P$ value $> 0.1 \Rightarrow$ little evidence against $H_0$

➡ $0.05 < P$ value $\leq 0.1 \Rightarrow$ some evidence against $H_0$

➡ $0.01 < P$ value $\leq 0.05 \Rightarrow$ moderate evidence against $H_0$

➡ $0.001 < P$ value $\leq 0.01 \Rightarrow$ strong evidence against $H_0$

➡ $P$ value $< 0.001 \Rightarrow$ very strong evidence against $H_0$

Using the above guide to make a decision, for **Examples 12-10** and **12-12,** we will not reject the null hypothesis, since both $P$ values are greater than 0.1. However, the $P$ value for **Example 12-11** is 0.0039, which indicates strong evidence against the null hypothesis. In this case, we would reject the null hypothesis and conclude that the alternative hypothesis is true at the 0.0039 level of significance.

## Quick Tip

Note that there is a clear distinction between the $\alpha$ value (level of significance) and the $P$ value. The $\alpha$ value is chosen before the statistical test is carried out, and the $P$ value is computed after an experiment is run and a sample statistic is computed.

Another approach to using the $P$ value to help in making a decision for a test is to compare it with a specified significance level $\alpha$. The following can be used:

- $P$ value $< \alpha \Rightarrow$ reject the null hypothesis
- $P$ value $\geq \alpha \Rightarrow$ do not reject the null hypothesis

Technology

### Technology Corner

All of the concepts discussed in this chapter can be computed and illustrated using most statistical software packages. All scientific and graphical calculators can be used for the computations. In addition, some of the newer calculators, like the TI-83, will allow you to do the test directly on the calculator. If you own a calculator, you should consult the manual to determine what statistical features are included.

Illustration: **Figure 12-7** shows the outputs computed by the MINITAB software for **Examples 12-1** and **12-7.** The MINITAB software does not allow you to directly do hypothesis tests for a population mean or for the difference of two means when you have summary data, as in **Examples 12-1** and **12-7.** The software will allow you to directly do tests for means when you have raw data. Note that the outputs also give the *P* values for the tests. **Figure 12-8** shows the outputs computed by the TI-83 calculator for **Examples 12-1, 12-5, 12-6,** and **12-9.** The TI-83 calculator allows you to use both summary data and raw data to do hypothesis tests for both means and proportions. The *P* values provided by both the MINITAB and TI-83 outputs can be used to perform the tests. Care should always be taken when using the formulas for computations in hypothesis testing. One can use other features of the technologies to illustrate other concepts discussed in the chapter.

```
Worksheet size: 100000 cells

Test and Confidence Interval for One Proportion--Example 12-1

Test of p = 0.6 vs p > 0.6

                                                      Exact
Sample      X     N  Sample p       95.0 % CI       P-Value
1          61   100  0.610000   (0.507314, 0.705990)  0.462


Test and Confidence Interval for Two Proportions--Example 12-7

Sample      X     N  Sample p
1          50    75  0.666667
2          45   100  0.450000

Estimate for p(1) - p(2):  0.216667
99% CI for p(1) - p(2):  (0.0267185, 0.406615)
Test for p(1) - p(2) = 0 (vs not = 0):  Z = 2.85  P-Value = 0.004
```

**Fig. 12-7:** MINITAB output for Examples 12-1 and 12-7

```
1-PropZTest        Z-Test           2-PropZTest        2-SampZTest
prop>.6            μ≠32             p1>p2              μ1>μ2
z=1.020620726      z=4.8            z=3.294345637      z=4.629100499
p=.1537170965      p=1.5887053e-6   p=4.9331431e-4     p=1.8382544e-6
p̂=.65              x̄=35             p̂1=.7              x̄1=85
n=100              n=64             p̂2=.4              x̄2=83
                                    ↓p̂=.6142857143     ↓n1=75

Example 12-1       Example 12-5      Example 12-6       Example 12-9
```

**Fig. 12-8:** TI-83 output for Examples 12-1, 12-5, 12-6, and 12-9

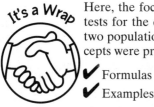

It's a Wrap

Here, the focus was on hypothesis tests for a single mean and a single proportion. Also, tests for the difference between two population proportions and the difference between two population means were addressed. In all tests, we assumed large samples. These concepts were presented through

✔ Formulas
✔ Examples

Test
Yourself

## True/False Questions

1. A claim or statement about a population parameter is classified as the null hypothesis.
2. A statement contradicting the claim in the null hypothesis about a population parameter is classified as the alternative hypothesis.
3. If we want to claim that a population parameter is different from a specified value, this situation can be considered as a one-tailed test.
4. The null hypothesis is considered correct until proven otherwise.
5. A Type I error is the error we make when we fail to reject an incorrect null hypothesis.
6. The probability of making a Type I error and the level of significance are equal or the same.
7. The range of $z$ values that indicates that there is a significant difference between the value of the sample statistic and the proposed parameter value is called the rejection region or the critical region.
8. If the sample size $n$ is less than 30, then a $z$ score will always be associated with any hypothesis that deals with the mean.
9. In the $P$-value approach to hypothesis testing, if the $P$ value is less than a specified significance level, we fail to reject the null hypothesis.
10. In the $P$-value approach to hypothesis testing, if $0.01 < P$ value $< 0.05$, there is insufficient evidence to reject the null hypothesis.
11. In the $P$-value approach to hypothesis testing, if $P$ value $< 0.001$, there is very strong evidence to reject the null hypothesis.
12. When large samples ($n \geq 30$) are associated with hypothesis tests for population proportions, the associated test statistic is a $z$ score.
13. The distribution of sample proportions from a single population is approximately normal provided that the sample size is large enough ($n \geq 30$).
14. The distribution of the difference between two sample means is approximately normal with variance $\dfrac{\sigma_1^2}{n_1} + \dfrac{\sigma_2^2}{n_2}$, where $n_1$ and $n_2$ are the sample sizes from populations 1 and 2, respectively, and $\sigma_1^2$ and $\sigma_2^2$ are the respective variances, if the sample sizes are both greater than or equal to 30.
15. In testing for the difference of two population means, if the population variances are unknown and the sample sizes from the populations are both greater than or equal to 30, the associated test statistic is approximately a $z$ score.
16. In making inferences on the difference of two population proportions, the calculated (pooled) proportion is given by $\hat{p} = \dfrac{x_1 + n_1}{x_2 + n_2}$, where $x_1$ and $x_2$ are the respective numbers of successes from populations 1 and 2, and $n_1$ and $n_2$ are the respective sample sizes.
17. If the null hypothesis is rejected, this means that the null hypothesis is not true.
18. When performing hypothesis tests on two population means, it is necessary to assume that the populations are normally distributed.

19. The $P$ value of a hypothesis test can be computed without the value of the test statistic.
20. The $P$ value of a hypothesis test is the smallest level of significance at which the null hypothesis can be rejected.
21. In hypothesis testing, the alternative hypothesis is assumed to be true.
22. In hypothesis testing, if the null hypothesis is rejected, the alternative hypothesis must also be rejected.

## Completion Questions

1. The two broad areas of statistical inference are hypothesis testing and (point, interval) _____ estimation.
2. Rejecting a true null hypothesis is classified as a (Type I, Type II) _____ error.
3. Failing to reject a false null hypothesis is classified as a _____ error.
4. In hypothesis testing, the cutoff values which define the rejection region are known as ($P$, critical, $\alpha$) _____ values.
5. In a hypothesis test, if the computed $P$ value is less than 0.001, there is very strong evidence to (fail to reject, reject) _____ the null hypothesis.
6. In a hypothesis test, if the computed $P$ value is greater than a specified level of significance, then we (reject, fail to reject) _____ the null hypothesis.
7. The level of significance for a hypothesis test is equal to the probability of a (Type I, Type II) _____ error.
8. If we reject the (null, alternative) _____ hypothesis, the (null, alternative) _____ hypothesis may be accepted.
9. The point estimate for the difference of two population means $\mu_1 - \mu_2$ can be represented by ($\hat{p}_1 - \hat{p}_2, \bar{x}_1 - \bar{x}_2, \mu_1 - \mu_2$) _____, where the subscripts represent the corresponding populations.
10. In a hypothesis test for means, it is necessary to assume that the sample was selected from a(n) _____ population.
11. The value of the level of significance lies between _____ and _____.
12. If a hypothesis test is performed at the 0.02 level of significance and the computed $P$ value is 0.01, you will (reject, fail to reject) _____ the null hypothesis.
13. The level of significance for a hypothesis test is the probability of rejecting a (true, false) _____ null hypothesis.
14. When conducting a hypothesis test for a single population mean, the test statistic is assumed to have a normal distribution if the sample size is (small, large) _____.
15. The area under a curve that leads to the rejection of the null hypothesis is also known as the (critical or rejection, do not reject) _____ region.
16. The number that separates the rejection region from the do not reject region is called a (critical, noncritical) _____ value of the test.
17. If we are performing a right-tailed test and the computed $z$ value is 2.99, the $P$ value will be _____.
18. If we are performing a two-tailed test for the population mean when the population standard deviation is known, and if the test statistic value is 2.79, the $P$ value will be _____.
19. The $P$ value of a hypothesis test is the smallest level of significance at which the null hypothesis can (be rejected, fail to be rejected) _____.

20. In the *P*-value approach to hypothesis testing, if the *P* value is less than a specified significance level α, then we (reject, do not reject) _____ the null hypothesis.

## Multiple-Choice Questions

1. The calculated numerical value that is compared to a table value in a hypothesis test is called the
   (a) level of significance.
   (b) critical value.
   (c) population parameter.
   (d) test statistic.

2. A right-tailed test is conducted with α = 0.0582. If the *z* tables are used, the critical value will be
   (a) −1.57.
   (b) 1.57.
   (c) −0.15.
   (d) 0.15.

3. A right-tailed test is performed, with the test statistic having a standard normal distribution. If the computed test statistic is 3.00, the *P* value for this test is
   (a) 0.4996.
   (b) 0.9996.
   (c) 0.0013.
   (d) 0.0500.

4. New software is being integrated into the teaching of a course with the hope that it will help to improve the overall average score for this course. The historical average score for this course is 72. If a statistical test is done for this situation, the alternative hypothesis will be
   (a) $H_1: \mu \neq 72$.
   (b) $H_1: \mu < 72$.
   (c) $H_1: \mu = 72$.
   (d) $H_1: \mu > 72$.

5. Dr. J claims that 40 percent of his College Algebra class (a very large section) will drop his course by midterm. To test his claim, he selected 45 names at random and discovered that 20 of them had already dropped long before midterm. The test statistic value for his hypothesis test is
   (a) 0.6086.
   (b) 0.3704.
   (c) 8.3333.
   (d) 0.6847.

6. In testing a hypothesis, the hypothesis that is assumed to be true is
   (a) the alternative hypothesis.
   (b) the null hypothesis.
   (c) the null or the alternative hypothesis.
   (d) neither the null nor the alternative hypothesis.

7. A Type I error is defined to be the probability of
   (a) failing to reject a true null hypothesis.
   (b) failing to reject a false null hypothesis.
   (c) rejecting a false null hypothesis.
   (d) rejecting a true null hypothesis.

8. A Type II error is defined to be the probability of
   (a) failing to reject a true null hypothesis.
   (b) failing to reject a false null hypothesis.
   (c) rejecting a false null hypothesis.
   (d) rejecting a true null hypothesis.

9. In hypothesis testing, the level of significance is the probability of
   (a) failing to reject a true null hypothesis.
   (b) failing to reject a false null hypothesis.
   (c) rejecting a false null hypothesis.
   (d) rejecting a true null hypothesis.

10. The level of significance can be any
    (a) $z$ value.
    (b) parameter value.
    (c) value between 0 and 1, inclusive.
    (d) $\alpha$ value.

11. If you fail to reject the null hypothesis in the testing of a hypothesis, then
    (a) a Type I error has definitely occurred.
    (b) a Type II error has definitely occurred.
    (c) the computed test statistic is incorrect.
    (d) there is insufficient evidence to claim that the alternative hypothesis is true.

12. If we were testing the hypotheses $H_0$: $\mu = \mu_0$ vs. $H_1$: $\mu > \mu_0$ (where $\mu_0$ is a specified value of $\mu$) at a given significance level $\alpha$, with large samples and unknown population variance, then $H_0$ will be rejected if the computed test statistic is
    (a) $z > z_\alpha$.
    (b) $z < -z_\alpha$.
    (c) $z > z_{\alpha/2}$.
    (d) $z < -z_{\alpha/2}$.

13. Which of the following general guidelines is used when using the $P$ value to perform hypothesis tests?
    (a) If the $P$ value $> 0.1$, there is little or no evidence to reject the null hypothesis.
    (b) If $0.01 < P$ value $\leq 0.05$, there is moderate evidence to reject the null hypothesis.
    (c) If the $P$ value $\leq 0.001$, there is very strong evidence to reject the null hypothesis.
    (d) All of the above.

14. When the $P$ value is used in testing a hypothesis, we will not reject the null hypothesis for a level of significance $\alpha$ when
    (a) $P$ value $< \alpha$.
    (b) $P$ value $\geq \alpha$.
    (c) $P$ value $= \alpha$.
    (d) $P$ value $\neq \alpha$.

15. A real estate agent claims that the average price for homes in a certain subdivision is $150,000. You believe that the average price is lower. If you plan to test his claim by taking a random sample of the prices of the homes in the subdivision, the formulated set of hypotheses will be

   (a) $H_0: \mu \leq 150,000$ vs. $H_1: \mu > 150,000$.
   (b) $H_0: \mu = 150,000$ vs. $H_1: \mu \neq 150,000$.
   (c) $H_0: \mu < 150,000$ vs. $H_1: \mu \geq 150,000$.
   (d) $H_0: \mu \geq 150,000$ vs. $H_1: \mu < 150,000$.

16. A statistics student was not pleased with his final grade in his statistics course, so he decided to appeal his grade. He believes that the average score on the final examination was less than 69 (out of a possible 100 points), so he believes that it was an unfair examination. He thinks that he should have made at least a grade of B in the course. He decided to test his claim about the average of the final examination. If he knows his "statistics," the correct set of hypothesis he will set up to test his claim is

   (a) $H_0: \mu \leq 69$ vs. $H_1: \mu > 69$.
   (b) $H_0: \mu \geq 69$ vs. $H_1: \mu < 69$.
   (c) $H_0: \mu = 69$ vs. $H_1: \mu \neq 69$.
   (d) $H_0: \mu \neq 69$ vs. $H_1: \mu = 69$.

17. An advertisement on the TV claims that a certain brand of tire has an average lifetime of 50,000 miles. Suppose you plan to test this claim by taking a sample of tires and putting them on test. The correct set of hypotheses to set up is

   (a) $H_0: \mu \leq 50,000$ vs. $H_1: \mu > 50,000$.
   (b) $H_0: \mu \geq 50,000$ vs. $H_1: \mu < 50,000$.
   (c) $H_0: \mu = 50,000$ vs. $H_1: \mu \neq 50,000$.
   (d) $H_0: \mu \neq 50,000$ vs. $H_1: \mu = 50,000$.

18. The local newspaper reported that at least 25 percent of the population in a university community works at the university. You believe that the proportion is lower. If you selected a random sample to test this claim, the appropriate set of hypotheses would be

   (a) $H_0: p \leq 0.25$ vs. $H_1: p > 0.25$.
   (b) $H_0: p = 0.25$ vs. $H_1: p \neq 0.25$.
   (c) $H_0: p < 0.25$ vs. $H_1: p \geq 0.25$.
   (d) $H_0: p \geq 0.25$ vs. $H_1: p < 0.25$.

19. The local newspaper claims that 15 percent of the residents of the community play the state lottery. If you plan to test the claim by taking a random sample from the community, the appropriate set of hypotheses is

   (a) $H_0: p \geq 0.15$ vs. $H_1: p < 0.15$.
   (b) $H_0: p \leq 0.15$ vs. $H_1: p > 0.15$.
   (c) $H_0: p = 0.15$ vs. $H_1: p \neq 0.15$.
   (d) $H_0: p \neq 0.15$ vs. $H_1: p = 0.15$.

20. The local newspaper claims that no more than 5 percent of the residents of the community are on welfare. If you plan to test the claim by taking a random sample from the community, the appropriate set of hypotheses is

   (a) $H_0: p \leq 0.05$ vs. $H_1: p > 0.05$.
   (b) $H_0: p \geq 0.05$ vs. $H_1: p < 0.05$.
   (c) $H_0: p = 0.05$ vs. $H_1: p \neq 0.05$.
   (d) $H_0: p > 0.05$ vs. $H_1: p \leq 0.05$.

21. For the following information,

$$n = 16 \qquad \mu = 15 \qquad \bar{x} = 16 \qquad \sigma^2 = 16$$

assume that the population is normal and compute the test statistic : 1 if you were testing for a single population mean.

(a) $z = 1$
(b) $z = \frac{1}{4}$
(c) $z = 0$
(d) $z = -1$

22. For the following information

$$n = 16 \qquad \mu = 15 \qquad \bar{x} = 16 \qquad \sigma^2 = 16$$

assume that the population is normal. If you are performing a right-tail. e( 1 test for a single population mean, then

(a) $P$ value $= 0.3413$.
(b) $P$ value $< 0.05$.
(c) $P$ value $= 0.1587$.
(d) $P$ value $= 0.0794$.

23. For the following information,

$$n = 16 \qquad \mu = 15 \qquad \bar{x} = 16 \qquad \sigma^2 = 16$$

assume that the population is normal. If you are performing a right-tailed t( 2: st for a single population mean, then you

(a) will reject the null hypothesis if $\alpha = 0.1$.
(b) will not reject the null hypothesis if $\alpha = 0.1$.
(c) will not be able to do the test, since more information is needed.
(d) need the hypotheses to be given.

24. For the following information,

$$n = 16 \qquad \mu = 15 \qquad \bar{x} = 16 \qquad \sigma^2 = 16$$

assume that the population is normal. If you are performing a left-tailed test f or ' a single population mean, then you

(a) will reject the null hypothesis if $\alpha = 0.2$.
(b) will not reject the null hypothesis if $\alpha = 0.2$.
(c) will not be able to do the test, since more information is needed.
(d) need the hypotheses to be given.

25. If a null hypothesis is rejected at the 0.05 level of significance for a two-tailed test. , : you
(a) will always reject it at the 99 percent level of confidence.
(b) will always reject it at the 90 percent level of confidence.
(c) will always not reject it at the 99 percent level of confidence.
(d) will always not reject it at the 96 percent level of confidence.

26. If a null hypothesis is rejected at the 5 percent significance level for a right-tailed te si t, you
(a) will always reject it at the 0.1 level of significance.
(b) will always reject it at the 0.01 level of significance.
(c) will always not reject it at the 0.01 level of significance.
(d) will sometimes reject it at the 0.06 level of significance.

27. For a left-tailed test concerning the population proportion with sample size 203 and $\alpha = 0.05$, the null hypothesis will be rejected if the computed test statistic is
    (a) less than $-1.96$.
    (b) less than $-1.717$.
    (c) less than $-1.645$.
    (d) less than $-2.704$.

28. It was reported that a certain population had a mean of 27. To test this claim, you selected a random sample of size 100. The computed sample mean and sample standard deviation were 25 and 7, respectively. The appropriate set of hypotheses for this test is
    (a) $H_0: \mu \leq 27$ vs. $H_1: \mu > 27$.
    (b) $H_0: \mu = 27$ vs. $H_1: \mu \neq 27$.
    (c) $H_0: \mu \geq 25$ vs. $H_1: \mu < 25$.
    (d) $H_0: \mu \neq 25$ vs. $H_1: \mu = 25$.

29. It was reported that a certain population had a mean of 27. To test this claim, you selected a random sample of size 100. The computed sample mean and sample standard deviation were 25 and 7, respectively. The computed test statistic for the appropriate set of hypotheses is
    (a) $-4.0816$.
    (b) $-0.4082$.
    (c) $-28.5714$.
    (d) $-2.8571$.

30. It was reported that a certain population had a mean of 27. To test this claim, you selected a random sample of size 100. The computed sample mean and sample standard deviation were 25 and 7, respectively. The $P$ value for the appropriate set of hypotheses is
    (a) $0.0021$.
    (b) $0.9979$.
    (c) $0.0042$.
    (d) $-0.4979$.

31. It was reported that a certain population had a mean of 27. To test this claim, you selected a random sample of size 100. The computed sample mean and sample standard deviation were 25 and 7, respectively. At the 0.05 level of significance, you can claim that the average of this population is
    (a) not equal to 25.
    (b) equal to 25.
    (c) not equal to 27.
    (d) equal to 27.

32. For a highly publicized murder trial, it was estimated that 25 percent of the population watched the proceedings on TV. A statistics student felt that this estimate was too small for his community and decided to do a hypothesis test. He selected a random sample of 100 residents from the university community where he lives and found that 32 of them actually watched at least three hours of the proceedings. The appropriate set of hypotheses for the test is
    (a) $H_0: p \leq 0.32$ vs. $H_1: p > 0.32$.
    (b) $H_0: p \geq 0.25$ vs. $H_1: p < 0.25$.

(c) $H_0$: $p \geq 0.32$ vs. $H_1$: $p < 0.32$.

(d) $H_0$: $p \leq 0.25$ vs. $H_1$: $p > 0.25$.

33. For a highly publicized murder trial, it was estimated that 25 percent of the population watched the proceedings on TV. A statistics student felt that this estimate was too small for his community and decided to do a hypothesis test. He selected a random sample of 100 residents from the university community where he lives and found that 32 of them actually watched at least three hours of the proceedings. The computed test statistic for the test is

    (a) 1.6167.
    (b) 1.5006.
    (c) −1.6167.
    (d) −1.5006.

34. For a highly publicized murder trial, it was estimated that 25 percent of the population watched the proceedings on TV. A statistics student felt that this estimate was too small for his community and decided to do a hypothesis test. He selected a random sample of 100 residents from the university community where he lives and found that 32 of them actually watched at least three hours of the proceedings. The $P$ value for the test is

    (a) 0.4332.
    (b) 0.0526.
    (c) 0.0668.
    (d) 0.4474.

35. For a highly publicized murder trial, it was estimated that 25 percent of the population watched the proceedings on TV. A statistics student felt that this estimate was too small for his community and decided to do a hypothesis test. He selected a random sample of 100 residents from the university community where he lives and found that 32 of them actually watched at least three hours of the proceedings. At the 10 percent significance level, you can claim that the proportion of viewers in this community was

    (a) significantly greater than 32 percent.
    (b) significantly smaller than 32 percent.
    (c) significantly smaller than 25 percent.
    (d) significantly greater than 25 percent.

36. For a highly publicized murder trial, it was estimated that 25 percent of the population watched the proceedings on TV. A statistics student felt that this estimate was too small for his community and decided to do a hypothesis test. He selected a random sample of 100 residents from the university community where he lives and found that 32 of them actually watched at least three hours of the proceedings. The standard deviation for the distribution of the sample proportion is

    (a) 0.19.
    (b) 4.67.
    (c) 4.33.
    (d) 0.23.

37. If two large samples are selected independently from two different populations, the sampling distribution of the difference of the sample means

    (a) has a mean that is the sum of the two population means.
    (b) has a variance that is the difference of the two variances for the two populations.

(c) has a distribution that is approximately normal.

(d) has a mean and variance that are the average of the two population means and variances, respectively.

38. If we are trying to establish that the mean of population 1 is greater than the mean of population 2, the appropriate set of hypotheses is
    (a) $H_0: \mu_2 - \mu_1 \le 0$ vs. $H_1: \mu_2 - \mu_1 > 0$
    (b) $H_0: \mu_1 - \mu_2 \ge 0$ vs. $H_1: \mu_1 - \mu_2 < 0$
    (c) $H_0: \mu_1 - \mu_2 = 0$ vs. $H_1: \mu_1 - \mu_2 \ne 0$
    (d) $H_0: \mu_1 - \mu_2 \le 0$ vs. $H_1: \mu_1 - \mu_2 > 0$

39. If we are trying to establish that the mean of population 1 is not the same as the mean of population 2, the appropriate set of hypotheses is
    (a) $H_0: \mu_1 - \mu_2 \ne 0$ vs. $H_1: \mu_1 - \mu_2 = 0$.
    (b) $H_0: \mu_1 - \mu_2 = 0$ vs. $H_1: \mu_1 - \mu_2 \ne 0$.
    (c) $H_0: \mu_1 - \mu_2 \ge 0$ vs. $H_1: \mu_1 - \mu_2 < 0$.
    (d) $H_0: \mu_1 - \mu_2 \le 0$ vs. $H_1: \mu_1 - \mu_2 > 0$.

40. In performing hypothesis tests for the difference of two population proportions, if $n_1$ and $n_2$ are the respective sample sizes and $x_1$ and $x_2$ are the respective successes, then the pooled estimate of the difference of the population proportions is given by
    (a) $\dfrac{x_1 - x_2}{n_1 + n_2}$.

    (b) $\dfrac{x_1 - x_2}{n_1 - n_2}$.

    (c) $\dfrac{x_1 + x_2}{n_1 + n_2}$.

    (d) $\dfrac{x_1 + x_2}{n_1 - n_2}$.

41. Two machines are used to fill 50-lb bags of dog food. Sample information for these two machines is given in the table.

|                       | MACHINE A | MACHINE B |
| --------------------- | --------- | --------- |
| Sample size           | 81        | 64        |
| Sample mean (pounds)  | 51        | 48        |
| Sample variance       | 16        | 12        |

The point estimate for the difference between the two population means ($\mu_A - \mu_B$) is
    (a) 17.
    (b) 3.
    (c) 4.
    (d) –4.

42. Two machines are used to fill 50-lb bags of dog food. Sample information for these two machines is given in the table.

|                      | MACHINE A | MACHINE B |
| -------------------- | --------- | --------- |
| Sample size          | 81        | 64        |
| Sample mean (pounds) | 51        | 48        |
| Sample variance      | 16        | 12        |

The standard deviation (standard error) for the distribution of differences of sample means $(\bar{x}_A - \bar{x}_B)$ is

(a) 0.6205.
(b) 0.1931.
(c) 0.3850.
(d) 0.3217.

43. Two machines are used to fill 50-lb bags of dog food. Sample information for these two machines is given in the table.

|                      | MACHINE A | MACHINE B |
| -------------------- | --------- | --------- |
| Sample size          | 81        | 64        |
| Sample mean (pounds) | 51        | 48        |
| Sample variance      | 16        | 12        |

If you are to conduct a test to determine whether the average amount dispensed by machine A is significantly more than the average amount dispensed by machine B, the appropriate set of hypotheses is

(a) $H_0: \mu_B - \mu_A \leq 0$ vs. $H_1: \mu_B - \mu_A > 0$.
(b) $H_0: \mu_A - \mu_B = 0$ vs. $H_1: \mu_A - \mu_B \neq 0$.
(c) $H_0: \mu_A - \mu_B \geq 0$ vs. $H_1: \mu_A - \mu_B < 0$.
(d) $H_0: \mu_A - \mu_B \leq 0$ vs. $H_1: \mu_A - \mu_B > 0$.

44. Two machines are used to fill 50-lb bags of dog food. Sample information for these two machines is given in the table.

|                      | MACHINE A | MACHINE B |
| -------------------- | --------- | --------- |
| Sample size          | 81        | 64        |
| Sample mean (pounds) | 51        | 48        |
| Sample variance      | 16        | 12        |

If you are to conduct a test to determine whether the average amount dispensed by machine A is significantly more than the average amount dispensed by machine B, the computed test statistic for this test is

(a) 7.7918.
(b) –7.7918.
(c) 4.8348.
(d) 2.1988.

45. Two machines are used to fill 50-lb bags of dog food. Sample information for these two machines is given in the table.

| | MACHINE A | MACHINE B |
|---|---|---|
| Sample size | 81 | 64 |
| Sample mean (pounds) | 51 | 48 |
| Sample variance | 16 | 12 |

If you are to conduct a test to determine whether the average amount dispensed by machine A is significantly more than the average amount dispensed by machine B, the *P* value for the test is

(a) approximately 0.5.
(b) approximately 0.0.
(c) approximately 1.0.
(d) none of the above answers.

46. Two machines are used to fill 50-lb bags of dog food. Sample information for these two machines is given in the table.

| | MACHINE A | MACHINE B |
|---|---|---|
| Sample size | 81 | 64 |
| Sample mean (pounds) | 51 | 48 |
| Sample variance | 16 | 12 |

If you are to conduct a test at the 0.01 significance level to determine whether the average amount dispensed by machine A is significantly more than the average amount dispensed by machine B, the correct decision is

(a) do not reject the null hypothesis.
(b) reject the null hypothesis.
(c) reject the alternative hypothesis.
(d) do not reject the alternative hypothesis.

47. A study was conducted to determine whether remediation in mathematics enabled students to be more successful in college algebra. Success here means that a student received a grade of C or better, and remediation was for one year (students took an equivalent of one year of high school algebra). The following table shows the results of this study.

| | REMEDIAL | NONREMEDIAL |
|---|---|---|
| Sample size | $n_1 = 34$ | $n_2 = 150$ |
| Number of successes | $x_1 = 20$ | $x_2 = 104$ |

If we assume that the two population proportions are both equal to $p$, then a point estimate for $p$ is

(a) 0.4565.
(b) 0.6739.
(c) 0.4078.
(d) 0.7241.

48. A study was conducted to determine whether remediation in mathematics enabled students to be more successful in college algebra. Success here means that a student received a grade of C or better, and remediation was for one year (students took an equivalent of one year of high school algebra). The following table shows the results of this study.

| | REMEDIAL | NONREMEDIAL |
|---|---|---|
| Sample size | $n_1 = 34$ | $n_2 = 150$ |
| Number of successes | $x_1 = 20$ | $x_2 = 104$ |

If we assume that the two population proportions are both equal to $p$, then an estimate for the standard deviation for the distribution of differences of sample proportions is approximately

(a) 0.0079.
(b) 0.1898.
(c) 0.0360.
(d) 0.0890.

49. A study was conducted to determine whether remediation in mathematics enabled students to be more successful in college algebra. Success here means that a student received a grade of C or better, and remediation was for one year (students took an equivalent of one year of high school algebra). The following table shows the results of this study.

| | REMEDIAL | NONREMEDIAL |
|---|---|---|
| Sample size | $n_1 = 34$ | $n_2 = 150$ |
| Number of successes | $x_1 = 20$ | $x_2 = 104$ |

If $p_r$ and $p_n$ are the population proportions for the remedial and nonremedial groups, respectively, the appropriate set of hypotheses for this situation is

(a) $H_0: p_r - p_n > 0$ vs. $H_0: p_r - p_n \leq 0$.
(b) $H_0: p_r - p_n = 0$ vs. $H_0: p_r - p_n \neq 0$.
(c) $H_0: p_r - p_n \geq 0$ vs. $H_0: p_r - p_n < 0$.
(d) $H_0: p_r - p_n \leq 0$ vs. $H_0: p_r - p_n > 0$.

50. A study was conducted to determine whether remediation in mathematics enabled students to be more successful in college algebra. Success here means that a student received a grade of C or better, and remediation was for one year (students took an equivalent of one year of high school algebra). The following table shows the results of this study.

| | REMEDIAL | NONREMEDIAL |
|---|---|---|
| Sample size | $n_1 = 34$ | $n_2 = 150$ |
| Number of successes | $x_1 = 20$ | $x_2 = 104$ |

If $p_r$ and $p_n$ are the population proportions for the remedial and nonremedial groups, respectively, the computed test statistic for the appropriate test is

(a) 3.6426.
(b) −1.1803.
(c) −13.2685.
(d) 1.3931.

51. A study was conducted to determine whether remediation in mathematics enabled students to be more successful in college algebra. Success here means that a student received a grade of C or better, and remediation was for one year (students took an equivalent of one year of high school algebra). The following table shows the results of this study.

| | REMEDIAL | NONREMEDIAL |
|---|---|---|
| Sample size | $n_1 = 34$ | $n_2 = 150$ |
| Number of successes | $x_1 = 20$ | $x_2 = 104$ |

If $p_r$ and $p_n$ are the population proportions for the remedial and nonremedial groups, respectively, the $P$ value for the appropriate test is

(a) −0.1190.
(b) 0.1190.
(c) 0.8810.
(d) 0.3810.

52. A study was conducted to determine whether remediation in mathematics enabled students to be more successful in college algebra. Success here means that a student received a grade of C or better, and remediation was for one year (students took an equivalent of one year of high school algebra). The following table shows the results of this study.

| | REMEDIAL | NONREMEDIAL |
|---|---|---|
| Sample size | $n_1 = 34$ | $n_2 = 150$ |
| Number of successes | $x_1 = 20$ | $x_2 = 104$ |

If $p_r$ and $p_n$ are the population proportions for the remedial and nonremedial groups, respectively, the correct decision for the appropriate test at the 10 percent level of significance is

(a) do not reject the null hypothesis.
(b) reject the null hypothesis.
(c) reject the alternative hypothesis.
(d) do not reject the alternative hypothesis.

## Further Exercises

If possible, you can use any technology help available to solve the following questions.

1. A new shampoo is being test-marketed. A large number of 16-ounce bottles were mailed out at random to potential customers in the hope that the customers will return an enclosed questionnaire. Out of the 1,000 returned questionnaires, 575 indicated that they like the shampoo and will consider buying it when it becomes available on the market.

    Perform a hypothesis test to determine if the proportion of potential customers is at most 50 percent. Use a level of significance of 0.05.

2. In a test to compare the performance of two models of cars, the Bullet and the Speeding Bullet, 75 cars of each model were driven on the same speedway with a full tank of gas in each car (same size tanks). The mean number of miles for the Bullet was 540 miles with a standard deviation of 20 miles; the mean number of miles for the Speeding Bullet was 600 with a standard deviation of 38.

    (a) What is the point estimate of the difference of the means for the populations (the Speeding Bullet – the Bullet)?

    (b) At the 2 percent level of significance, are the two models of cars significantly different?

3. A local politician ran on the issue of whether alcohol should be sold on Sundays in two adjoining counties. The following results were obtained by his staff:

| COUNTY | SURVEYED VOTERS | VOTERS FAVORING THE PROPOSAL |
|--------|-----------------|------------------------------|
| 1 | 600 | 220 |
| 2 | 400 | 160 |

    (a) What is the point estimate for the difference of the population proportions for county 1 and county 2?

    (b) At the 5 percent level of significance, test the hypothesis that the proportion of voters favoring the proposal was the same in both counties.

    (c) What is the $P$ value for the test in (b)?

4. For a highly publicized murder trial, a CNN poll showed that 68 out of every 100 whites surveyed said that the defendant was guilty, while 15 out of every 100 blacks surveyed said that the defendant was guilty.

    (a) Compute the point estimate for the difference of the two population proportions (blacks – whites).

    (b) At the 5 percent level of significance, can you conclude that the proportion of whites who thought that the defendant was guilty was significantly different from the proportion of blacks who thought that the defendant was guilty?

    (c) Compute a $P$ value for the test in part (b).

# ANSWER KEY

## True/False Questions

1. T   2. T   3. F   4. T   5. F   6. T   7. T   8. F   9. F   10. F   11. T
12. T   13. T   14. T   15. T   16. F   17. F   18. T   19. F   20. T   21. F   22. F

## Completion Questions

1. interval   2. Type I   3. Type II   4. critical   5. reject   6. fail to reject
7. Type I   8. null, alternative   9. $\bar{x}_1 - \bar{x}_2$   10. normally distributed   11. 0, 1
12. reject   13. true   14. large   15. critical or rejection   16. critical   17. 0.0014
18. 0.0052   19. be rejected   20. reject

## Multiple-Choice Questions

1. (d)   2. (b)   3. (c)   4. (d)   5. (a)   6. (b)   7. (d)   8. (b)   9. (d)
10. (c)   11. (d)   12. (a)   13. (d)   14. (b)   15. (d)   16. (b)   17. (c)   18. (d)
19. (c)   20. (a)   21. (a)   22. (c)   23. (b)   24. (a)   25. (b)   26. (a)
27. (c)   28. (b)   29. (d)   30. (c)   31. (c)   32. (d)   33. (a)   34. (b)   35. (d)
36. (c)   37. (c)   38. (d)   39. (b)   40. (c)   41. (b)   42. (a)   43. (d)
44. (c)   45. (b)   46. (b)   47. (b)   48. (d)   49. (d)   50. (b)   51. (c)   52. (a)

# CHAPTER 13

# Confidence Intervals and Hypothesis Tests— Small Samples

**Do I Need to Read This Chapter?**

You should read this chapter if you need to review or need to learn about

→ The *t* distribution

→ Confidence intervals for the population mean—small samples

→ Hypothesis tests for a population mean—small samples

→ Confidence intervals for the difference between two population means—small independent samples

→ Hypothesis tests for the difference between two population means—small independent samples

→ Confidence intervals for the difference between two population means—small dependent samples

→ Hypothesis tests for the difference between two population means—small dependent samples

---

**Get Started**

Here we will focus on small-sample confidence intervals and hypothesis tests for the mean. We will consider both single- and two-population confidence intervals and hypothesis tests for the mean. In addition, for the two-population case, we will consider both independent and dependent samples.

---

## 13-1 The *t* Distribution

Recall that for the confidence interval for the mean, we assumed that either the population standard deviation was known or, when it was unknown, the sample size was large ($n \geq 30$). In the latter case, we replaced the population standard deviation with the sample standard deviation. In both cases, the standard normal distribution was used to find the confidence interval for the mean, and the variable of interest was assumed to be normal. In many cases, however, the population standard deviation is unknown and the sample size is small ($n < 30$). Again we can replace the population standard deviation with the sample standard deviation, but in this case we use the *t distribution* and not the standard normal distribution.

The *t* distribution has some properties that are similar to and some that are different from the properties of the standard normal distribution. Properties of the *t* distribution are listed below.

**➤➤ Properties that are similar to those of the *z* distribution:**

- It is bell-shaped.
- It is symmetrical about the mean.
- The mean, median, and mode are all equal to 0.
- The curve never touches the *x* axis (horizontal axis).

**➤➤ Properties that are different from those of the *z* distribution:**

- The variance is greater than 1.
- The shape of the distribution depends on the sample size or on the concept of degrees of freedom (*df*).
- As the sample size gets larger, the *t* distribution converges to the *z* distribution.

**Note:** The **degrees of freedom** (*df*) are the number of values that are free to vary after a statistic is computed from a set of data values. They tell us which *t* distribution we should use. For example, if the mean of 10 values is 3, then 9 of the 10 values are free to vary. However, once 9 values have been selected, the 10th value must be a specific number. It must be the number such that the sum of all the numbers is $10 \times 3 = 30$. Thus, the degrees of freedom are $10 - 1 = 9$. This tells us which *t* distribution to use.

**Figure 13-1** displays some of these properties.

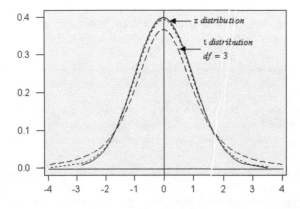

**Fig. 13-1:** Comparison between the standard normal distribution and the *t* distribution with $df = 3$

**Note:** The middle curve in **Fig. 13-1** is a *t* distribution with 25 degrees of freedom. Observe that it is almost the same as the *z* distribution.

## Quick Tip

Extensive tables of critical *t* values are available for use in solving confidence interval and hypothesis testing problems for small samples.

In order to state formulas that can be used to compute the small-sample confidence intervals and hypothesis tests for a population mean, we need to be familiar with the notation $t_{\alpha, n-1}$ (read as "*t* sub alpha with $n-1$ degrees of freedom").

**Explanation of the notation—$t_{\alpha, n-1}$:** $t_{\alpha, n-1}$ is a *t* score with $n-1$ degrees of freedom such that $\alpha$ area is to the right of the *t*-score value.

The diagram in **Fig. 13-2** explains the notation.

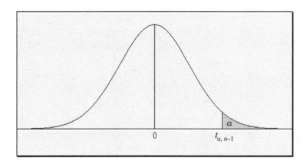

**Fig. 13-2:** Diagram explaining the notation $t_{\alpha, n-1}$

Values for *t* scores with the appropriate degrees of freedom can be obtained from the *t* tables in the Appendix.

***Example 13-1:*** What is the value of $t_{0.05, 10}$?

**Solution:** From the table in the appendix, $t_{0.05, 10} = 1.812$.

***Example 13-2:*** What is the value of $t_{0.9, 16}$?

**Solution:** Since the *t* distribution is symmetrical about 0, based on the definition of $t_{\alpha, n-1}$, the value of $t_{0.9, 16}$ is equivalent to $-t_{0.1, 16}$. From the table, we have $t_{0.9, 16} = -1.337$.

## 13-2 Small-Sample Confidence Interval for a Mean

The general equation used in constructing a $(1 - \alpha) \times 100$ percent confidence interval for the population mean when the population standard deviation is unknown and the sample size $n < 30$ is given by

$$\bar{x} \pm t_{\alpha/2, n-1} \frac{s}{\sqrt{n}}$$

**Note:** The margin of error is given by $E = \pm t_{\alpha/2, n-1} \frac{s}{\sqrt{n}}$.

***Example 13-3:*** A random sample of 16 public school teachers in a particular state has a mean salary of \$33,000 with a standard deviation of \$1,000. Construct a 99 percent confidence interval estimate for the true mean salary for public school teachers for the given state.

**Solution:** Given $\alpha = 0.01$ (1 percent), $\alpha/2 = 0.005$, $n = 16$, $df = 16 - 1 = 15$, $t_{0.005, 15} = 2.947$, $\bar{x} = 33,000$, $s = 1,000$, and $\dfrac{s}{\sqrt{n}} = 250$. Thus, the 99 percent confidence interval estimate for the mean salary is $33,000 \pm 2.947 \times 250 = 33,000 \pm 736.75$. That is, we are 99 percent confident that the average salary for public school teachers for the given state will lie between \$32,263.25 and \$33,736.75.

***Example 13-4:*** The president of a small community college wishes to estimate the average distance commuting students travel to the campus. A sample of 12 students was randomly selected and yielded the following distances in miles: 27, 35, 33, 30, 39, 25, 38, 22, 27, 37, 33, 40. Construct a 95 percent confidence interval estimate for the true mean distance commuting students travel to the campus.

**Solution:** Since raw data are given, we need to find the sample mean and the sample standard deviation. These values are $\bar{x} = 32.1667$ and $s = 5.9365$. (Verify that these values are correct.) Also, we are given $\alpha = 0.05$, $\alpha/2 = 0.025$, $n = 12$, $df = 12 - 1 = 11$, $t_{0.025, 11} = 2.201$, and $\dfrac{s}{\sqrt{n}} = 1.7137$. Thus, the 95 percent confidence interval estimate for the mean distance is $32.1667 \pm 2.201 \times 1.7137 = 32.1667 \pm 3.7719$. That is, we are 95 percent confident that the average distance commuting students travel to the campus will lie between 28.3948 and 35.9386 mi.

### 13-3 Small-Sample Test for a Mean

We refer to tests based on the statistic $t = \dfrac{\bar{x} - \mu_0}{s/\sqrt{n}}$ as small-sample tests because we are assuming that the sampling distribution for the sample means has a $t$ distribution. The test requires a sample size $n < 30$ and requires that the population standard deviation be unknown. We also assume that the sampling distribution is normal.

Following is a summary of the tests for a population mean under these conditions.

### Summary of Hypothesis Tests

**(a) One-tailed (right-tailed)**

$H_0$: $\mu \leq \mu_0$ (where $\mu_0$ is a specified value of the population mean)

$H_1$: $\mu > \mu_0$

T.S.: $t = \dfrac{\bar{x} - \mu_0}{s/\sqrt{n}}$, for $\sigma$ unknown and $n < 30$

D.R.: For a specified significance level $\alpha$, reject the null hypothesis if the computed test statistic value $t$ is greater than $+t_{\alpha, n-1}$.

Conclusion: . . .

**Note:** This is a right-tailed test because the direction of the inequality sign in the alternative hypothesis is to the right.

**(b) One-tailed (left-tailed)**

$H_0$: $\mu \geq \mu_0$ (where $\mu_0$ is a specified value of the population mean)

$H_1$: $\mu < \mu_0$

T.S.: $t = \dfrac{\bar{x} - \mu_0}{s/\sqrt{n}}$, for $\sigma$ unknown and $n < 30$

D.R.: For a specified significance level $\alpha$, reject the null hypothesis if the computed test statistic value $t$ is less than $-t_{\alpha, n-1}$.

Conclusion: . . .

**Note:** This is a left-tailed test because the direction of the inequality sign in the alternative hypothesis is to the left.

**(c) Two-tailed**

$H_0$: $\mu = \mu_0$ (where $\mu_0$ is a specified value of the population mean)

$H_1$: $\mu \neq \mu_0$

T.S.: $t = \dfrac{\bar{x} - \mu_0}{s/\sqrt{n}}$, for $\sigma$ unknown and $n < 30$

D.R.: For a specified significance level $\alpha$, reject the null hypothesis if the computed test statistic value $t$ is less than $-t_{\alpha/2, n-1}$ or if it is greater than $+t_{\alpha/2, n-1}$.

Conclusion: . . .

**Note:** This is a two-tailed test because of the not-equal-to symbol in the alternative hypothesis. Also, note that the level of significance is shared equally when finding the critical $t$ value $(t_{\alpha/2, n-1})$.

***Example 13-5:*** A teachers' union would like to establish that the average salary for high school teachers in a particular state is less than $35,500. A random sample of 25 public high school teachers in the particular state has a mean salary of $34,578 with a standard deviation of $910. Test to establish whether the union's claim is correct at the 5 percent level of significance.

**Solution:** This is a left-tailed test. Why? Given $\alpha = 0.05$, $n = 25$, $df = 25 - 1 = 24$, $t_{0.05, 24} = 1.711$, $\bar{x} = 34{,}578$, $s = 910$, $\mu_0 = 35{,}500$, and $\dfrac{s}{\sqrt{n}} = \dfrac{910}{\sqrt{25}} = 182$.

Thus,

$H_0$: $\mu \geq 35{,}500$

$H_1$: $\mu < 35{,}500$

T.S.: $t = \dfrac{\bar{x} - \mu_0}{s/\sqrt{n}} = \dfrac{34{,}578 - 35{,}500}{182} = -5.0659$

D.R.: For a significance level of $\alpha = 0.05$, reject the null hypothesis if the computed test statistic value $t = -5.0659 < -t_{0.05, 24} = -1.711$.

Conclusion: Since $-5.0659 < -1.711$, reject $H_0$. There is sufficient sample evidence to support the claim that the average salary for high school teachers in the state is less than $35,500 at the 5 percent level of significance. That is, there is a significant difference between the sample mean and the postulated value of the population mean.

**Figure 13-3** depicts the rejection region.

***Example 13-6:*** The dean of students of a private college claims that the average distance commuting students travel to the campus is 35 mi. The commuting students feel otherwise. A sample

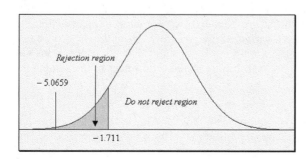

**Fig. 13-3:** Diagram depicting the rejection region for Example 13-5

of 16 students was randomly selected and yielded a mean of 36 miles and a standard deviation of 5 miles. Test the dean's claim at the 5 percent level of significance.

**Solution:** This is a two-tailed test, since the students do not believe the claim. The students are not concerned whether the average distance traveled is less than 35 mi or more than 35 mi. Given $\alpha = 0.05$, $\alpha/2 = 0.025$, $n = 16$, $df = n - 1 = 16 - 1 = 15$, $t_{0.025,\,15} = 2.131$, $\bar{x} = 36$, $s = 5$, $\mu_0 = 35$, and $\dfrac{s}{\sqrt{n}} = 1.25$. Thus,

$H_0$: $\mu = 35$

$H_1$: $\mu \neq 35$

T.S.: $t = \dfrac{\bar{x} - \mu_0}{s/\sqrt{n}} = \dfrac{36 - 35}{1.25} = 0.8$

D.R.: For a significance level $\alpha = 0.05$, reject the null hypothesis if the computed test statistic value, $t = 0.8 < -t_{0.025,\,15} = -2.131$ or if $t = 0.8 > t_{0.025,\,15} = 2.131$.

Conclusion: Since neither of the conditions is satisfied, do not reject $H_0$. There is insufficient sample evidence to refute the dean's claim at the 5 percent level of significance. That is, there is not a significant difference between the sample mean and the postulated value of the population mean of 36 mi.

**Figure 13-4** depicts the rejection region.

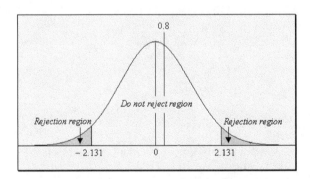

**Fig. 13-4:** Diagram depicting the rejection region for Example 13-6

## 13-4 Independent Small-Sample Confidence Interval for the Difference between Two Population Means

When we have independent samples, there is a procedure that we can use to construct confidence intervals for the difference between two population means when the sample sizes are small and the population variances are unknown. Here the procedure assumes that the samples are obtained from normal populations and that the populations have equal variances. Thus, we have that the distribution of the differences of the sample means will have

a standard deviation $\sigma_{\bar{x}_1 - \bar{x}_2} = \sqrt{\dfrac{\sigma_1^2}{n_1} + \dfrac{\sigma_2^2}{n_2}} = \sqrt{\dfrac{\sigma^2}{n_1} + \dfrac{\sigma^2}{n_2}} = \sigma\sqrt{\dfrac{1}{n_1} + \dfrac{1}{n_2}}$. Since $\sigma$ is unknown,

the question is, with what should we estimate it? We use the pooled standard deviation $s_p$ as the estimate for $\sigma$. The equation for the pooled variance is given by $s_p^2 = \dfrac{(n_1 - 1)s_1^2 + (n_2 - 1)s_2^2}{n_1 + n_2 - 2}$. Note that the subscripts refer to the populations. Thus we can write

$$\sigma_{\bar{x}_1 - \bar{x}_2} \approx \sqrt{\frac{(n_1 - 1)s_1^2 + (n_2 - 1)s_2^2}{n_1 + n_2 - 2}} \times \sqrt{\frac{1}{n_1} + \frac{1}{n_2}}. \text{ The degrees of freedom for this situation are}$$

$df = n_1 + n_2 - 2$.

These properties can aid us in the construction of a $(1 - \alpha) \times 100$ percent confidence interval for the difference of two population means. The general equation used in constructing a $(1 - \alpha) \times 100$ percent confidence interval for the difference between two population means for small samples is given by

$$(\bar{x}_1 - \bar{x}_2) \pm t_{\alpha/2, \, df} \times s_p \sqrt{\frac{1}{n_1} + \frac{1}{n_2}}$$

**Example 13-7:** Two methods were used to teach a biostatistics course for a nurse practitioner program. A sample of 16 scores was selected for method 1, and a sample of 25 scores was selected for method 2, with the summary results given in **Table 13-1.**

**Table 13-1**

|  | METHOD 1 | METHOD 2 |
|---|---|---|
| Sample mean | 88 | 81 |
| Sample standard deviation | 2.5 | 1.8 |

Construct a 99 percent confidence interval for the difference in the mean scores for the two methods. Assume that the scores are normally distributed.

**Solution:** From the information given, we have $n_1 = 16$, $n_2 = 25$, $\bar{x}_1 = 88$, $\bar{x}_2 = 81$, $s_1 = 2.5$, $s_2 = 1.8$, $s_p^2 = \dfrac{(n_1 - 1)s_1^2 + (n_2 - 1)s_2^2}{n_1 + n_2 - 2} = \dfrac{(16 - 1) \times 2.5^2 + (25 - 1) \times 1.8^2}{16 + 25 - 2} = 4.3977$, $s_p = 2.0971$, $\alpha = 0.01$, $\alpha/2 = 0.005$, $df = 16 + 25 - 2 = 39$, $t_{0.005, \, 39} = 2.576$, and $\sqrt{\dfrac{1}{n_1} + \dfrac{1}{n_2}} = \sqrt{\dfrac{1}{16} + \dfrac{1}{25}} = 0.3202$.

Thus, the 99 percent confidence interval estimate for the difference of the means for the two methods (method 1 − method 2) is $(88 - 81) \pm 2.576 \times 2.0971 \times 0.3202 = 7 \pm 1.7295$. That is, we are 99 percent confident that the difference between the mean scores for the two teaching methods will lie between 5.2705 and 8.7295. Since both limits are positive, one may conclude that method 1 seems to be the better of the two methods. Thus, we can also say that the average score for method 1 will be between 5.2705 and 8.7295 more than that of method 2.

**Example 13-8:** A male instructor claims that male and female instructors in his department, on average, wait the same amount of time for promotion to full professor. He collected data on 10 male and 10 female instructors. The number of years to full professor are given in **Table 13-2.**

**Table 13-2**

| Male | 11 | 17 | 14 | 10 | 13 | 11 | 17 | 10 | 19 | 14 |
|---|---|---|---|---|---|---|---|---|---|---|
| Female | 10 | 13 | 12 | 12 | 18 | 14 | 14 | 17 | 17 | 10 |

Construct a 95 percent confidence interval for the difference in the mean time for promotion to full professor. Assume that the years to promotion are normally distributed.

**Solution:** Since raw data are given, we will have to compute the sample means and sample variances. From the information given, we have $n_1 = 10$, $n_2 = 10$, $\bar{x}_1 = 13.6$, $\bar{x}_2 = 13.7$, $s_1 = 3.2$, $s_2 = 2.869$, $\alpha = 0.05$, $\alpha/2 = 0.025$, $df = 10 + 10 - 2 = 18$, $t_{0.025,\,18} = 2.101$,

$$s_p^2 = \frac{(n_1 - 1)s_1^2 + (n_2 - 1)s_2^2}{n_1 + n_2 - 2} = \frac{(10 - 1) \times 3.2^2 + (10 - 1) \times 2.869^2}{10 + 10 - 2} = 9.2356,$$

$s_p = 3.0390$, and $\sqrt{\dfrac{1}{n_1} + \dfrac{1}{n_2}} = \sqrt{\dfrac{1}{10} + \dfrac{1}{10}} = 0.4472$. Thus, the 95 percent confidence interval estimate for the difference of the means for the waiting times to full professor is $(13.6 - 13.7) \pm 2.101 \times 3.0390 \times 0.4472 = -0.1 \pm 2.8553$. That is, we are 95 percent confident that the difference between the mean waiting times to full professor for male and female instructors in the given department will lie between $-2.95534$ and $2.7553$. Observe that the lower limit is negative while the upper limit is positive. Thus, zero is included in the interval. This would imply that we cannot say that the means are different. That is, based on the confidence interval, we cannot refute the instructor's claim.

### 13-5 Independent Small-Sample Tests for the Difference between Two Population Means

There may be problems in which one must decide whether the observed difference between two small-sample means is due to chance or whether the difference is due to the fact that the corresponding population means are not the same. Here we will discuss such problems.

Below is a summary of the tests for the difference between two population means when the samples are small, the population variances are unknown but *equal,* and the sampling populations are normally distributed.

### Summary of Hypothesis Tests

**(a) One-tailed (right-tailed)**

$H_0$: $\mu_1 \le \mu_2$

$H_1$: $\mu_1 > \mu_2$

T.S.: $t = \dfrac{\bar{x}_1 - \bar{x}_2}{s_p \sqrt{\dfrac{1}{n_1} + \dfrac{1}{n_2}}}$, where $s_p$ is the pooled standard deviation

D.R.: For a specified significance level $\alpha$, reject the null hypothesis if the computed test statistic value $t$ is greater than $+t_{\alpha,\,df}$, where $df = n_1 + n_2 - 2$.

Conclusion: . . .

**Note:** This is a right-tailed test because the direction of the inequality sign in the alternative hypothesis is to the right.

**(b) One-tailed (left-tailed)**

$H_0$: $\mu_1 \ge \mu_2$

$H_1$: $\mu_1 < \mu_2$

T.S.: $t = \dfrac{\bar{x}_1 - \bar{x}_2}{s_p \sqrt{\dfrac{1}{n_1} + \dfrac{1}{n_2}}}$, where $s_p$ is the pooled standard deviation

D.R.: For a specified significance level $\alpha$, reject the null hypothesis if the computed test statistic value $t$ is less than $-t_{\alpha, df}$, where $df = n_1 + n_2 - 2$.

Conclusion: . . .

**Note:** This is a left-tailed test because the direction of the inequality sign in the alternative hypothesis is to the left.

**(c) Two-tailed**

$H_0$: $\mu_1 = \mu_2$

$H_1$: $\mu_1 \neq \mu_2$

T.S.: $t = \dfrac{\bar{x}_1 - \bar{x}_2}{s_p \sqrt{\dfrac{1}{n_1} + \dfrac{1}{n_2}}}$, where $s_p$ is the pooled standard deviation

D.R.: For a specified significance level $\alpha$, reject the null hypothesis if the computed test statistic value $t$ is less than $-t_{\alpha/2, df}$ or if it is greater than $+t_{\alpha/2, df}$, where $df = n_1 + n_2 - 2$.

Conclusion: . . .

**Note:** This is a two-tailed test because of the not-equal-to symbol in the alternative hypothesis. Also, note that the level of significance is shared equally when finding the critical $t$ value ($t_{\alpha/2, df}$).

***Example 13-9:*** A researcher claims that the starting salary for male nurse-practitioners is more than that for female nurse-practitioners. A random sample of 12 male and 9 female nurse-practitioners yielded the information given in **Table 13-3.** Is there enough sample evidence to support the researcher's claim. Test at the 1% level of significance.

**Table 13-3**

|  | MALE | FEMALE |
|---|---|---|
| Sample size | 12 | 9 |
| Sample mean | $71,000 | $69,500 |
| Sample standard deviation | $1,000 | $1,500 |

**Solution:** This is a right-tailed test. Why? From the information given, we have $n_1 = 12$, $n_2 = 9$, $\bar{x}_1 = 71,000$, $\bar{x}_2 = 69,500$, $s_1 = 1,000$, $s_2 = 1,500$, $\alpha = 0.01$, $df = 12 + 9 - 2 = 19$, $t_{0.01, 19} = 2.539$,

$$s_p^2 = \frac{(n_1 - 1)s_1^2 + (n_2 - 1)s_2^2}{n_1 + n_2 - 2} = \frac{(12 - 1) \times 1,000^2 + (9 - 1) \times 1,500^2}{12 + 9 - 2} = 1,526,315, \; s_p = 1,235.4415,$$

and $\sqrt{\dfrac{1}{n_1} + \dfrac{1}{n_2}} = \sqrt{\dfrac{1}{12} + \dfrac{1}{9}} = 0.4410.$

$H_0$: $\mu_1 \leq \mu_2$

$H_1$: $\mu_1 > \mu_2$

T.S.: $t = \dfrac{\bar{x}_1 - \bar{x}_2}{s_p \sqrt{\dfrac{1}{n_1} + \dfrac{1}{n_2}}} = \dfrac{71{,}000 - 69{,}500}{1{,}235.4415 \times 0.4410} = 2.7532$

D.R.: For a significance level $\alpha = 0.01$, reject the null hypothesis if the computed test statistic value $t = 2.7532 > t_{0.01,\,19} = 2.539$.

Conclusion:  Since $2.7532 > 2.539$, reject the null hypothesis. At the 1 percent significance level, we can conclude that the average starting salary for male nurse-practitioners is greater than that for the female nurse-practitioners. That is, the difference of the sample means is significantly different from zero.

### 13-6  Dependent Small-Sample Confidence Interval for the Difference between Two Population Means

In this section, the $t$ test is used when the samples are *dependent*. Samples are considered to be dependent when they are paired or matched in some way. For example, an instructor may give a test at the beginning of the semester to determine the basic math skill level of the students in a course. At the end of the semester, the instructor will give the same test to determine the basic math skill level of the students again. Although we have two different sets of data, they were obtained from the same set of students (assuming that all students remain in the course). Thus, we say that the data are dependent, since the same experimental units (students) were used. Another situation in which we may have dependent samples is, for example, when patients are matched or paired according to some variable of interest. Patients may then be assigned to two different groups. For instance, patients may be paired according to their age (blood pressure, etc.). That is, two patients with the same age will be paired, then one will be assigned to one sample group and the other to another sample group. Care should be taken when matching experimental units. In this example we matched by age, but this does not eliminate the influence of other variables.

In constructing confidence intervals for dependent data, we use the difference of the before-and-after values or the difference of the values of the matched pairs. By doing this, we will have a single sample of differences with which to construct a confidence interval.

The general equation used in constructing a $(1 - \alpha) \times 100$ percent confidence interval for the differences is given by

$$\bar{d} \pm t_{\alpha/2,\,n-1} \dfrac{s_d}{\sqrt{n}}$$

Here, $\bar{d}$ is the mean of the sample differences, $s_d$ is the standard deviation of the differences, $n$ is the number of pairs, and the $df$ for the $t$ distribution is $n - 1$.

**Note:**  The margin of error is given by $E = \pm t_{\alpha/2,\,n-1} \dfrac{s_d}{\sqrt{n}}$.

***Example 13-10:***  An instructor wanted to measure the basic math skills of his students before and after his college algebra course. A skills test was administered at the beginning of the semester, and the scores were recorded. At the end of the semester, he administered the same test and recorded the scores. **Table 13-4** shows the before-and-after scores for the test for the students who remained in the course until the end of the semester. The maximum possible score on the test was 100 points.

**Table 13-4**

| STUDENT # | 1 | 2 | 3 | 4 | 5 | 6 | 7 | 8 | 9 |
|---|---|---|---|---|---|---|---|---|---|
| Before | 61 | 58 | 79 | 69 | 62 | 71 | 25 | 48 | 53 |
| After | 68 | 62 | 83 | 65 | 62 | 74 | 31 | 52 | 51 |

Construct a 95 percent confidence interval for this set of dependent data.

**Solution:** First we need to find the differences (before – after) and use these differences as the raw data. The differences are given in **Table 13-5.**

**Table 13-5**

| STUDENT # | 1 | 2 | 3 | 4 | 5 | 6 | 7 | 8 | 9 |
|---|---|---|---|---|---|---|---|---|---|
| Difference | –7 | –4 | –4 | 4 | 0 | –3 | –6 | –4 | 2 |

For the differences, we have $\bar{d} = -2.4444$, $s_d = 3.6780$, and $n = 9$. Also, $\alpha = 0.05$, $\alpha/2 = 0.025$, $df = 9 - 1 = 8$, $t_{0.025, 8} = 2.306$, and $\dfrac{s_d}{\sqrt{n}} = 1.226$. Thus, the 95 percent confidence interval for the differences is $-2.4444 \pm 2.306 \times 1.226$ or $-2.4444 \pm 2.8272$. That is, we are 95 percent confident that the true mean difference will lie between $-5.2716$ and $0.3828$. Since 0 is contained in the interval, we cannot say that the course significantly improved the basic math skills of the students at this significance level.

## 13-7 Dependent Small-Sample Tests for the Difference between Two Population Means

We refer to tests based on the statistic $t = \dfrac{\bar{d} - \mu_d}{s_d/\sqrt{n}}$ as small-sample tests because we are assuming that the sampling distribution for the sample means of the differences has a $t$ distribution and that $\mu_d$ is the mean of the population of differences. The test requires that the population standard deviation of the differences be unknown and that the sampling distribution of the differences be normal.

Below is a summary of the tests for a population mean.

### Summary of Hypothesis Tests

**(a) One-tailed (right-tailed)**

$H_0$: $\mu_d \le 0$

$H_1$: $\mu_d > 0$

T.S.: $t = \dfrac{\bar{d} - \mu_d}{s_d/\sqrt{n}}$

D.R.: For a specified significance level $\alpha$, reject the null hypothesis if the computed test statistic value $t$ is greater than $+t_{\alpha, n-1}$.

Conclusion: . . .

**Note:** This is a right-tailed test because the direction of the inequality sign in the alternative hypothesis is to the right.

**(b) One-tailed (left-tailed)**

$H_0$: $\mu_d \geq 0$

$H_1$: $\mu_d < 0$

T.S.: $t = \dfrac{\bar{d} - \mu_d}{s_d/\sqrt{n}}$

D.R.: For a specified significance level $\alpha$, reject the null hypothesis if the computed test statistic value $t$ is less than $-t_{\alpha, n-1}$.

Conclusion: . . .

**Note:** This is a left-tailed test because the direction of the inequality sign in the alternative hypothesis is to the left.

**(c) Two-tailed**

$H_0$: $\mu_d = 0$

$H_1$: $\mu_d \neq 0$

T.S.: $t = \dfrac{\bar{d} - \mu_d}{s_d/\sqrt{n}}$

D.R.: For a specified significance level $\alpha$, reject the null hypothesis if the computed test statistic value $t$ is less than $-t_{\alpha/2, n-1}$ or if it is greater than $+t_{\alpha/2, n-1}$.

Conclusion: . . .

**Note:** This is a two-tailed test because of the not-equal-to symbol in the alternative hypothesis. Also, note that the level of significance is shared equally when finding the critical $t$ value ($t_{\alpha/2, n-1}$).

***Example 13-11:*** Test at the 5 percent level of significance whether the college algebra course in **Example 13-10** improved the basic math skills of the students.

**Solution:** We will use the information given in **Example 13-10.** In order for the course to improve the basic math skills of the students, the "after" scores should be significantly greater than the "before" scores. That is, we would like to establish whether the average of the difference (before – after) is less than zero. Thus,

$H_0$: $\mu_d \geq 0$

$H_1$: $\mu_d < 0$

T.S.: $t = \dfrac{\bar{d} - \mu_d}{s_d/\sqrt{n}} = \dfrac{-2.4444}{3.6780/\sqrt{9}} = -1.9938$

D.R.: For a significance level of $\alpha = 0.05$, reject the null hypothesis if the computed test statistic value $t = -1.9938 < -t_{0.05, 8} = -1.86$.

Conclusion: Since $-1.9938 < -1.86$, reject the null hypothesis and claim at the 5 percent level of significance that the average of the differences is less than zero. That is, the average of the "after" scores is significantly larger than the average of the "before" scores. Thus, one can conclude that the course indeed improved the basic math skills of the students who took it.

## Technology Corner

All of the concepts discussed in this chapter can be computed and illustrated using most statistical software packages. All scientific and graphical calculators can be used for the computations. In addition, some of the newer calculators, like the TI-83, will allow you to do the test directly on the calculator. If you own a calculator, you should consult the manual to determine what statistical features are included.

**Illustration:** **Figure 13-5** shows the outputs computed by the MINITAB software for **Examples 13-8, 13-10,** and **13-11.** Observe that the MINITAB output has rounded some of the answers. You can extract the information from the output to write up responses for the examples. Again recall that the MINITAB software does not allow you to do hypothesis tests and construct confidence intervals for population means directly when you have summary data. The software will allow you to directly do tests for means when you have raw data. Note that the outputs also give the *P* values for the tests. **Figure 13-6** shows the outputs computed by the TI-83 calculator for **Examples 13-8, 13-10,** and **13-11.** For **Examples**

```
Worksheet size: 100000 cells

Two Sample T-Test and Confidence Interval -- Example 13-8


Two sample T for Male vs Female

            N      Mean    StDev   SE Mean
Male       10     13.60     3.20      1.0
Female     10     13.70     2.87     0.91

95% CI for mu Male - mu Female: ( -3.0,  2.76)
T-Test mu Male = mu Female (vs not =): T = -0.07  P = 0.94  DF = 18
Both use Pooled StDev = 3.04


Paired T-Test and Confidence Interval -- Examples 13-10 and 13-11

Paired T for Before - After

              N      Mean    StDev   SE Mean
Before        9     58.44    15.69      5.23
After         9     60.89    15.00      5.00
Difference    9     -2.44     3.68      1.23

95% CI for mean difference: (-5.27, 0.38)
T-Test of mean difference = 0 (vs not = 0): T-Value = -1.99  P-Value = 0.081
```

**Fig. 13-5:** MINITAB output for Examples 13-8, 13-10, and 13-11

### Example 13-8

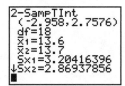

### Example 13-10

### Example 13-11

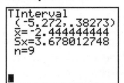

**Fig. 13-6:** TI-83 output for Examples 13-8, 13-10, and 13-11

**13-10** and **13-11,** when using the TI-83, you need to use the differences as one set of data values and use the calculator's *t*-test option. Again, the TI-83 calculator allows you to use both summary data and raw data in hypothesis tests. The *P* values provided by both the MINITAB and TI-83 outputs can be used to perform the tests. Care should always be taken when using the formulas for computations in hypothesis testing. One can use other features of the technologies to illustrate other concepts discussed in the chapter.

Here, we considered small-sample confidence intervals and small-sample hypothesis tests for a single mean. We also addressed independent small-sample confidence intervals and independent small-sample hypothesis tests for the difference between two means. In addition, dependent small-sample confidence intervals and dependent small-sample hypothesis tests for the difference between two means were considered. These concepts were presented through

✔ Formulas
✔ Examples

## True/False Questions

1. If the sample size $n$ is less than or equal to 30, a $z$ score will always be associated with any hypothesis that deals with the mean.
2. In the *P*-value approach to hypothesis testing, if the *P* value is less than a specified $\alpha$ value, then we will reject the null hypothesis.
3. In making inferences about the difference of two population means, if the variances of the two populations are assumed to be equal, the distribution of the differences of the sample means follows a $t$ distribution with degrees of freedom $n_1 + n_2 - 2$ for $n_1$ and/or $n_2$ less than 30.
4. In testing the difference of two population means, the pooled standard deviation (variance) is used when the underlying populations have unequal variances.
5. The matched-pair $t$ test is used to test the difference of two means when the two selected samples are independent.
6. When performing hypothesis tests on two population means using small samples, it is necessary to assume that the populations from which the samples are obtained are normally distributed.
7. The *P* value of a hypothesis test can be computed without the value of the test statistic.
8. The $t$ distribution is used in the construction of confidence intervals for the population mean when the population standard deviation is unknown, the sample size is small, and the sampling population is normal.
9. When data are obtained from matching or pairing, one should use the $z$ distribution in making inferences for the data.
10. The mean and the variance for the $z$ distribution and the $t$ distribution are the same.

## Completion Questions

1. The point estimate for the difference of two population means $\mu_1 - \mu_2$ is $(\hat{p}_1 - \hat{p}_2, \bar{x}_1 - \bar{x}_2, \mu_1 - \mu_2)$ _____, where the subscripts represent the corresponding populations.
2. If a small sample size is used in a hypothesis test for the mean, it is necessary to assume that the sample was selected from a (normal, $t$) _____ distribution.
3. When testing for the difference of two population means with small sample sizes, if we assume that the two population variances are equal and the populations are approximately normally distributed, the distribution of the differences of sample means will follow a (normal, $t$) _____ distribution.

4. We use a matched-pair $t$ test when the samples are (independent, dependent) _____ .

5. In conducting a hypothesis test for a single population mean, the test statistic is assumed to have a standard normal distribution if the sample size is (small, large) _____.

6. In computing the confidence interval estimate for the difference of two population means, the $t$ distribution (is, is not) _____ restricted to small samples.

7. In testing for the difference of two population means, when the sample variances of the two independent samples have to be combined, the resulting variance is called the _____ estimator of the variance.

8. When constructing a confidence interval for the population mean $\mu$ when the population standard deviation $\sigma$ is known, the correct distribution to use is the $(z, t)$ _____ distribution.

9. When using the pooled variance for inferences about the difference between two population means, we assume that the variances for the populations are (equal, not equal) _____ .

10. The variance for the $t$ distribution is (less than, equal to, greater than) _____ 1.

## Multiple-Choice Questions

1. Which of the following is true? The $t$ distribution should be used when
   (a) the sampling population is nonnormal.
   (b) the sampling population is unimodal.
   (c) the population standard deviation is unknown, the sample size is small, and the sampling distribution is normal.
   (d) the population standard deviation is known.

2. Commuter students at a certain college claim that the average distance they have to commute to campus is 26 miles per day. A random sample of 16 commuter students was surveyed and yielded an average distance of 31 mi and a variance of 64. The test statistic for this test is
   (a) $z = -2.5$.
   (b) $z = 0.0394$.
   (c) $t = 2.5$.
   (d) $t = 1.25$.

3. For a small-sample left-tailed test for the population mean, the sample size was 18 and $\alpha = 0.01$. The critical (table) value for this test is
   (a) $-2.878$.
   (b) $-2.552$.
   (c) $2.878$.
   (d) $-2.567$.

4. For the following information,
   $$n = 16 \qquad \mu_0 = 15 \qquad \bar{x} = 16 \qquad s^2 = 16$$
   assume that the population is normal. Compute the test statistic for testing for a population mean.

   (a) $z = 1$
   (b) $z = 1/4$

(c) $t = 1$

(d) $t = 1/4$

5. For the following information,

$$n = 16 \qquad \mu_0 = 14.239 \qquad \bar{x} = 16 \qquad s^2 = 16$$

assume that the population is normal. If you were performing a right-tailed test for the population mean, then the

(a) $P$ value $< 0.25$.

(b) $P$ value $= 0.05$.

(c) $P$ value $< 0.04$.

(d) $P$ value $< 0.025$.

6. An advertising agency would like to create an advertisement for a fast food restaurant claiming that the average waiting time from ordering to receiving your order at the restaurant is less than 5 min. The agency measured the time from ordering to delivery of order for 25 customers and found that the average time was 4.7 min with a standard deviation of 0.6 min. The test statistic that would be computed is

(a) –4.2.

(b) –12.5.

(c) –2.5.

(d) –20.8.

7. An advertising agency would like to create an advertisement for a fast food restaurant claiming that the average waiting time from ordering to receiving your order at the restaurant is less than 5 min. The agency measured the time from ordering to delivery of order for 25 customers and found that the average time was 4.7 min with a standard deviation of 0.6 min. The $P$ value for this test would be

(a) 0.100.

(b) 0.050.

(c) 0.025.

(d) 0.010.

8. An advertising agency would like to create an advertisement for a fast food restaurant claiming that the average waiting time from ordering to receiving your order at the restaurant is less than 5 min. The agency measured the time from ordering to delivery of order for 25 customers and found that the average time was 4.7 min with a standard deviation of 0.6 min. The appropriate set of hypotheses to be tested is

(a) $H_0$: $\mu \le 4.7$ vs. $H_1$: $\mu > 4.7$.

(b) $H_0$: $\mu \ge 4.7$ vs. $H_1$: $\mu < 4.7$.

(c) $H_0$: $\mu \ge 5$ vs. $H_1$: $\mu < 5$.

(d) $H_0$: $\mu \le 5$ vs. $H_1$: $\mu > 5$.

9. An advertising agency would like to create an advertisement for a fast food restaurant claiming that the average waiting time from ordering to receiving your order at the restaurant is less than 5 min. The agency measured the time from ordering to delivery of order for 25 customers and found that the average time was 4.7 min with a standard deviation of 0.6 min. At the 5 percent level of significance, we can claim that the average time between ordering and receiving the order is

(a) significantly greater than 4.7 min.

(b) significantly smaller than 4.7 min.

(c) significantly greater than 5 min.

(d) significantly smaller than 5 min.

10. If two small samples are selected independently from two different normal populations with equal variances, the sampling distribution of the difference of the sample means
    (a) has a mean that is the difference of the two sample means.
    (b) has a variance that is the difference of the two variances for the two populations.
    (c) has a distribution that is normal.
    (d) has a $t$ distribution.

11. If two small samples are selected independently from two different populations with equal variances, the combined variance that is associated with this situation is called
    (a) an estimate for the sample variance for the distribution of the differences.
    (b) the pooled estimate for the sample variance for the distribution of the differences.
    (c) the pooled estimate for the equal variances of the sampling populations.
    (d) none of the above.

12. If we are testing for the difference of two population means, the pooled variance is appropriate if
    (a) the populations are normally distributed.
    (b) the samples are small.
    (c) the population variances are unknown but assumed to be equal.
    (d) all of the above are true.

13. In constructing a confidence interval for the difference of two population means with the assumptions that the sample sizes $n_1$ and $n_2$ are small and the populations are normally distributed with equal variances, the degrees of freedom for the associated $t$ distribution are
    (a) $n_1 + n_2 - 1$.
    (b) $n_1 + n_2 - 2$.
    (c) $n_1 + n_2 + 1$.
    (d) $n_1 + n_2 + 2$.

14. In performing a large-sample test for the difference between two population means with known population variances, which of the following is not correct?
    (a) The test statistic has a $t$ distribution with $n_1 + n_2 - 2$ degrees of freedom.
    (b) The test statistic has a standard normal distribution.
    (c) Both of the sample sizes need not be greater than 30.
    (d) The population variances need not be equal to each other.

15. The matched-pair $t$ test is appropriate
    (a) when the samples are independent.
    (b) only when the population variances are equal.
    (c) when the samples are dependent.
    (d) in none of the above cases.

16. Two machines are used to fill 50-lb bags of dog food. Sample information for these two machines is given below.

| | MACHINE A | MACHINE B |
|---|---|---|
| Sample size | 81 | 64 |
| Sample mean (pounds) | 51 | 48 |
| Sample variance | 16 | 12 |

The 90 percent confidence interval for the difference between the two population means $(\mu_A - \mu_B)$ is

(a) $3 \pm 0.3850$.
(b) $3 \pm 0.0310$.
(c) $3 \pm 1.0207$.
(d) $3 \pm 1.4458$.

17. A mathematics professor wants to determine whether there is a difference in final averages between the past two semesters (semester I and semester II) of his business statistics classes. For a random sample of 16 students from semester I, the mean of the final averages was 75 with a standard deviation of 4. For a random sample of 9 students from semester II, the mean was 73 with a standard deviation of 6. If the final averages from semesters I and II are assumed to be normally distributed with equal variances, the point estimate for the difference between the means of the two populations (semester I – semester II) is

(a) $-2$.
(b) 2.
(c) 3.
(d) 7.

18. A mathematics professor wants to determine whether there is a difference in the final averages between the past two semesters (semester I and semester II) of his business statistics classes. For a random sample of 16 students from semester I, the mean of the final averages was 75 with a standard deviation of 4. For a random sample of 9 students from semester II, the mean was 73 with a standard deviation of 6. If the final averages from semesters I and II are assumed to be normally distributed with equal variances, the appropriate set of hypotheses is

(a) $H_0$: $\mu_I - \mu_{II} \neq 0$ vs. $H_1$: $\mu_I - \mu_{II} = 0$.
(b) $H_0$: $\mu_I - \mu_{II} = 0$ vs. $H_1$: $\mu_I - \mu_{II} \neq 0$.
(c) $H_0$: $\mu_I - \mu_{II} \geq 0$ vs. $H_1$: $\mu_I - \mu_{II} < 0$.
(d) $H_0$: $\mu_I - \mu_{II} \leq 0$ vs. $H_1$: $\mu_I - \mu_{II} > 0$.

19. A mathematics professor wants to determine whether there is a difference in the final averages between the past two semesters (semester I and semester II) of his business statistics classes. For a random sample of 16 students from semester I, the mean of the final averages was 75 with a standard deviation of 4. For a random sample of 9 students from semester II, the mean was 73 with a standard deviation of 6. If the final averages from semesters I and II are assumed to be normally distributed with equal variances, the computed test statistic for the appropriate test is

(a) $z = 1.9964$.
(b) $t = 0.5009$.
(c) $t = 1.0018$.
(d) $z = 0.5009$.

20. A mathematics professor wants to determine whether there is a difference in the final averages between the past two semesters (semester I and semester II) of his business statistics classes. For a random sample of 16 students from semester I, the mean of the final averages was 75 with a standard deviation of 4. For a random sample of 9 students from semester II, the mean was 73 with a standard deviation of 6. If the final averages from semesters I and II are assumed to be normally distributed with equal variances, an appropriate range for the $P$ value for the appropriate test is

(a) $0.3 < P$ value $< 0.4$.

(b) $0.4 < P$ value $< 0.5$.

(c) $0.2 < P$ value $< 0.3$.

(d) $0.1 < P$ value $< 0.2$.

21. A mathematics professor wants to determine whether there is a difference in the final averages between the past two semesters (semester I and semester II) of his business statistics classes. For a random sample of 16 students from semester I, the mean of the final averages was 75 with a standard deviation of 4. For a random sample of 9 students from semester II, the mean was 73 with a standard deviation of 6. If the final averages from semesters I and II are assumed to be normally distributed with equal variances, the correct decision for the appropriate test at a 0.05 level of significance when the population variances are assumed to be equal is

(a) do not reject the null hypothesis.

(b) reject the null hypothesis.

(c) reject the alternative hypothesis.

(d) do not reject the alternative hypothesis.

22. A mathematics professor wants to determine whether there is a difference in the final averages between the past two semesters (semester I and semester II) of his business statistics classes. For a random sample of 16 students from semester I, the mean of the final averages was 75 with a standard deviation of 4. For a random sample of 9 students from semester II, the mean was 73 with a standard deviation of 6. If the final averages from semesters I and II are assumed to be normally distributed with equal variances, the 90 percent confidence interval for the difference of the means (semester I – semester II) when the population variances are assumed to be equal is

(a) −1.4218 to 5.4218.

(b) −4.8313 to 8.8313.

(c) −2.991 to 6.991.

(d) −7.9640 to 11.9640.

23. A mathematics professor wants to determine whether there is a difference in the final averages between the past two semesters (semester I and semester II) of his business statistics classes. For a random sample of 16 students from semester I, the mean of the final averages was 75 with a standard deviation of 4. For a random sample of 9 students from semester II, the mean was 73 with a standard deviation of 6. If the final averages from semesters I and II are assumed to be normally distributed with equal variances, the standard error for the distribution of the differences of the sample means is

(a) 4.9910.

(b) 1.9964.

(c) 3.9856.

(d) 3.4737.

24. A mathematics professor wants to determine whether there is a difference in the final averages between the past two semesters (semester I and semester II) of his business statistics classes. For a random sample of 16 students from semester I, the mean of the final averages was 75 with a standard deviation of 4. For a random sample of 9 students from semester II, the mean was 73 with a standard deviation of 6. If the final averages from semesters I and II are assumed to be normally distributed with equal variances, the degrees of freedom for the appropriate test at a 0.05 level of significance are

(a) 25.

(b) 26.

(c) 24.

(d) 23.

25. A group of foreign students who would like to study in the United States registered for a special TOEFL (Test of English as a Foreign Language) preparatory course offered in their home country. They took a sample examination on the first day of classes and then retook it at the end of the course. The results for six of the students are given below.

| STUDENT | 1 | 2 | 3 | 4 | 5 | 6 |
|---------|-----|-----|-----|-----|-----|-----|
| Before | 325 | 495 | 525 | 480 | 525 | 480 |
| After | 375 | 520 | 510 | 515 | 550 | 490 |

Such sample data would be considered

(a) independent data.

(b) dependent data.

(c) not large enough data.

(d) none of the above.

26. A group of foreign students who would like to study in the United States registered for a special TOEFL (Test of English as a Foreign Language) preparatory course offered in their home country. They took a sample examination on the first day of classes and then retook it at the end of the course. The results for six of the students are given below.

| STUDENT | 1 | 2 | 3 | 4 | 5 | 6 |
|---------|-----|-----|-----|-----|-----|-----|
| Before | 325 | 495 | 525 | 480 | 525 | 480 |
| After | 375 | 520 | 510 | 515 | 550 | 490 |

Let $\mu_d$ represent the mean of the population of differences (after score – before score). If you want to determine whether the course helped to improve the students' scores, the appropriate set of hypotheses will be

(a) $H_0: \mu_d > 0$ vs. $H_1: \mu_d \leq 0$.

(b) $H_0: \mu_d = 0$ vs. $H_1: \mu_d \neq 0$.

(c) $H_0: \mu_d \geq 0$ vs. $H_1: \mu_d < 0$.

(d) $H_0: \mu_d \leq 0$ vs. $H_1: \mu_d > 0$.

27. A group of foreign students who would like to study in the United States registered for a special TOEFL (Test of English as a Foreign Language) preparatory course offered in their home country. They took a sample examination on the first day of classes and then retook it at the end of the course. The results for six of the students are given below.

| STUDENT | 1 | 2 | 3 | 4 | 5 | 6 |
|---------|-----|-----|-----|-----|-----|-----|
| Before | 325 | 495 | 525 | 480 | 525 | 480 |
| After | 375 | 520 | 510 | 515 | 550 | 490 |

If you want to determine whether the course helped to improve the students' scores, the computed test statistic for the appropriate test is

(a) $z = 2.3814$.
(b) $t = 0.0169$.
(c) $z = -0.0169$.
(d) $t = 2.3814$.

28. A group of foreign students who would like to study in the United States registered for a special TOEFL (Test of English as a Foreign Language) preparatory course offered in their home country. They took a sample examination on the first day of classes and then retook it at the end of the course. The results for six of the students are given below.

| STUDENT | 1 | 2 | 3 | 4 | 5 | 6 |
|---------|-----|-----|-----|-----|-----|-----|
| Before | 325 | 495 | 525 | 480 | 525 | 480 |
| After | 375 | 520 | 510 | 515 | 550 | 490 |

If you want to determine whether the course helped to improve the students' scores, the computed $P$ value for the appropriate hypothesis test is

(a) $P$ value $> 0.05$.
(b) $0.025 < P$ value $< 0.05$.
(c) $P$ value $< 0.025$.
(d) $P$ value $= 0.1$.

29. A group of foreign students who would like to study in the United States registered for a special TOEFL (Test of English as a Foreign Language) preparatory course offered in their home country. They took a sample examination on the first day of classes and then retook it at the end of the course. The results for six of the students are given below.

| STUDENT | 1 | 2 | 3 | 4 | 5 | 6 |
|---------|-----|-----|-----|-----|-----|-----|
| Before | 325 | 495 | 525 | 480 | 525 | 480 |
| After | 375 | 520 | 510 | 515 | 550 | 490 |

If you want to determine whether the course helped to improve the students' scores, the appropriate degrees of freedom for the appropriate test are

(a) 6.
(b) 7.
(c) 5.
(d) 4.

30. A group of foreign students who would like to study in the United States registered for a special TOEFL (Test of English as a Foreign Language) preparatory course offered in their home country. They took a sample examination on the first day of classes and then retook it at the end of the course. The results for six of the students are given below.

| STUDENT | 1 | 2 | 3 | 4 | 5 | 6 |
|---------|-----|-----|-----|-----|-----|-----|
| Before | 325 | 495 | 525 | 480 | 525 | 480 |
| After | 375 | 520 | 510 | 515 | 550 | 490 |

If you want to determine whether the course helped to improve the students' scores, the correct decision at the 5 percent level of significance is

(a) do not reject the null hypothesis.
(b) reject the null hypothesis.
(c) reject the alternative hypothesis.
(d) do not reject the alternative hypothesis.

31. A group of foreign students who would like to study in the United States registered for a special TOEFL (Test of English as a Foreign Language) preparatory course offered in their home country. They took a sample examination on the first day of classes and then retook it at the end of the course. The results for six of the students are given below.

| STUDENT | 1 | 2 | 3 | 4 | 5 | 6 |
|---------|-----|-----|-----|-----|-----|-----|
| Before | 325 | 495 | 525 | 480 | 525 | 480 |
| After | 375 | 520 | 510 | 515 | 550 | 490 |

The 99 percent confidence interval for the difference of the means (after − before) is

(a) −15.0173 to 58.3507.
(b) 6.6950 to 36.6429.
(c) −58.3507 to 15.0173.
(d) −36.6429 to −6.6950.

32. Independent random samples are taken to test the difference between two means. The sample sizes are 50 and 60. The sampling distribution for the difference of sample means has a(n)

(a) $t$ distribution with 110 degrees of freedom.
(b) $t$ distribution with 108 degrees of freedom.
(c) exact normal distribution.
(d) $t$ distribution with 112 degrees of freedom.

## Further Exercises

If possible, you can use any technology help available to solve the following questions.

1. You are given the following random sample from a normal population:

    25        39        59        32        46        49

    (a) Construct a 99 percent confidence interval for the population mean.
    (b) Test the hypothesis that the mean of the population is at most 50 at a 0.05 level of significance.

2. In a test to compare the performance of two models of cars, the Arrow and the Sparrow, 10 cars of each model were driven on the same speedway with a full tank of gas in each car (same size tanks). The mean number of miles for the Arrow was 550 miles with a standard deviation of 15 miles; the mean number of miles for the Sparrow was 600 with a standard deviation of 18.

   (a) What is the point estimate of the difference of the means for the populations (the Arrow – the Sparrow)?

   (b) Construct a 95 percent confidence interval for the difference of the two means (the Arrow – the Sparrow).

   (c) At the 2 percent level of significance, are the two models of cars significantly different?

3. To investigate whether one teaching method (method I) will yield better averages than the traditional method (method II) of teaching a specific course, two sections of the same course using the two teaching methods were offered by the same instructor. The table below gives a summary of some of the results

| | METHOD I | METHOD II |
|---|---|---|
| Sample size | 15 | 13 |
| Mean | 88.9 | 82.9 |
| Standard deviation | 3,200 | 20 |

Assume that the two population (methods I and II) variances are equal, and that the populations of scores are normally distributed.

   (a) What is the point estimate for the difference of the two population means (method I – method II)?

   (b) Construct a 95 percent confidence for the difference of the means (method I – method II).

   (c) Can you conclude that method I has improved the overall scores of the students? Test at the 5 percent level of significance.

4. In a manufacturing company, a specific group of workers are responsible for assembling a certain component of an item. The floor manager was not satisfied with the rate at which they worked, so he decided to offer an incentive. The following is the number of components assembled per hour by five workers before and after the incentive was offered.

| WORKER | BEFORE | AFTER |
|---|---|---|
| 1 | 6 | 9 |
| 2 | 10 | 11 |
| 3 | 9 | 10 |
| 4 | 7 | 11 |
| 5 | 6 | 8 |

At the 5 percent level of significance, test to see whether the incentive improved production.

## ANSWER KEY

### True/False Questions

1. F   2. T   3. T   4. F   5. F   6. T   7. F   8. T   9. F   10. F

### Completion Questions

1. $\bar{x}_1 - \bar{x}_2$   2. normal   3. $t$   4. dependent   5. large   6. is   7. pooled   8. $z$
9. equal   10. greater than

### Multiple-Choice Questions

1. (c)   2. (c)   3. (d)   4. (c)   5. (b)   6. (c)   7. (d)   8. (c)   9. (d)
10. (d)   11. (c)   12. (d)   13. (b)   14. (a)   15. (c)   16. (c)   17. (b)
18. (b)   19. (c)   20. (a)   21. (a)   22. (a)   23. (b)   24. (d)   25. (b)
26. (d)   27. (d)   28. (b)   29. (c)   30. (b)   31. (a)   32. (b)

# CHAPTER 14

# Chi-Square Procedures

**Do I Need to Read This Chapter?**

**Y**ou should read this chapter if you need to review or need to learn about

➡ Properties of the chi-square distribution
➡ The chi-square test for goodness of fit
➡ The chi-square test for independence

---

**Get Started**

Here we will focus on the chi-square distribution and how it is used to test for goodness of fit and independence.

---

## 14-1 The Chi-Square Distribution

Here we will present some properties of the chi-square ($\chi^2$) distribution.

### Properties of the $\chi^2$ Distribution

- It is a continuous distribution.
- It is not symmetrical.
- It is skewed to the right.
- The distribution depends on the degrees of freedom $df = n - 1$, where $n$ is the sample size.
- The value of a $\chi^2$ random variable is always nonnegative.
- There are infinitely many $\chi^2$ distributions, since each is uniquely defined by its degrees of freedom.

- For small sample size, the $\chi^2$ distribution is very skewed to the right.
- As $n$ increases, the $\chi^2$ distribution becomes more and more symmetrical.

**Figure 14-1** displays some of the properties of the $\chi^2$ distribution.

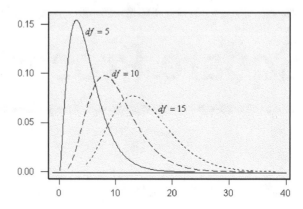

**Fig. 14-1:** Diagram of a family of $\chi^2$ distributions

Since we will be using the $\chi^2$ distribution for the tests in this chapter, we will need to be able to find critical values associated with the distribution.

**Quick Tip**

Extensive tables of critical $\chi^2$ values are available for use in solving confidence intervals and hypothesis testing problems that are associated with the $\chi^2$ distribution.

In order to perform the hypothesis tests in this chapter, we need to be familiar with the notation $\chi^2_{\alpha,\, n-1}$ (read as "chi-square sub alpha with $n - 1$ degrees of freedom").

**Explanation of the notation—$\chi^2_{\alpha,\, n-1}$:** $\chi^2_{\alpha,\, n-1}$ is a $\chi^2$ value with $n - 1$ degrees of freedom such that $\alpha$ area is to the right of the corresponding $\chi^2$ value.

The diagram in **Fig. 14-2** shows a diagram that explains the notation of $\chi^2_{\alpha,\, n-1}$.

Values for the $\chi^2$ random variable with the appropriate degrees of freedom can be obtained from the $\chi^2$ tables in the appendix.

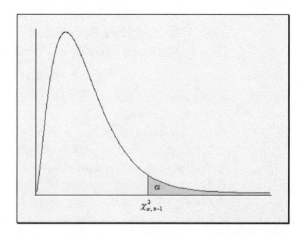

**Fig. 14-2:** Diagram explaining the notation $\chi^2_{\alpha,\, n-1}$

***Example 14-1:*** What is the value of $\chi^2_{0.05, 10}$?

**Solution:** From **Table 4** in the appendix, $\chi^2_{0.05, 10} = 18.307$. (Verify.)

***Example 14-2:*** What is the value of $\chi^2_{0.95, 20}$?

**Solution:** From **Table 4** in the appendix, $\chi^2_{0.95, 20} = 10.851$. (Verify.)

### 14-2 The Chi-Square Test for Goodness of Fit

Have you ever wondered whether a sample of observed data (frequency distribution or proportions) fits some pattern or distribution? We should not expect the pattern to exactly fit a given distribution, so we can look for differences and make conclusions as to the *goodness of fit* of the data.

We can assume that a good fit exists. That is, we can propose a hypothesis that a specified theoretical distribution is appropriate to model the pattern. This, of course, will be your null hypothesis. Since this sample is one of the many possible samples, we can investigate the chance of obtaining this sample with its differences from the model when we assume that the null hypothesis is true. If the chance is small, we can reject the null hypothesis and claim that the model is not appropriate.

How should one go about deciding the significance of the observed differences? To do this, we use a statistic composed of the weighted differences of the frequencies. This statistics has a chi-square distribution with $n - 1$ degrees of freedom, where $n$ is the number of (frequency) categories, and is given by

$$\chi^2 = \sum \frac{(\text{obs} - \text{exp})^2}{\text{exp}}$$

In the equation, obs represents the observed frequencies and exp represents the expected frequencies.

Below is a summary of the tests for goodness of fit.

#### Summary of Hypothesis Test

$H_0$: (statement indicating that the observed data fit some pattern or distribution)

$H_1$: (statement indicating that the observed data do not fit the pattern or distribution indicated in the null hypothesis)

T.S.: $\chi^2 = \sum \dfrac{(\text{obs} - \text{exp})^2}{\text{exp}}$

D.R.: For a specified significance level $\alpha$, reject the null hypothesis if the computed test statistic value $\chi^2$ is greater than $\chi^2_{\alpha, n-1}$.

Conclusion: . . .

**Quick Tip**

The chi-square goodness-of-fit test is always a right-tailed test.

***Example 14-3:*** At an international student organization function, the president of the organization believed that the students who were present consisted of 10 percent Africans, 25 percent Indians, 40 percent Asians, and 25 percent Americans. The students present consisted of 25 Africans, 15 Indians, 80 Asians, and 20 Americans. At the 5 percent level of significance, test the president's belief.

**Solution:** Given $\alpha = 0.05$, $n = 4$ (number of categories), $df = 4 - 1 = 3$, $\chi^2_{0.05,3} = 7.815$. Observe that there are a total of 140 students. The expected values will then be $0.10 \times 140 = 14$, $0.25 \times 140 = 35$, $0.40 \times 140 = 56$, and $0.25 \times 140 = 35$. The observed and expected frequencies are given in **Table 14-1.**

**Table 14-1**

| Observed | 25 | 15 | 80 | 20 |
|----------|----|----|----|----|
| Expected | 14 | 35 | 56 | 35 |

From the table, $\chi^2 = \dfrac{(25 - 14)^2}{14} + \dfrac{(15 - 35)^2}{35} + \dfrac{(80 - 56)^2}{56} + \dfrac{(20 - 35)^2}{35} = 36.7857$. Thus, we can write up the test as

$H_0$:  The composition of the students at the function was 10 percent Africans, 25 percent Indians, 40 percent Asians, and 25 percent Americans

$H_1$:  The distribution stated in the null hypothesis is not the same.

T.S.:  $\chi^2 = \sum \dfrac{(\text{obs} - \text{exp})^2}{\text{exp}} = 36.7857$

D.R.:  For a significance level 0.05, reject the null hypothesis if the computed test statistic value $\chi^2 = 36.7857 > \chi^2_{0.05,3} = 7.815$.

Conclusion:  Since $36.7857 > 7.815$, reject the null hypothesis. That is, at the 5 percent level of significance, there is enough sample evidence to reject the belief of the president of the international student organization.

**Quick Tip**

For the chi-square goodness-of-fit test, the expected frequencies should be at least 5. When the expected frequency of a class or category is less than 5, this class or category can be combined with another class or category so that the expected frequency is at least 5.

***Example 14-4:*** There are 4 TV sets in the student center of a large university. At a particular time each day, four different soap operas (1, 2, 3, and 4) are viewed on these TV sets. The percentages of the audience captured by these shows during one semester were 25 percent, 30 percent, 25 percent, and 20 percent, respectively. During the first week of the following semester, 300 students are surveyed.

(a) If the viewing pattern has not changed, what number of students is expected to watch each soap opera?

**Solution:**  Based on the information, the expected values will be $0.25 \times 300 = 75$, $0.30 \times 300 = 90$, $0.25 \times 300 = 75$, and $0.20 \times 300 = 60$. These expected values are shown in **Table 14-2.**

**Table 14-2**

|  | SOAP OPERA | | | |
|---|---|---|---|---|
|  | 1 | 2 | 3 | 4 |
| Expected Number | 75 | 90 | 75 | 60 |

(b) Suppose that the actual observed numbers of students viewing the soap operas are those given in **Table 14-3.**

**Table 14-3**

|  | SOAP OPERA | | | |
|---|---|---|---|---|
|  | 1 | 2 | 3 | 4 |
| Observed Number | 80 | 88 | 79 | 53 |

Test whether these numbers indicate a change at the 1 percent level of significance.

**Solution:** Given $\alpha = 0.01$, $n = 4$, $df = 4 - 1 = 3$, $\chi^2_{0.01,\,3} = 11.345$. The observed and expected frequencies are given in **Table 14-4.**

**Table 14-4**

| Observed | 80 | 88 | 79 | 53 |
|---|---|---|---|---|
| Expected | 75 | 90 | 75 | 60 |

From **Table 14-4,** $\chi^2 = \dfrac{(80-75)^2}{75} + \dfrac{(88-90)^2}{90} + \dfrac{(79-75)^2}{75} + \dfrac{(53-60)^2}{60} = 1.4978$. Thus, we can write up the test as

$H_0$: The proportions of students who watched the soap operas (1, 2, 3, and 4) were 25 percent, 30 percent, 25 percent, and 20 percent, respectively.

$H_1$: The distribution stated in the null hypothesis is not correct.

T.S.: $\chi^2 = \sum \dfrac{(\text{obs} - \text{exp})^2}{\text{exp}} = 1.4978$

D.R.: For a significance level 0.01, reject the null hypothesis if the computed test statistic value $\chi^2 = 1.4978 > \chi^2_{0.01,\,3} = 11.345$.

Conclusion: Since $1.4978 < 11.345$, do not reject the null hypothesis. That is, at the 1 percent level of significance, there is not enough sample evidence to reject the postulated distribution of students who watch the four soap operas.

## 14-3 The Chi-Square Test for Independence

The chi-square *independence test* can be used to test for the independence of two variables.

***Example 14-5:*** A survey was done by a car manufacturer concerning a particular make and model. A group of 500 potential customers were asked whether they purchased their current car because of its appearance, its performance rating, or its fixed price (no negotiating). The results, broken down by gender, are given in **Table 14-5.**

**Table 14-5**

| | OBSERVED FREQUENCIES | | |
| --- | --- | --- | --- |
| | APPEARANCE | PERFORMANCE | COST |
| Male | 100 | 50 | 35 |
| Female | 80 | 170 | 65 |

*Question:* Do females feel differently from males about the three different criteria used in choosing a car, or do they feel basically the same?

One way of answering this question is to determine whether the criterion used in buying a car is independent of gender. That is, we can do a test for independence. Thus the null hypothesis will be that the criterion used is independent of gender, while the alternative hypothesis will be that the criterion used is dependent on gender.

When data are arranged in tabular form for the chi-square independence test, the table is called a *contingency table*. Here **Table 14-5** has 2 rows and 3 columns, so we say that we have a 2 by 3 ($2 \times 3$) contingency table. The degrees of freedom for any contingency table are given by (number of rows – 1) × (number of columns – 1). In this example, $df = (2 - 1) \times (3 - 1) = 2$.

In order to test for independence using the chi-square independence test, we must compute expected values under the assumption that the null hypothesis is true. To find these expected values, we need to compute the row totals and the column totals. **Table 14-6** shows the observed frequencies with the row and column totals. These row and column totals are called *marginal totals*.

**Table 14-6**

| | OBSERVED FREQUENCIES WITH MARGINAL TOTALS | | | |
| --- | --- | --- | --- | --- |
| | APPEARANCE | PERFORMANCE | COST | TOTAL |
| Male | 100 | 50 | 35 | 185 |
| Female | 80 | 170 | 65 | 315 |
| Total | 180 | 220 | 100 | Grand total = 500 |

The total for the first row (*male*) is 185, and the total for the first column (*appearance*) is 180. The expected value for the cell in the table where the first row (*male*) and the first column (*appearance*) intersect will be $\dfrac{185 \times 180}{500} = 66.6$. Recall that the grand total of the observed values was 500. The expected value for the cell corresponding to the intersection of the second row (*female*) and the third column (*cost*) is $\dfrac{315 \times 100}{500} = 63$. We can continue in this manner to obtain the expected values for the rest of the cells. **Table 14-7** gives the expected frequencies. Check to see that the entries are correct.

**Table 14-7**

| | EXPECTED FREQUENCIES | | | |
| --- | --- | --- | --- | --- |
| | APPEARANCE | PERFORMANCE | COST | TOTAL |
| Male | 66.6 | 81.4 | 37 | 185 |
| Female | 113.4 | 138.6 | 63 | 315 |
| Total | 180 | 220 | 100 | Grand total = 500 |

The $\chi^2$ test statistic is computed in the same manner as for the goodness-of-fit test. A test for the above situation is given next.

**Solution:** Let us use $\alpha = 0.01$. So $df = (2-1)(3-1) = 2$ and $\chi^2_{0.01, 2} = 9.210$.

From the previous computations,

$$\chi^2 = \frac{(100-66.6)^2}{66.6} + \frac{(50-81.4)^2}{81.4} + \frac{(35-37)^2}{37} + \frac{(80-113.4)^2}{113.4} + $$

$$\frac{(170-138.6)^2}{138.6} + \frac{(65-63)^2}{63} = 45.9536$$

Thus, we can write up the test as

$H_0$: The criterion used in purchasing a car is independent of gender.

$H_1$: The criterion used in purchasing a car is dependent on gender.

T.S.: $\chi^2 = \sum \dfrac{(\text{obs}-\text{exp})^2}{\text{exp}} = 45.9536$

D.R.: For a significance level 0.01, reject the null hypothesis if the computed test statistic value $\chi^2 = 45.9536 > \chi^2_{0.01, 3} = 9.210$.

Conclusion: Since $45.9536 > 9.210$, reject the null hypothesis. That is, at the 1 percent level of significance, there is enough sample evidence to claim that the criterion used in purchasing a car is dependent on gender.

## Technology Corner

All of the concepts discussed in this chapter can be computed and illustrated using most statistical software packages. All scientific and graphical calculators can be used for the computations. In addition, some of the newer calculators, like the TI-83, will allow you to do the test directly on the calculator. If you own a calculator, you should consult the manual to determine what statistical features are included.

Illustration: **Figure 14-3** shows the output computed by the MINITAB software for **Example 14-5.** You can extract the information from the output to help in writing up the test. Note that the output also gives the $P$ value for the test. **Figure 14-4** shows the output computed by the TI-83 calculator for **Example 14-5.** The $P$ values provided by both the MINITAB and TI-83 outputs can be used to write up the conclusion for the test. Care should always be taken when using the formulas for computations in hypothesis testing. One can use other features of the technologies to illustrate other concepts discussed in the chapter.

```
Worksheet size: 100000 cells

Chi-Square Test -- Example 14-5

Expected counts are printed below observed counts

            C1        C2        C3     Total
    1       100        50        35       185
          66.60     81.40     37.00

    2        80       170        65       315
         113.40    138.60     63.00

Total       180       220       100       500

Chi-Sq = 16.750 + 12.113 +   0.108 +
          9.837 +  7.114 +   0.063 = 45.985
DF = 2, P-Value = 0.000
```

**Fig. 14-3:** MINITAB output for Example 14-5

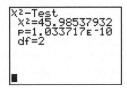

**Fig. 14-4:** TI-83 output for Example 14-5

Here, the focus was on chi-square tests for goodness of fit and for independence between two variables. Sometimes, these tests are called nonparametric tests because no assumptions are made about the distribution of the sampling populations. The concepts in the chapter were illustrated through

✔ Formulas
✔ Examples

Test
Yourself

## True/False Questions

1. The goodness-of-fit test provides us with a procedure for determining whether a set of data is a good fit to a theoretical model.
2. In general, the number of degrees of freedom for any goodness-of-fit test is given by (number of frequency categories – 1).
3. The expected frequencies in a contingency table are the actual numbers observed or recorded in each cell of the table.
4. The observed frequencies in a contingency table represent the theoretical expected outcomes assuming that the null hypothesis is true.
5. In a contingency table, the expected frequency for any cell is computed from (row total + column total)/(grand total).
6. The degrees of freedom for an $r \times c$ contingency table is given by $(r-1)(c-1)$, where $r$ represents the number of rows and $c$ represents the number of columns.
7. Values of the $\chi^2$ random variable can sometimes assume negative values.
8. The sampling distribution for a goodness-of-fit test is the $\chi^2$ distribution.
9. The $\chi^2$ goodness-of-fit test is always demonstrated as a left-tailed test.
10. The number of degrees of freedom for a contingency table with 12 rows and 12 columns is 144.
11. One of the criteria used when performing the $\chi^2$ goodness-of-fit test is that the expected frequency in each cell be less than 5.
12. The shape of the $\chi^2$ distribution depends on the sample size.
13. The $\chi^2$ distribution is used to test hypotheses concerning the population mean.
14. The $\chi^2$ goodness-of-fit test is always a right-tailed test.
15. In a contingency table, you should pool categories whenever the observed frequency in any cell is less than 5.
16. The difference between the observed and expected frequencies in a contingency table is measured by a $\chi^2$ statistic.

## Completion Questions

1. The $\chi^2$ goodness-of-fit test is always a (left, right, two) _____-tailed test.
2. In an $r \times c$ contingency table, where $r$ is the number of rows and $c$ is the number of columns, the degrees of freedom will be $\{(r-1)(c+1), (r-1)(c-1), (r+1)(c-1)\}$ _____.
3. The values of the $\chi^2$ random variable can never assume (negative, positive, zero) _____ values.
4. The number of degrees of freedom for a $\chi^2$ goodness-of-fit test is the (number of categories, sample size, total expected frequency) _____ – 1.

5. When doing tests that involve contingency tables and the $\chi^2$ distribution, it is recommended that the expected frequencies in each cell be greater than or equal to _____.

6. In a $\chi^2$ goodness-of-fit test, the expected values for the cells in the contingency table are computed by assuming that the null hypothesis is (true, false) _____.

7. In a $5 \times 7$ contingency table, the degrees of freedom are (35, 28, 30, 24) _____.

8. For the $\chi^2$ goodness-of-fit test, the $P$ value will be the area to the (right, left) _____ of the computed $\chi^2$ test statistic with the appropriate degrees of freedom.

9. A cross-classification of values in tabular form is called a(n) _____.

10. The number of cells that will result in a cross-classification of the two variables grade (A, B, C, D, E) and number of days absent from class (1, 2, 3, 4, 5) is (20, 16, 25) _____.

11. If the computed test statistic for the $\chi^2$ goodness-of-fit test is large, this will tend to support the (null, alternative) _____ hypothesis.

12. In a contingency table, the expected cell frequency is computed by multiplying the row total and the column total, and dividing by the (row, column, grand) _____ total.

13. Frequencies obtained from a sample are called (observed, expected) _____ frequencies.

14. The $\chi^2$ procedure is a procedure defined for testing how well frequencies of categories in a sample represent frequencies of categories in the (sample, population) _____ .

## Multiple-Choice Questions

1. The number of cells for a $5 \times 7$ contingency table is
   (a) 35.
   (b) 24.
   (c) 48.
   (d) 28.

2. A cross-classification of two categorical variables in tabular form is called a
   (a) frequency distribution table.
   (b) probability distribution table.
   (c) rows and columns table.
   (d) contingency table.

3. Consider the table below, formed by cross-classifying age group and brand of cola consumed.

| | UNDER AGE 15 | AGE 15–25 | AGE 25–35 |
|---|---|---|---|
| Cola 1 | 150 | 100 | 200 |
| Cola 2 | 300 | 125 | 200 |
| Cola 3 | 300 | 200 | 300 |

The *under age 15* relative frequency is

   (a) 0.4000.
   (b) 0.2400.
   (c) 0.3733.
   (d) 0.4267.

4. Consider the table below, formed by cross-classifying age group and brand of cola consumed.

|  | UNDER AGE 15 | AGE 15–25 | AGE 25–35 |
|---|---|---|---|
| Cola 1 | 150 | 100 | 200 |
| Cola 2 | 300 | 125 | 200 |
| Cola 3 | 300 | 200 | 300 |

The observed cell frequency for *age 15–25 cola 3* consumers is

(a) 300.

(b) 200.

(c) 125.

(d) 100.

5. Consider the table below, formed by cross-classifying age group and brand of cola consumed.

|  | UNDER AGE 15 | AGE 15–25 | AGE 25–35 |
|---|---|---|---|
| Cola 1 | 150 | 100 | 200 |
| Cola 2 | 300 | 125 | 200 |
| Cola 3 | 300 | 200 | 300 |

The expected cell frequency for *age 15–25 cola 3* consumers is

(a) 180.

(b) 250.

(c) 320.

(d) 750.

6. Consider the table below, formed by cross-classifying age group and brand of cola consumed.

|  | UNDER AGE 15 | AGE 15–25 | AGE 25–35 |
|---|---|---|---|
| Cola 1 | 150 | 100 | 200 |
| Cola 2 | 300 | 125 | 200 |
| Cola 3 | 300 | 200 | 300 |

If you were to test whether there is any difference in the proportions of people consuming the different brands based on age, the test statistic will be

(a) 31.029.

(b) 26.035.

(c) 30.996.

(d) 31.035.

7. Consider the table below, formed by cross-classifying age group and brand of cola consumed.

|  | UNDER AGE 15 | AGE 15–25 | AGE 25–35 |
|---|---|---|---|
| Cola 1 | 150 | 100 | 200 |
| Cola 2 | 300 | 125 | 200 |
| Cola 3 | 300 | 200 | 300 |

If you were to test whether there is any difference in the proportions of people consuming the different brands based on age, the degrees of freedom for the distribution of the test statistic would be

(a) 4.
(b) 9.
(c) 16.
(d) 12.

8. Consider the table below, formed by cross-classifying age group and brand of cola consumed.

|  | UNDER AGE 15 | AGE 15–25 | AGE 25–35 |
|---|---|---|---|
| Cola 1 | 150 | 100 | 200 |
| Cola 2 | 300 | 125 | 200 |
| Cola 3 | 300 | 200 | 300 |

If you were to test at the 5 percent level of significance whether there is any difference in the proportions of people consuming the different brands based on age, the rejection region would be

(a) $\chi^2 > 16.919$.
(b) $\chi^2 > 9.488$.
(c) $\chi^2 > 26.296$.
(d) $\chi^2 > 21.026$.

9. Consider the table below, formed by cross-classifying age group and brand of cola consumed.

|  | UNDER AGE 15 | AGE 15–25 | AGE 25–35 |
|---|---|---|---|
| Cola 1 | 150 | 100 | 200 |
| Cola 2 | 300 | 125 | 200 |
| Cola 3 | 300 | 200 | 300 |

If you were to test at the 5 percent level of significance whether there is any difference in the proportions of people consuming the different brands based on age, the appropriate null hypothesis for the test would be

(a) there is some difference among the proportions of people consuming the different brands of cola based on age.

(b) there is a small difference among the proportions of people consuming the different brands of cola based on age.

(c) there is no difference among the proportions of people consuming the different brands of cola based on age.

(d) there is a large difference among the proportions of people consuming the different brands of cola based on age.

10. Consider the table below, formed by cross-classifying age group and brand of cola consumed.

| | UNDER AGE 15 | AGE 15–25 | AGE 25–35 |
|---|---|---|---|
| Cola 1 | 150 | 100 | 200 |
| Cola 2 | 300 | 125 | 200 |
| Cola 3 | 300 | 200 | 300 |

If you were to test at the 5 percent level of significance whether there is any difference in the proportions of people consuming the different brands based on age, the decision for the test will be

(a) do not reject the null hypothesis.
(b) reject the alternative hypothesis.
(c) do not reject the alternative hypothesis.
(d) reject the null hypothesis.

11. A six-sided die with faces numbered $1, 2, \ldots, 6$ is rolled 300 times, with the following outcomes:

| OUTCOMES | 1 | 2 | 3 | 4 | 5 | 6 |
|---|---|---|---|---|---|---|
| Observed frequencies | 44 | 48 | 47 | 53 | 52 | 56 |

For an outcome of 4, the relative frequency is

(a) 53 percent.
(b) 17.67 percent.
(c) 50 percent.
(d) 14.2857 percent.

12. A six-sided die with faces numbered $1, 2, \ldots, 6$ is rolled 300 times, with the following outcomes:

| OUTCOMES | 1 | 2 | 3 | 4 | 5 | 6 |
|---|---|---|---|---|---|---|
| Observed frequencies | 44 | 48 | 47 | 53 | 52 | 56 |

The observed frequency for the outcome value of 5 is

(a) 260.
(b) 52/300.
(c) 52/5.
(d) 52.

13. A six-sided die with faces numbered 1, 2, . . . , 6 is rolled 300 times, with the following outcomes:

| OUTCOMES | 1 | 2 | 3 | 4 | 5 | 6 |
|---|---|---|---|---|---|---|
| Observed frequencies | 44 | 48 | 47 | 53 | 52 | 56 |

The expected frequency for the outcome value of 5 is (*Hint:* The expected value for each category will be the same, since the probability is ⅙ of observing a 1, 2, 3, 4, 5, or 6.)

(a) 52.
(b) 0.1733.
(c) 50.
(d) 5.

14. A six-sided die with faces numbered 1, 2, . . . , 6 is rolled 300 times, with the following outcomes:

| OUTCOMES | 1 | 2 | 3 | 4 | 5 | 6 |
|---|---|---|---|---|---|---|
| Observed frequencies | 44 | 48 | 47 | 53 | 52 | 56 |

If you are testing whether the die is fair, the alternative hypothesis will be that

(a) the die is fair.
(b) the die is not fair.
(c) the die is sometimes fair.
(d) the die is almost never fair.

15. A fair six-sided die with faces numbered 1, 2, . . . , 6 is rolled 300 times, with the following outcomes:

| OUTCOMES | 1 | 2 | 3 | 4 | 5 | 6 |
|---|---|---|---|---|---|---|
| Observed frequencies | 44 | 48 | 47 | 53 | 52 | 56 |

If you are testing whether the die is fair, the computed test statistic will be

(a) 1.96.
(b) 1.80.
(c) 1.76.
(d) 2.00.

16. A fair six-sided die with faces numbered 1, 2, . . . , 6 is rolled 300 times, with the following outcomes:

| OUTCOMES | 1 | 2 | 3 | 4 | 5 | 6 |
|---|---|---|---|---|---|---|
| Observed frequencies | 44 | 48 | 47 | 53 | 52 | 56 |

If you are testing whether the die is fair, the degrees of freedom for the distribution of the test statistic will be

(a) 6.

(b) 12.

(c) 7.

(d) 5.

17. A six-sided die with faces numbered 1, 2, ..., 6 is rolled 300 times, with the following outcomes:

| OUTCOMES | 1 | 2 | 3 | 4 | 5 | 6 |
|---|---|---|---|---|---|---|
| Observed frequencies | 44 | 48 | 47 | 53 | 52 | 56 |

If you are testing at the 5 percent level of significance whether the die is fair, the rejection region will be

(a) $\chi^2 > 12.597$.

(b) $\chi^2 > 11.070$.

(c) $\chi^2 > 21.026$.

(d) $\chi^2 > 14.067$.

18. A six-sided die with faces numbered 1, 2, ..., 6 is rolled 300 times, with the following outcomes:

| OUTCOMES | 1 | 2 | 3 | 4 | 5 | 6 |
|---|---|---|---|---|---|---|
| Observed frequencies | 44 | 48 | 47 | 53 | 52 | 56 |

If you are testing at the 5 percent level of significance whether the die is fair, your decision will be

(a) do not reject the null hypothesis.

(b) reject the alternative hypothesis.

(c) do not reject the alternative hypothesis.

(d) reject the null hypothesis.

## Further Exercises

If possible, you can use any technology help available to solve the following questions.

1. A survey was done by a car manufacturer concerning a particular make and model. A group of 500 individuals were asked whether they purchased their car because of its appearance, its performance ratings, or its fixed price (no negotiating). The results are given in the following table:

| OWNER/CRITERIA | APPEARANCE | PERFORMANCE | COST |
|---|---|---|---|
| Male | 100 | 50 | 35 |
| Female | 80 | 170 | 65 |

(a) Compute the expected frequency for each entry.
(b) State the null and alternative hypotheses if you are testing whether there is any difference in the proportions using the three different criteria to purchase the given car based on gender.
(c) Compute the test statistic for the goodness-of-fit test.
(d) If you test at the 5 percent level of significance, what will be the rejection region?
(e) Perform a goodness-of-fit test at the 5 percent level of significance for the hypotheses stated in part (b).
(f) Estimate the $P$ value for this test.

2. A regular six-sided die was tossed 440 times, and the individual outcomes were recorded in the table below.

| OUTCOMES | 1 | 2 | 3 | 4 | 5 | 6 |
|---|---|---|---|---|---|---|
| Observed frequencies | 61 | 83 | 57 | 64 | 85 | 90 |

Test at the 5 percent level of significance to determine whether the die is fair.

# ANSWER KEY
## True/False Questions

1. T   2. T   3. F   4. F   5. F   6. T   7. F   8. T   9. F   10. F   11. F
12. T   13. F   14. T   15. F   16. T

## Completion Questions

1. right   2. $(r-1)(c-1)$   3. negative   4. number of categories   5. five (5)
6. true   7. 24   8. right   9. contingency table   10. twenty five (25)
11. alternative   12. grand   13. observed   14. population

## Multiple-Choice Questions

1. (a)   2. (d)   3. (a)   4. (b)   5. (c)   6. (d)   7. (a)   8. (b)   9. (c)   10. (d)
11. (b)   12. (d)   13. (c)   14. (b)   15. (a)   16. (d)   17. (b)   18. (a)

# Appendix

✔ Table 1—The Binomial Distribution
✔ Table 2—The Standard Normal Distribution
✔ Table 3—The *t* Distribution
✔ Table 4—The Chi-Square Distribution

**Table 1 The Binomial Distribution**

| | | | | | | $p$ | | | | | | |
|---|---|---|---|---|---|---|---|---|---|---|---|---|
| $n$ | $x$ | 0.05 | 0.1 | 0.2 | 0.3 | 0.4 | 0.5 | 0.6 | 0.7 | 0.8 | 0.9 | 0.95 |
| 2 | 0 | 0.9025 | 0.81 | 0.64 | 0.49 | 0.36 | 0.25 | 0.16 | 0.09 | 0.04 | 0.01 | 0.002 |
| | 1 | 0.0950 | 0.18 | 0.32 | 0.42 | 0.48 | 0.50 | 0.48 | 0.42 | 0.32 | 0.18 | 0.095 |
| | 2 | 0.0025 | 0.01 | 0.04 | 0.09 | 0.16 | 0.25 | 0.36 | 0.49 | 0.64 | 0.81 | 0.902 |
| 3 | 0 | 0.857 | 0.729 | 0.512 | 0.343 | 0.216 | 0.125 | 0.064 | 0.027 | 0.008 | 0.001 | 0.000 |
| | 1 | 0.135 | 0.243 | 0.384 | 0.441 | 0.432 | 0.375 | 0.288 | 0.189 | 0.096 | 0.027 | 0.007 |
| | 2 | 0.007 | 0.027 | 0.096 | 0.189 | 0.288 | 0.375 | 0.432 | 0.441 | 0.384 | 0.243 | 0.135 |
| | 3 | 0.000 | 0.001 | 0.008 | 0.027 | 0.064 | 0.125 | 0.216 | 0.343 | 0.512 | 0.729 | 0.857 |
| 4 | 0 | 0.815 | 0.656 | 0.410 | 0.240 | 0.130 | 0.063 | 0.026 | 0.008 | 0.002 | 0.000 | 0.000 |
| | 1 | 0.171 | 0.292 | 0.410 | 0.412 | 0.346 | 0.250 | 0.154 | 0.076 | 0.026 | 0.004 | 0.000 |
| | 2 | 0.014 | 0.049 | 0.154 | 0.265 | 0.346 | 0.375 | 0.346 | 0.265 | 0.154 | 0.049 | 0.014 |
| | 3 | 0.000 | 0.004 | 0.026 | 0.076 | 0.154 | 0.250 | 0.346 | 0.412 | 0.410 | 0.292 | 0.171 |
| | 4 | 0.000 | 0.000 | 0.002 | 0.008 | 0.026 | 0.063 | 0.130 | 0.240 | 0.410 | 0.656 | 0.815 |

**Table 1 The Binomial Distribution** *(continued)*

| | | | | | | p | | | | | | |
|---|---|---|---|---|---|---|---|---|---|---|---|---|
| *n* | *x* | 0.05 | 0.1 | 0.2 | 0.3 | 0.4 | 0.5 | 0.6 | 0.7 | 0.8 | 0.9 | 0.95 |
| 5 | 0 | 0.774 | 0.590 | 0.328 | 0.168 | 0.078 | 0.031 | 0.010 | 0.002 | 0.000 | 0.000 | 0.000 |
| | 1 | 0.204 | 0.328 | 0.410 | 0.360 | 0.259 | 0.156 | 0.077 | 0.028 | 0.006 | 0.000 | 0.000 |
| | 2 | 0.021 | 0.073 | 0.205 | 0.309 | 0.346 | 0.313 | 0.230 | 0.132 | 0.051 | 0.008 | 0.001 |
| | 3 | 0.001 | 0.008 | 0.051 | 0.132 | 0.230 | 0.313 | 0.346 | 0.309 | 0.205 | 0.073 | 0.021 |
| | 4 | 0.000 | 0.000 | 0.006 | 0.028 | 0.077 | 0.156 | 0.259 | 0.360 | 0.410 | 0.328 | 0.204 |
| | 5 | 0.000 | 0.000 | 0.000 | 0.002 | 0.010 | 0.031 | 0.078 | 0.168 | 0.328 | 0.590 | 0.774 |
| 6 | 0 | 0.735 | 0.531 | 0.262 | 0.118 | 0.047 | 0.016 | 0.004 | 0.001 | 0.000 | 0.000 | 0.000 |
| | 1 | 0.232 | 0.354 | 0.393 | 0.303 | 0.187 | 0.094 | 0.037 | 0.010 | 0.002 | 0.000 | 0.000 |
| | 2 | 0.031 | 0.098 | 0.246 | 0.324 | 0.311 | 0.234 | 0.138 | 0.060 | 0.015 | 0.001 | 0.000 |
| | 3 | 0.002 | 0.015 | 0.082 | 0.185 | 0.276 | 0.313 | 0.276 | 0.185 | 0.082 | 0.015 | 0.002 |
| | 4 | 0.000 | 0.001 | 0.015 | 0.060 | 0.138 | 0.234 | 0.311 | 0.324 | 0.246 | 0.098 | 0.031 |
| | 5 | 0.000 | 0.000 | 0.002 | 0.010 | 0.037 | 0.094 | 0.187 | 0.303 | 0.393 | 0.354 | 0.232 |
| | 6 | 0.000 | 0.000 | 0.000 | 0.001 | 0.004 | 0.016 | 0.047 | 0.118 | 0.262 | 0.531 | 0.735 |
| 7 | 0 | 0.698 | 0.478 | 0.210 | 0.082 | 0.028 | 0.008 | 0.002 | 0.000 | 0.000 | 0.000 | 0.000 |
| | 1 | 0.257 | 0.372 | 0.367 | 0.247 | 0.131 | 0.055 | 0.017 | 0.004 | 0.000 | 0.000 | 0.000 |
| | 2 | 0.041 | 0.124 | 0.275 | 0.318 | 0.261 | 0.164 | 0.077 | 0.025 | 0.004 | 0.000 | 0.000 |
| | 3 | 0.004 | 0.023 | 0.115 | 0.227 | 0.290 | 0.273 | 0.194 | 0.097 | 0.029 | 0.003 | 0.000 |
| | 4 | 0.000 | 0.003 | 0.029 | 0.097 | 0.194 | 0.273 | 0.290 | 0.227 | 0.115 | 0.023 | 0.004 |
| | 5 | 0.000 | 0.000 | 0.004 | 0.025 | 0.077 | 0.164 | 0.261 | 0.318 | 0.275 | 0.124 | 0.041 |
| | 6 | 0.000 | 0.000 | 0.000 | 0.004 | 0.017 | 0.055 | 0.131 | 0.247 | 0.367 | 0.372 | 0.257 |
| | 7 | 0.000 | 0.000 | 0.000 | 0.000 | 0.002 | 0.008 | 0.028 | 0.082 | 0.210 | 0.478 | 0.698 |
| 8 | 0 | 0.663 | 0.430 | 0.168 | 0.058 | 0.017 | 0.004 | 0.001 | 0.000 | 0.000 | 0.000 | 0.000 |
| | 1 | 0.279 | 0.383 | 0.336 | 0.198 | 0.090 | 0.031 | 0.008 | 0.001 | 0.000 | 0.000 | 0.000 |
| | 2 | 0.051 | 0.149 | 0.294 | 0.296 | 0.209 | 0.109 | 0.041 | 0.010 | 0.001 | 0.000 | 0.000 |
| | 3 | 0.005 | 0.033 | 0.147 | 0.254 | 0.279 | 0.219 | 0.124 | 0.047 | 0.009 | 0.000 | 0.000 |
| | 4 | 0.000 | 0.005 | 0.046 | 0.136 | 0.232 | 0.273 | 0.232 | 0.136 | 0.046 | 0.005 | 0.000 |
| | 5 | 0.000 | 0.000 | 0.009 | 0.047 | 0.124 | 0.219 | 0.279 | 0.254 | 0.147 | 0.033 | 0.005 |
| | 6 | 0.000 | 0.000 | 0.001 | 0.010 | 0.041 | 0.109 | 0.209 | 0.296 | 0.294 | 0.149 | 0.051 |
| | 7 | 0.000 | 0.000 | 0.000 | 0.001 | 0.008 | 0.031 | 0.090 | 0.198 | 0.336 | 0.383 | 0.279 |
| | 8 | 0.000 | 0.000 | 0.000 | 0.000 | 0.001 | 0.004 | 0.017 | 0.058 | 0.168 | 0.430 | 0.663 |
| 9 | 0 | 0.630 | 0.387 | 0.134 | 0.040 | 0.010 | 0.002 | 0.000 | 0.000 | 0.000 | 0.000 | 0.000 |
| | 1 | 0.299 | 0.387 | 0.302 | 0.156 | 0.060 | 0.018 | 0.004 | 0.000 | 0.000 | 0.000 | 0.000 |
| | 2 | 0.063 | 0.172 | 0.302 | 0.267 | 0.161 | 0.070 | 0.021 | 0.004 | 0.000 | 0.000 | 0.000 |
| | 3 | 0.008 | 0.045 | 0.176 | 0.267 | 0.251 | 0.164 | 0.074 | 0.021 | 0.003 | 0.000 | 0.000 |
| | 4 | 0.001 | 0.007 | 0.066 | 0.172 | 0.251 | 0.246 | 0.167 | 0.074 | 0.017 | 0.001 | 0.000 |
| | 5 | 0.000 | 0.001 | 0.017 | 0.074 | 0.167 | 0.246 | 0.251 | 0.172 | 0.066 | 0.007 | 0.001 |
| | 6 | 0.000 | 0.000 | 0.003 | 0.021 | 0.074 | 0.164 | 0.251 | 0.267 | 0.176 | 0.045 | 0.008 |
| | 7 | 0.000 | 0.000 | 0.000 | 0.004 | 0.021 | 0.070 | 0.161 | 0.267 | 0.302 | 0.172 | 0.063 |
| | 8 | 0.000 | 0.000 | 0.000 | 0.000 | 0.004 | 0.018 | 0.060 | 0.156 | 0.302 | 0.387 | 0.299 |
| | 9 | 0.000 | 0.000 | 0.000 | 0.000 | 0.000 | 0.002 | 0.010 | 0.040 | 0.134 | 0.387 | 0.630 |

| | | | | | | $p$ | | | | | | |
|---|---|---|---|---|---|---|---|---|---|---|---|---|
| $n$ | $x$ | 0.05 | 0.1 | 0.2 | 0.3 | 0.4 | 0.5 | 0.6 | 0.7 | 0.8 | 0.9 | 0.95 |
| 10 | 0 | 0.599 | 0.349 | 0.107 | 0.028 | 0.006 | 0.001 | 0.000 | 0.000 | 0.000 | 0.000 | 0.000 |
| | 1 | 0.315 | 0.387 | 0.268 | 0.121 | 0.040 | 0.010 | 0.002 | 0.000 | 0.000 | 0.000 | 0.000 |
| | 2 | 0.075 | 0.194 | 0.302 | 0.233 | 0.121 | 0.044 | 0.011 | 0.001 | 0.000 | 0.000 | 0.000 |
| | 3 | 0.010 | 0.057 | 0.201 | 0.267 | 0.215 | 0.117 | 0.042 | 0.009 | 0.001 | 0.000 | 0.000 |
| | 4 | 0.001 | 0.011 | 0.088 | 0.200 | 0.251 | 0.205 | 0.111 | 0.037 | 0.006 | 0.000 | 0.000 |
| | 5 | 0.000 | 0.001 | 0.026 | 0.103 | 0.201 | 0.246 | 0.201 | 0.103 | 0.026 | 0.001 | 0.000 |
| | 6 | 0.000 | 0.000 | 0.006 | 0.037 | 0.111 | 0.205 | 0.251 | 0.200 | 0.088 | 0.011 | 0.001 |
| | 7 | 0.000 | 0.000 | 0.001 | 0.009 | 0.042 | 0.117 | 0.215 | 0.267 | 0.201 | 0.057 | 0.010 |
| | 8 | 0.000 | 0.000 | 0.000 | 0.001 | 0.011 | 0.044 | 0.121 | 0.233 | 0.302 | 0.194 | 0.075 |
| | 9 | 0.000 | 0.000 | 0.000 | 0.000 | 0.002 | 0.010 | 0.040 | 0.121 | 0.268 | 0.387 | 0.315 |
| | 10 | 0.000 | 0.000 | 0.000 | 0.000 | 0.000 | 0.001 | 0.006 | 0.028 | 0.107 | 0.349 | 0.599 |
| 11 | 0 | 0.569 | 0.314 | 0.086 | 0.020 | 0.004 | 0.000 | 0.000 | 0.000 | 0.000 | 0.000 | 0.000 |
| | 1 | 0.329 | 0.384 | 0.236 | 0.093 | 0.027 | 0.005 | 0.001 | 0.000 | 0.000 | 0.000 | 0.000 |
| | 2 | 0.087 | 0.213 | 0.295 | 0.200 | 0.089 | 0.027 | 0.005 | 0.001 | 0.000 | 0.000 | 0.000 |
| | 3 | 0.014 | 0.071 | 0.221 | 0.257 | 0.177 | 0.081 | 0.023 | 0.004 | 0.000 | 0.000 | 0.000 |
| | 4 | 0.001 | 0.016 | 0.111 | 0.220 | 0.236 | 0.161 | 0.070 | 0.017 | 0.002 | 0.000 | 0.000 |
| | 5 | 0.000 | 0.002 | 0.039 | 0.132 | 0.221 | 0.226 | 0.147 | 0.057 | 0.010 | 0.000 | 0.000 |
| | 6 | 0.000 | 0.000 | 0.010 | 0.057 | 0.147 | 0.226 | 0.221 | 0.132 | 0.039 | 0.002 | 0.000 |
| | 7 | 0.000 | 0.000 | 0.002 | 0.017 | 0.070 | 0.161 | 0.236 | 0.220 | 0.111 | 0.016 | 0.001 |
| | 8 | 0.000 | 0.000 | 0.000 | 0.004 | 0.023 | 0.081 | 0.177 | 0.257 | 0.221 | 0.071 | 0.014 |
| | 9 | 0.000 | 0.000 | 0.000 | 0.001 | 0.005 | 0.027 | 0.089 | 0.200 | 0.295 | 0.213 | 0.087 |
| | 10 | 0.000 | 0.000 | 0.000 | 0.000 | 0.001 | 0.005 | 0.027 | 0.093 | 0.236 | 0.384 | 0.329 |
| | 11 | 0.000 | 0.000 | 0.000 | 0.000 | 0.000 | 0.000 | 0.004 | 0.020 | 0.086 | 0.314 | 0.569 |
| 12 | 0 | 0.540 | 0.282 | 0.069 | 0.014 | 0.002 | 0.000 | 0.000 | 0.000 | 0.000 | 0.000 | 0.000 |
| | 1 | 0.341 | 0.377 | 0.206 | 0.071 | 0.017 | 0.003 | 0.000 | 0.000 | 0.000 | 0.000 | 0.000 |
| | 2 | 0.099 | 0.230 | 0.283 | 0.168 | 0.064 | 0.016 | 0.002 | 0.000 | 0.000 | 0.000 | 0.000 |
| | 3 | 0.017 | 0.085 | 0.236 | 0.240 | 0.142 | 0.054 | 0.012 | 0.001 | 0.000 | 0.000 | 0.000 |
| | 4 | 0.002 | 0.021 | 0.133 | 0.231 | 0.213 | 0.121 | 0.042 | 0.008 | 0.001 | 0.000 | 0.000 |
| | 5 | 0.000 | 0.004 | 0.053 | 0.158 | 0.227 | 0.193 | 0.101 | 0.029 | 0.003 | 0.000 | 0.000 |
| | 6 | 0.000 | 0.000 | 0.016 | 0.079 | 0.177 | 0.226 | 0.177 | 0.079 | 0.016 | 0.000 | 0.000 |
| | 7 | 0.000 | 0.000 | 0.003 | 0.029 | 0.101 | 0.193 | 0.227 | 0.158 | 0.053 | 0.004 | 0.000 |
| | 8 | 0.000 | 0.000 | 0.001 | 0.008 | 0.042 | 0.121 | 0.213 | 0.231 | 0.133 | 0.021 | 0.002 |
| | 9 | 0.000 | 0.000 | 0.000 | 0.001 | 0.012 | 0.054 | 0.142 | 0.240 | 0.236 | 0.085 | 0.017 |
| | 10 | 0.000 | 0.000 | 0.000 | 0.000 | 0.002 | 0.016 | 0.064 | 0.168 | 0.283 | 0.230 | 0.099 |
| | 11 | 0.000 | 0.000 | 0.000 | 0.000 | 0.000 | 0.003 | 0.017 | 0.071 | 0.206 | 0.377 | 0.341 |
| | 12 | 0.000 | 0.000 | 0.000 | 0.000 | 0.000 | 0.000 | 0.002 | 0.014 | 0.069 | 0.282 | 0.540 |
| 13 | 0 | 0.513 | 0.254 | 0.055 | 0.010 | 0.001 | 0.000 | 0.000 | 0.000 | 0.000 | 0.000 | 0.000 |
| | 1 | 0.351 | 0.367 | 0.179 | 0.054 | 0.011 | 0.002 | 0.000 | 0.000 | 0.000 | 0.000 | 0.000 |
| | 2 | 0.111 | 0.245 | 0.268 | 0.139 | 0.045 | 0.010 | 0.001 | 0.000 | 0.000 | 0.000 | 0.000 |
| | 3 | 0.021 | 0.100 | 0.246 | 0.218 | 0.111 | 0.035 | 0.006 | 0.001 | 0.000 | 0.000 | 0.000 |
| | 4 | 0.003 | 0.028 | 0.154 | 0.234 | 0.184 | 0.087 | 0.024 | 0.003 | 0.000 | 0.000 | 0.000 |

**Table 1 The Binomial Distribution** *(continued)*

| | | \multicolumn{11}{c}{*p*} |
| *n* | *x* | 0.05 | 0.1 | 0.2 | 0.3 | 0.4 | 0.5 | 0.6 | 0.7 | 0.8 | 0.9 | 0.95 |
|---|---|---|---|---|---|---|---|---|---|---|---|---|
| 13 | 5 | 0.000 | 0.006 | 0.069 | 0.180 | 0.221 | 0.157 | 0.066 | 0.014 | 0.001 | 0.000 | 0.000 |
| | 6 | 0.000 | 0.001 | 0.023 | 0.103 | 0.197 | 0.209 | 0.131 | 0.044 | 0.006 | 0.000 | 0.000 |
| | 7 | 0.000 | 0.000 | 0.006 | 0.044 | 0.131 | 0.209 | 0.197 | 0.103 | 0.023 | 0.001 | 0.000 |
| | 8 | 0.000 | 0.000 | 0.001 | 0.014 | 0.066 | 0.157 | 0.221 | 0.180 | 0.069 | 0.006 | 0.000 |
| | 9 | 0.000 | 0.000 | 0.000 | 0.003 | 0.024 | 0.087 | 0.184 | 0.234 | 0.154 | 0.028 | 0.003 |
| | 10 | 0.000 | 0.000 | 0.000 | 0.001 | 0.006 | 0.035 | 0.111 | 0.218 | 0.246 | 0.100 | 0.021 |
| | 11 | 0.000 | 0.000 | 0.000 | 0.000 | 0.001 | 0.010 | 0.045 | 0.139 | 0.268 | 0.245 | 0.111 |
| | 12 | 0.000 | 0.000 | 0.000 | 0.000 | 0.000 | 0.002 | 0.011 | 0.054 | 0.179 | 0.367 | 0.351 |
| | 13 | 0.000 | 0.000 | 0.000 | 0.000 | 0.000 | 0.000 | 0.001 | 0.010 | 0.055 | 0.254 | 0.513 |
| 14 | 0 | 0.488 | 0.229 | 0.044 | 0.007 | 0.001 | 0.000 | 0.000 | 0.000 | 0.000 | 0.000 | 0.000 |
| | 1 | 0.359 | 0.356 | 0.154 | 0.041 | 0.007 | 0.001 | 0.000 | 0.000 | 0.000 | 0.000 | 0.000 |
| | 2 | 0.123 | 0.257 | 0.250 | 0.113 | 0.032 | 0.006 | 0.001 | 0.000 | 0.000 | 0.000 | 0.000 |
| | 3 | 0.026 | 0.114 | 0.250 | 0.194 | 0.085 | 0.022 | 0.003 | 0.000 | 0.000 | 0.000 | 0.000 |
| | 4 | 0.004 | 0.035 | 0.172 | 0.229 | 0.155 | 0.061 | 0.014 | 0.001 | 0.000 | 0.000 | 0.000 |
| | 5 | 0.000 | 0.008 | 0.086 | 0.196 | 0.207 | 0.122 | 0.041 | 0.007 | 0.000 | 0.000 | 0.000 |
| | 6 | 0.000 | 0.001 | 0.032 | 0.126 | 0.207 | 0.183 | 0.092 | 0.023 | 0.002 | 0.000 | 0.000 |
| | 7 | 0.000 | 0.000 | 0.009 | 0.062 | 0.157 | 0.209 | 0.157 | 0.062 | 0.009 | 0.000 | 0.000 |
| | 8 | 0.000 | 0.000 | 0.002 | 0.023 | 0.092 | 0.183 | 0.207 | 0.126 | 0.032 | 0.001 | 0.000 |
| | 9 | 0.000 | 0.000 | 0.000 | 0.007 | 0.041 | 0.122 | 0.207 | 0.196 | 0.086 | 0.008 | 0.000 |
| | 10 | 0.000 | 0.000 | 0.000 | 0.001 | 0.014 | 0.061 | 0.155 | 0.229 | 0.172 | 0.035 | 0.004 |
| | 11 | 0.000 | 0.000 | 0.000 | 0.000 | 0.003 | 0.022 | 0.085 | 0.194 | 0.250 | 0.114 | 0.026 |
| | 12 | 0.000 | 0.000 | 0.000 | 0.000 | 0.001 | 0.006 | 0.032 | 0.113 | 0.250 | 0.257 | 0.123 |
| | 13 | 0.000 | 0.000 | 0.000 | 0.000 | 0.000 | 0.001 | 0.007 | 0.041 | 0.154 | 0.356 | 0.359 |
| | 14 | 0.000 | 0.000 | 0.000 | 0.000 | 0.000 | 0.000 | 0.001 | 0.007 | 0.044 | 0.229 | 0.488 |
| 15 | 0 | 0.463 | 0.206 | 0.035 | 0.005 | 0.000 | 0.000 | 0.000 | 0.000 | 0.000 | 0.000 | 0.000 |
| | 1 | 0.366 | 0.343 | 0.132 | 0.031 | 0.005 | 0.000 | 0.000 | 0.000 | 0.000 | 0.000 | 0.000 |
| | 2 | 0.135 | 0.267 | 0.231 | 0.092 | 0.022 | 0.003 | 0.000 | 0.000 | 0.000 | 0.000 | 0.000 |
| | 3 | 0.031 | 0.129 | 0.250 | 0.170 | 0.063 | 0.014 | 0.002 | 0.000 | 0.000 | 0.000 | 0.000 |
| | 4 | 0.005 | 0.043 | 0.188 | 0.219 | 0.127 | 0.042 | 0.007 | 0.001 | 0.000 | 0.000 | 0.000 |
| | 5 | 0.001 | 0.010 | 0.103 | 0.206 | 0.186 | 0.092 | 0.024 | 0.003 | 0.000 | 0.000 | 0.000 |
| | 6 | 0.000 | 0.002 | 0.043 | 0.147 | 0.207 | 0.153 | 0.061 | 0.012 | 0.001 | 0.000 | 0.000 |
| | 7 | 0.000 | 0.000 | 0.014 | 0.081 | 0.177 | 0.196 | 0.118 | 0.035 | 0.003 | 0.000 | 0.000 |
| | 8 | 0.000 | 0.000 | 0.003 | 0.035 | 0.118 | 0.196 | 0.177 | 0.081 | 0.014 | 0.000 | 0.000 |
| | 9 | 0.000 | 0.000 | 0.001 | 0.012 | 0.061 | 0.153 | 0.207 | 0.147 | 0.043 | 0.002 | 0.000 |
| | 10 | 0.000 | 0.000 | 0.000 | 0.003 | 0.024 | 0.092 | 0.186 | 0.206 | 0.103 | 0.010 | 0.001 |
| | 11 | 0.000 | 0.000 | 0.000 | 0.001 | 0.007 | 0.042 | 0.127 | 0.219 | 0.188 | 0.043 | 0.005 |
| | 12 | 0.000 | 0.000 | 0.000 | 0.000 | 0.002 | 0.014 | 0.063 | 0.170 | 0.250 | 0.129 | 0.031 |
| | 13 | 0.000 | 0.000 | 0.000 | 0.000 | 0.000 | 0.003 | 0.022 | 0.092 | 0.231 | 0.267 | 0.135 |
| | 14 | 0.000 | 0.000 | 0.000 | 0.000 | 0.000 | 0.000 | 0.005 | 0.031 | 0.132 | 0.343 | 0.366 |
| | 15 | 0.000 | 0.000 | 0.000 | 0.000 | 0.000 | 0.000 | 0.000 | 0.005 | 0.035 | 0.206 | 0.463 |

| | | | | | | $p$ | | | | | | |
|---|---|---|---|---|---|---|---|---|---|---|---|---|
| $n$ | $x$ | 0.05 | 0.1 | 0.2 | 0.3 | 0.4 | 0.5 | 0.6 | 0.7 | 0.8 | 0.9 | 0.95 |
| 16 | 0 | 0.440 | 0.185 | 0.028 | 0.003 | 0.000 | 0.000 | 0.000 | 0.000 | 0.000 | 0.000 | 0.000 |
| | 1 | 0.371 | 0.329 | 0.113 | 0.023 | 0.003 | 0.000 | 0.000 | 0.000 | 0.000 | 0.000 | 0.000 |
| | 2 | 0.146 | 0.275 | 0.211 | 0.073 | 0.015 | 0.002 | 0.000 | 0.000 | 0.000 | 0.000 | 0.000 |
| | 3 | 0.036 | 0.142 | 0.246 | 0.146 | 0.047 | 0.009 | 0.001 | 0.000 | 0.000 | 0.000 | 0.000 |
| | 4 | 0.006 | 0.051 | 0.200 | 0.204 | 0.101 | 0.028 | 0.004 | 0.000 | 0.000 | 0.000 | 0.000 |
| | 5 | 0.001 | 0.014 | 0.120 | 0.210 | 0.162 | 0.067 | 0.014 | 0.001 | 0.000 | 0.000 | 0.000 |
| | 6 | 0.000 | 0.003 | 0.055 | 0.165 | 0.198 | 0.122 | 0.039 | 0.006 | 0.000 | 0.000 | 0.000 |
| | 7 | 0.000 | 0.000 | 0.020 | 0.101 | 0.189 | 0.175 | 0.084 | 0.019 | 0.001 | 0.000 | 0.000 |
| | 8 | 0.000 | 0.000 | 0.006 | 0.049 | 0.142 | 0.196 | 0.142 | 0.049 | 0.006 | 0.000 | 0.000 |
| | 9 | 0.000 | 0.000 | 0.001 | 0.019 | 0.084 | 0.175 | 0.189 | 0.101 | 0.020 | 0.000 | 0.000 |
| | 10 | 0.000 | 0.000 | 0.000 | 0.006 | 0.039 | 0.122 | 0.198 | 0.165 | 0.055 | 0.003 | 0.000 |
| | 11 | 0.000 | 0.000 | 0.000 | 0.001 | 0.014 | 0.067 | 0.162 | 0.210 | 0.120 | 0.014 | 0.001 |
| | 12 | 0.000 | 0.000 | 0.000 | 0.000 | 0.004 | 0.028 | 0.101 | 0.204 | 0.200 | 0.051 | 0.006 |
| | 13 | 0.000 | 0.000 | 0.000 | 0.000 | 0.001 | 0.009 | 0.047 | 0.146 | 0.246 | 0.142 | 0.036 |
| | 14 | 0.000 | 0.000 | 0.000 | 0.000 | 0.000 | 0.002 | 0.015 | 0.073 | 0.211 | 0.275 | 0.146 |
| | 15 | 0.000 | 0.000 | 0.000 | 0.000 | 0.000 | 0.000 | 0.003 | 0.023 | 0.113 | 0.329 | 0.371 |
| | 16 | 0.000 | 0.000 | 0.000 | 0.000 | 0.000 | 0.000 | 0.000 | 0.003 | 0.028 | 0.185 | 0.440 |
| 17 | 0 | 0.418 | 0.167 | 0.023 | 0.002 | 0.000 | 0.000 | 0.000 | 0.000 | 0.000 | 0.000 | 0.000 |
| | 1 | 0.374 | 0.315 | 0.096 | 0.017 | 0.002 | 0.000 | 0.000 | 0.000 | 0.000 | 0.000 | 0.000 |
| | 2 | 0.158 | 0.280 | 0.191 | 0.058 | 0.010 | 0.001 | 0.000 | 0.000 | 0.000 | 0.000 | 0.000 |
| | 3 | 0.041 | 0.156 | 0.239 | 0.125 | 0.034 | 0.005 | 0.000 | 0.000 | 0.000 | 0.000 | 0.000 |
| | 4 | 0.008 | 0.060 | 0.209 | 0.187 | 0.080 | 0.018 | 0.002 | 0.000 | 0.000 | 0.000 | 0.000 |
| | 5 | 0.001 | 0.017 | 0.136 | 0.208 | 0.138 | 0.047 | 0.008 | 0.001 | 0.000 | 0.000 | 0.000 |
| | 6 | 0.000 | 0.004 | 0.068 | 0.178 | 0.184 | 0.094 | 0.024 | 0.003 | 0.000 | 0.000 | 0.000 |
| | 7 | 0.000 | 0.001 | 0.027 | 0.120 | 0.193 | 0.148 | 0.057 | 0.009 | 0.000 | 0.000 | 0.000 |
| | 8 | 0.000 | 0.000 | 0.008 | 0.064 | 0.161 | 0.185 | 0.107 | 0.028 | 0.002 | 0.000 | 0.000 |
| | 9 | 0.000 | 0.000 | 0.002 | 0.028 | 0.107 | 0.185 | 0.161 | 0.064 | 0.008 | 0.000 | 0.000 |
| | 10 | 0.000 | 0.000 | 0.000 | 0.009 | 0.057 | 0.148 | 0.193 | 0.120 | 0.027 | 0.001 | 0.000 |
| | 11 | 0.000 | 0.000 | 0.000 | 0.003 | 0.024 | 0.094 | 0.184 | 0.178 | 0.068 | 0.004 | 0.000 |
| | 12 | 0.000 | 0.000 | 0.000 | 0.001 | 0.008 | 0.047 | 0.138 | 0.208 | 0.136 | 0.017 | 0.001 |
| | 13 | 0.000 | 0.000 | 0.000 | 0.000 | 0.002 | 0.018 | 0.080 | 0.187 | 0.209 | 0.060 | 0.008 |
| | 14 | 0.000 | 0.000 | 0.000 | 0.000 | 0.000 | 0.005 | 0.034 | 0.125 | 0.239 | 0.156 | 0.041 |
| | 15 | 0.000 | 0.000 | 0.000 | 0.000 | 0.000 | 0.001 | 0.010 | 0.058 | 0.191 | 0.280 | 0.158 |
| | 16 | 0.000 | 0.000 | 0.000 | 0.000 | 0.000 | 0.000 | 0.002 | 0.017 | 0.096 | 0.315 | 0.374 |
| | 17 | 0.000 | 0.000 | 0.000 | 0.000 | 0.000 | 0.000 | 0.000 | 0.002 | 0.023 | 0.167 | 0.418 |
| 18 | 0 | 0.397 | 0.150 | 0.018 | 0.002 | 0.000 | 0.000 | 0.000 | 0.000 | 0.000 | 0.000 | 0.000 |
| | 1 | 0.376 | 0.300 | 0.081 | 0.013 | 0.001 | 0.000 | 0.000 | 0.000 | 0.000 | 0.000 | 0.000 |
| | 2 | 0.168 | 0.284 | 0.172 | 0.046 | 0.007 | 0.001 | 0.000 | 0.000 | 0.000 | 0.000 | 0.000 |
| | 3 | 0.047 | 0.168 | 0.230 | 0.105 | 0.025 | 0.003 | 0.000 | 0.000 | 0.000 | 0.000 | 0.000 |
| | 4 | 0.009 | 0.070 | 0.215 | 0.168 | 0.061 | 0.012 | 0.001 | 0.000 | 0.000 | 0.000 | 0.000 |

**Table 1 The Binomial Distribution** *(continued)*

| n | x | 0.05 | 0.1 | 0.2 | 0.3 | 0.4 | 0.5 | 0.6 | 0.7 | 0.8 | 0.9 | 0.95 |
|---|---|------|-----|-----|-----|-----|-----|-----|-----|-----|-----|------|
| 18 | 5 | 0.001 | 0.022 | 0.151 | 0.202 | 0.115 | 0.033 | 0.004 | 0.000 | 0.000 | 0.000 | 0.000 |
|  | 6 | 0.000 | 0.005 | 0.082 | 0.187 | 0.166 | 0.071 | 0.015 | 0.001 | 0.000 | 0.000 | 0.000 |
|  | 7 | 0.000 | 0.001 | 0.035 | 0.138 | 0.189 | 0.121 | 0.037 | 0.005 | 0.000 | 0.000 | 0.000 |
|  | 8 | 0.000 | 0.000 | 0.012 | 0.081 | 0.173 | 0.167 | 0.077 | 0.015 | 0.001 | 0.000 | 0.000 |
|  | 9 | 0.000 | 0.000 | 0.003 | 0.039 | 0.128 | 0.185 | 0.128 | 0.039 | 0.003 | 0.000 | 0.000 |
|  | 10 | 0.000 | 0.000 | 0.001 | 0.015 | 0.077 | 0.167 | 0.173 | 0.081 | 0.012 | 0.000 | 0.000 |
|  | 11 | 0.000 | 0.000 | 0.000 | 0.005 | 0.037 | 0.121 | 0.189 | 0.138 | 0.035 | 0.001 | 0.000 |
|  | 12 | 0.000 | 0.000 | 0.000 | 0.001 | 0.015 | 0.071 | 0.166 | 0.187 | 0.082 | 0.005 | 0.000 |
|  | 13 | 0.000 | 0.000 | 0.000 | 0.000 | 0.004 | 0.033 | 0.115 | 0.202 | 0.151 | 0.022 | 0.001 |
|  | 14 | 0.000 | 0.000 | 0.000 | 0.000 | 0.001 | 0.012 | 0.061 | 0.168 | 0.215 | 0.070 | 0.009 |
|  | 15 | 0.000 | 0.000 | 0.000 | 0.000 | 0.000 | 0.003 | 0.025 | 0.105 | 0.230 | 0.168 | 0.047 |
|  | 16 | 0.000 | 0.000 | 0.000 | 0.000 | 0.000 | 0.001 | 0.007 | 0.046 | 0.172 | 0.284 | 0.168 |
|  | 17 | 0.000 | 0.000 | 0.000 | 0.000 | 0.000 | 0.000 | 0.001 | 0.013 | 0.081 | 0.300 | 0.376 |
|  | 18 | 0.000 | 0.000 | 0.000 | 0.000 | 0.000 | 0.000 | 0.000 | 0.002 | 0.018 | 0.150 | 0.397 |
| 19 | 0 | 0.377 | 0.135 | 0.014 | 0.001 | 0.000 | 0.000 | 0.000 | 0.000 | 0.000 | 0.000 | 0.000 |
|  | 1 | 0.377 | 0.285 | 0.068 | 0.009 | 0.001 | 0.000 | 0.000 | 0.000 | 0.000 | 0.000 | 0.000 |
|  | 2 | 0.179 | 0.285 | 0.154 | 0.036 | 0.005 | 0.000 | 0.000 | 0.000 | 0.000 | 0.000 | 0.000 |
|  | 3 | 0.053 | 0.180 | 0.218 | 0.087 | 0.017 | 0.002 | 0.000 | 0.000 | 0.000 | 0.000 | 0.000 |
|  | 4 | 0.011 | 0.080 | 0.218 | 0.149 | 0.047 | 0.007 | 0.001 | 0.000 | 0.000 | 0.000 | 0.000 |
|  | 5 | 0.002 | 0.027 | 0.164 | 0.192 | 0.093 | 0.022 | 0.002 | 0.000 | 0.000 | 0.000 | 0.000 |
|  | 6 | 0.000 | 0.007 | 0.095 | 0.192 | 0.145 | 0.052 | 0.008 | 0.001 | 0.000 | 0.000 | 0.000 |
|  | 7 | 0.000 | 0.001 | 0.044 | 0.153 | 0.180 | 0.096 | 0.024 | 0.002 | 0.000 | 0.000 | 0.000 |
|  | 8 | 0.000 | 0.000 | 0.017 | 0.098 | 0.180 | 0.144 | 0.053 | 0.008 | 0.000 | 0.000 | 0.000 |
|  | 9 | 0.000 | 0.000 | 0.005 | 0.051 | 0.146 | 0.176 | 0.098 | 0.022 | 0.001 | 0.000 | 0.000 |
|  | 10 | 0.000 | 0.000 | 0.001 | 0.022 | 0.098 | 0.176 | 0.146 | 0.051 | 0.005 | 0.000 | 0.000 |
|  | 11 | 0.000 | 0.000 | 0.000 | 0.008 | 0.053 | 0.144 | 0.180 | 0.098 | 0.017 | 0.000 | 0.000 |
|  | 12 | 0.000 | 0.000 | 0.000 | 0.002 | 0.024 | 0.096 | 0.180 | 0.153 | 0.044 | 0.001 | 0.000 |
|  | 13 | 0.000 | 0.000 | 0.000 | 0.001 | 0.008 | 0.052 | 0.145 | 0.192 | 0.095 | 0.007 | 0.000 |
|  | 14 | 0.000 | 0.000 | 0.000 | 0.000 | 0.002 | 0.022 | 0.093 | 0.192 | 0.164 | 0.027 | 0.002 |
|  | 15 | 0.000 | 0.000 | 0.000 | 0.000 | 0.001 | 0.007 | 0.047 | 0.149 | 0.218 | 0.080 | 0.011 |
|  | 16 | 0.000 | 0.000 | 0.000 | 0.000 | 0.000 | 0.002 | 0.017 | 0.087 | 0.218 | 0.180 | 0.053 |
|  | 17 | 0.000 | 0.000 | 0.000 | 0.000 | 0.000 | 0.000 | 0.005 | 0.036 | 0.154 | 0.285 | 0.179 |
|  | 18 | 0.000 | 0.000 | 0.000 | 0.000 | 0.000 | 0.000 | 0.001 | 0.009 | 0.068 | 0.285 | 0.377 |
|  | 19 | 0.000 | 0.000 | 0.000 | 0.000 | 0.000 | 0.000 | 0.000 | 0.001 | 0.014 | 0.135 | 0.377 |
| 20 | 0 | 0.358 | 0.122 | 0.012 | 0.001 | 0.000 | 0.000 | 0.000 | 0.000 | 0.000 | 0.000 | 0.000 |
|  | 1 | 0.377 | 0.270 | 0.058 | 0.007 | 0.000 | 0.000 | 0.000 | 0.000 | 0.000 | 0.000 | 0.000 |
|  | 2 | 0.189 | 0.285 | 0.137 | 0.028 | 0.003 | 0.000 | 0.000 | 0.000 | 0.000 | 0.000 | 0.000 |
|  | 3 | 0.060 | 0.190 | 0.205 | 0.072 | 0.012 | 0.001 | 0.000 | 0.000 | 0.000 | 0.000 | 0.000 |
|  | 4 | 0.013 | 0.090 | 0.218 | 0.130 | 0.035 | 0.005 | 0.000 | 0.000 | 0.000 | 0.000 | 0.000 |
|  | 5 | 0.002 | 0.032 | 0.175 | 0.179 | 0.075 | 0.015 | 0.001 | 0.000 | 0.000 | 0.000 | 0.000 |

| | | | | | | p | | | | | | |
|---|---|---|---|---|---|---|---|---|---|---|---|---|
| n | x | 0.05 | 0.1 | 0.2 | 0.3 | 0.4 | 0.5 | 0.6 | 0.7 | 0.8 | 0.9 | 0.95 |
| 20 | 6 | 0.000 | 0.009 | 0.109 | 0.192 | 0.124 | 0.037 | 0.005 | 0.000 | 0.000 | 0.000 | 0.000 |
| | 7 | 0.000 | 0.002 | 0.055 | 0.164 | 0.166 | 0.074 | 0.015 | 0.001 | 0.000 | 0.000 | 0.000 |
| | 8 | 0.000 | 0.000 | 0.022 | 0.114 | 0.180 | 0.120 | 0.035 | 0.004 | 0.000 | 0.000 | 0.000 |
| | 9 | 0.000 | 0.000 | 0.007 | 0.065 | 0.160 | 0.160 | 0.071 | 0.012 | 0.000 | 0.000 | 0.000 |
| | 10 | 0.000 | 0.000 | 0.002 | 0.031 | 0.117 | 0.176 | 0.117 | 0.031 | 0.002 | 0.000 | 0.000 |
| | 11 | 0.000 | 0.000 | 0.000 | 0.012 | 0.071 | 0.160 | 0.160 | 0.065 | 0.007 | 0.000 | 0.000 |
| | 12 | 0.000 | 0.000 | 0.000 | 0.004 | 0.035 | 0.120 | 0.180 | 0.114 | 0.022 | 0.000 | 0.000 |
| | 13 | 0.000 | 0.000 | 0.000 | 0.001 | 0.015 | 0.074 | 0.166 | 0.164 | 0.055 | 0.002 | 0.000 |
| | 14 | 0.000 | 0.000 | 0.000 | 0.000 | 0.005 | 0.037 | 0.124 | 0.192 | 0.109 | 0.009 | 0.000 |
| | 15 | 0.000 | 0.000 | 0.000 | 0.000 | 0.001 | 0.015 | 0.075 | 0.179 | 0.175 | 0.032 | 0.002 |
| | 16 | 0.000 | 0.000 | 0.000 | 0.000 | 0.000 | 0.005 | 0.035 | 0.130 | 0.218 | 0.090 | 0.013 |
| | 17 | 0.000 | 0.000 | 0.000 | 0.000 | 0.000 | 0.001 | 0.012 | 0.072 | 0.205 | 0.190 | 0.060 |
| | 18 | 0.000 | 0.000 | 0.000 | 0.000 | 0.000 | 0.000 | 0.003 | 0.028 | 0.137 | 0.285 | 0.189 |
| | 19 | 0.000 | 0.000 | 0.000 | 0.000 | 0.000 | 0.000 | 0.000 | 0.007 | 0.058 | 0.270 | 0.377 |
| | 20 | 0.000 | 0.000 | 0.000 | 0.000 | 0.000 | 0.000 | 0.000 | 0.001 | 0.012 | 0.122 | 0.358 |

**Table 2:   The Standard Normal Distribution**

| $z$ | 0.00 | 0.01 | 0.02 | 0.03 | 0.04 | 0.05 | 0.06 | 0.07 | 0.08 | 0.09 |
|-----|------|------|------|------|------|------|------|------|------|------|
| 0.0 | 0.0000 | 0.0040 | 0.0080 | 0.0120 | 0.0160 | 0.0199 | 0.0239 | 0.0279 | 0.0319 | 0.0359 |
| 0.1 | 0.0398 | 0.0438 | 0.0478 | 0.0517 | 0.0557 | 0.0596 | 0.0636 | 0.0675 | 0.0714 | 0.0753 |
| 0.2 | 0.0793 | 0.0832 | 0.0871 | 0.0910 | 0.0948 | 0.0987 | 0.1026 | 0.1064 | 0.1103 | 0.1141 |
| 0.3 | 0.1179 | 0.1217 | 0.1255 | 0.1293 | 0.1331 | 0.1368 | 0.1406 | 0.1443 | 0.1480 | 0.1517 |
| 0.4 | 0.1554 | 0.1591 | 0.1628 | 0.1664 | 0.1700 | 0.1736 | 0.1772 | 0.1808 | 0.1844 | 0.1879 |
| 0.5 | 0.1915 | 0.1950 | 0.1985 | 0.2019 | 0.2054 | 0.2088 | 0.2123 | 0.2157 | 0.2190 | 0.2224 |
| 0.6 | 0.2257 | 0.2291 | 0.2324 | 0.2357 | 0.2389 | 0.2422 | 0.2454 | 0.2486 | 0.2517 | 0.2549 |
| 0.7 | 0.2580 | 0.2611 | 0.2642 | 0.2673 | 0.2703 | 0.2734 | 0.2764 | 0.2794 | 0.2823 | 0.2852 |
| 0.8 | 0.2881 | 0.2910 | 0.2939 | 0.2967 | 0.2995 | 0.3023 | 0.3051 | 0.3078 | 0.3106 | 0.3133 |
| 0.9 | 0.3159 | 0.3186 | 0.3212 | 0.3238 | 0.3264 | 0.3289 | 0.3315 | 0.3340 | 0.3365 | 0.3389 |
| 1.0 | 0.3413 | 0.3438 | 0.3461 | 0.3485 | 0.3508 | 0.3531 | 0.3554 | 0.3577 | 0.3599 | 0.3621 |
| 1.1 | 0.3643 | 0.3665 | 0.3686 | 0.3708 | 0.3729 | 0.3749 | 0.3770 | 0.3790 | 0.3810 | 0.3830 |
| 1.2 | 0.3849 | 0.3869 | 0.3888 | 0.3907 | 0.3925 | 0.3944 | 0.3962 | 0.3980 | 0.3997 | 0.4015 |
| 1.3 | 0.4032 | 0.4049 | 0.4066 | 0.4082 | 0.4099 | 0.4115 | 0.4131 | 0.4147 | 0.4162 | 0.4177 |
| 1.4 | 0.4192 | 0.4207 | 0.4222 | 0.4236 | 0.4251 | 0.4265 | 0.4279 | 0.4292 | 0.4306 | 0.4319 |
| 1.5 | 0.4332 | 0.4345 | 0.4357 | 0.4370 | 0.4382 | 0.4394 | 0.4406 | 0.4418 | 0.4429 | 0.4441 |
| 1.6 | 0.4452 | 0.4463 | 0.4474 | 0.4484 | 0.4495 | 0.4505 | 0.4515 | 0.4525 | 0.4535 | 0.4545 |
| 1.7 | 0.4554 | 0.4564 | 0.4573 | 0.4582 | 0.4591 | 0.4599 | 0.4608 | 0.4616 | 0.4625 | 0.4633 |
| 1.8 | 0.4641 | 0.4649 | 0.4656 | 0.4664 | 0.4671 | 0.4678 | 0.4686 | 0.4693 | 0.4699 | 0.4706 |
| 1.9 | 0.4713 | 0.4719 | 0.4726 | 0.4732 | 0.4738 | 0.4744 | 0.4750 | 0.4756 | 0.4761 | 0.4767 |
| 2.0 | 0.4772 | 0.4778 | 0.4783 | 0.4788 | 0.4793 | 0.4798 | 0.4803 | 0.4808 | 0.4812 | 0.4817 |
| 2.1 | 0.4821 | 0.4826 | 0.4830 | 0.4834 | 0.4838 | 0.4842 | 0.4846 | 0.4850 | 0.4854 | 0.4857 |
| 2.2 | 0.4861 | 0.4864 | 0.4868 | 0.4871 | 0.4875 | 0.4878 | 0.4881 | 0.4884 | 0.4887 | 0.4890 |
| 2.3 | 0.4893 | 0.4896 | 0.4898 | 0.4901 | 0.4904 | 0.4906 | 0.4909 | 0.4911 | 0.4913 | 0.4916 |
| 2.4 | 0.4918 | 0.4920 | 0.4922 | 0.4925 | 0.4927 | 0.4929 | 0.4931 | 0.4932 | 0.4934 | 0.4936 |
| 2.5 | 0.4938 | 0.4940 | 0.4941 | 0.4943 | 0.4945 | 0.4946 | 0.4948 | 0.4949 | 0.4951 | 0.4952 |
| 2.6 | 0.4953 | 0.4955 | 0.4956 | 0.4957 | 0.4959 | 0.4960 | 0.4961 | 0.4962 | 0.4963 | 0.4964 |
| 2.7 | 0.4965 | 0.4966 | 0.4967 | 0.4968 | 0.4969 | 0.4970 | 0.4971 | 0.4972 | 0.4973 | 0.4974 |
| 2.8 | 0.4974 | 0.4975 | 0.4976 | 0.4977 | 0.4977 | 0.4978 | 0.4979 | 0.4979 | 0.4980 | 0.4981 |
| 2.9 | 0.4981 | 0.4982 | 0.4982 | 0.4983 | 0.4984 | 0.4984 | 0.4985 | 0.4985 | 0.4986 | 0.4986 |
| 3.0 | 0.4987 | 0.4987 | 0.4987 | 0.4988 | 0.4988 | 0.4989 | 0.4989 | 0.4989 | 0.4990 | 0.4990 |

Note: The table values are the area between 0 and a given $z$-score value.

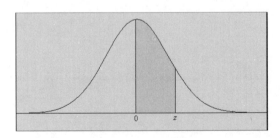

**Fig. 1:** Area under standard normal curve between 0 and $z$

## Table 3: The *t* Distribution

| df | Confidence Intervals | 80% | 90% | 95% | 98% | 99% |
|---|---|---|---|---|---|---|
| | One-Tailed, α | 0.10 | 0.05 | 0.025 | 0.01 | 0.005 |
| df | Two-Tailed, α | 0.20 | 0.10 | 0.05 | 0.02 | 0.01 |
| 1 | | 3.078 | 6.314 | 12.706 | 31.821 | 63.657 |
| 2 | | 1.886 | 2.920 | 4.303 | 6.965 | 9.925 |
| 3 | | 1.638 | 2.353 | 3.182 | 4.541 | 5.841 |
| 4 | | 1.533 | 2.132 | 2.776 | 3.747 | 4.604 |
| 5 | | 1.476 | 2.015 | 2.571 | 3.365 | 4.032 |
| 6 | | 1.440 | 1.943 | 2.447 | 3.143 | 3.707 |
| 7 | | 1.415 | 1.895 | 2.365 | 2.998 | 3.499 |
| 8 | | 1.397 | 1.860 | 2.306 | 2.896 | 3.355 |
| 9 | | 1.383 | 1.833 | 2.262 | 2.821 | 3.250 |
| 10 | | 1.372 | 1.812 | 2.228 | 2.764 | 3.169 |
| 11 | | 1.363 | 1.796 | 2.201 | 2.718 | 3.106 |
| 12 | | 1.356 | 1.782 | 2.179 | 2.681 | 3.055 |
| 13 | | 1.350 | 1.771 | 2.160 | 2.650 | 3.012 |
| 14 | | 1.345 | 1.761 | 2.145 | 2.624 | 2.977 |
| 15 | | 1.341 | 1.753 | 2.131 | 2.602 | 2.947 |
| 16 | | 1.337 | 1.746 | 2.120 | 2.583 | 2.921 |
| 17 | | 1.333 | 1.740 | 2.110 | 2.567 | 2.898 |
| 18 | | 1.330 | 1.734 | 2.101 | 2.552 | 2.878 |
| 19 | | 1.328 | 1.729 | 2.093 | 2.539 | 2.861 |
| 20 | | 1.325 | 1.725 | 2.086 | 2.528 | 2.845 |
| 21 | | 1.323 | 1.721 | 2.080 | 2.518 | 2.831 |
| 22 | | 1.321 | 1.717 | 2.074 | 2.508 | 2.819 |
| 23 | | 1.319 | 1.714 | 2.069 | 2.500 | 2.807 |
| 24 | | 1.318 | 1.711 | 2.064 | 2.492 | 2.797 |
| 25 | | 1.316 | 1.708 | 2.060 | 2.485 | 2.787 |
| 26 | | 1.315 | 1.706 | 2.056 | 2.479 | 2.779 |
| 27 | | 1.314 | 1.703 | 2.052 | 2.473 | 2.771 |
| 28 | | 1.313 | 1.701 | 2.048 | 2.467 | 2.763 |
| (z) ∞ | | 1.282[a] | 1.645 | 1.960 | 2.326[b] | 2.576[c] |

[a] $1.282 \approx 1.28$
[b] $2.326 \approx 2.33$
[c] $2.576 \approx 2.58$

**Table 4:   The Chi-Square Distribution**

| DEGREES OF FREEDOM | | | | | α | | | | | |
|---|---|---|---|---|---|---|---|---|---|---|
| | 0.005 | 0.01 | 0.025 | 0.05 | 0.1 | 0.9 | 0.95 | 0.975 | 0.99 | 0.995 |
| 1 | 7.879 | 6.635 | 5.024 | 3.841 | 2.706 | 0.016 | 0.004 | 0.001 | 0.000 | 0.000 |
| 2 | 10.597 | 9.210 | 7.378 | 5.991 | 4.605 | 0.211 | 0.103 | 0.051 | 0.020 | 0.010 |
| 3 | 12.838 | 11.345 | 9.348 | 7.815 | 6.251 | 0.584 | 0.352 | 0.216 | 0.115 | 0.072 |
| 4 | 14.860 | 13.277 | 11.143 | 9.488 | 7.779 | 1.064 | 0.711 | 0.484 | 0.297 | 0.207 |
| 5 | 16.750 | 15.086 | 12.833 | 11.070 | 9.236 | 1.610 | 1.145 | 0.831 | 0.554 | 0.412 |
| 6 | 18.548 | 16.812 | 14.449 | 12.592 | 10.645 | 2.204 | 1.635 | 1.237 | 0.872 | 0.676 |
| 7 | 20.278 | 18.475 | 16.013 | 14.067 | 12.017 | 2.833 | 2.167 | 1.690 | 1.239 | 0.989 |
| 8 | 21.955 | 20.090 | 17.535 | 15.507 | 13.362 | 3.490 | 2.733 | 2.180 | 1.646 | 1.344 |
| 9 | 23.589 | 21.666 | 19.023 | 16.919 | 14.684 | 4.168 | 3.325 | 2.700 | 2.088 | 1.735 |
| 10 | 25.188 | 23.209 | 20.483 | 18.307 | 15.987 | 4.865 | 3.940 | 3.247 | 2.558 | 2.156 |
| 11 | 26.757 | 24.725 | 21.920 | 19.675 | 17.275 | 5.578 | 4.575 | 3.816 | 3.053 | 2.603 |
| 12 | 28.300 | 26.217 | 23.337 | 21.026 | 18.549 | 6.304 | 5.226 | 4.404 | 3.571 | 3.074 |
| 13 | 29.819 | 27.688 | 24.736 | 22.362 | 19.812 | 7.042 | 5.892 | 5.009 | 4.107 | 3.565 |
| 14 | 31.319 | 29.141 | 26.119 | 23.685 | 21.064 | 7.790 | 6.571 | 5.629 | 4.660 | 4.075 |
| 15 | 32.801 | 30.578 | 27.488 | 24.996 | 22.307 | 8.547 | 7.261 | 6.262 | 5.229 | 4.601 |
| 16 | 34.267 | 32.000 | 28.845 | 26.296 | 23.542 | 9.312 | 7.962 | 6.908 | 5.812 | 5.142 |
| 17 | 35.718 | 33.409 | 30.191 | 27.587 | 24.769 | 10.085 | 8.672 | 7.564 | 6.408 | 5.697 |
| 18 | 37.156 | 34.805 | 31.526 | 28.869 | 25.989 | 10.865 | 9.390 | 8.231 | 7.015 | 6.265 |
| 19 | 38.582 | 36.191 | 32.852 | 30.144 | 27.204 | 11.651 | 10.117 | 8.907 | 7.633 | 6.844 |
| 20 | 39.997 | 37.566 | 34.170 | 31.410 | 28.412 | 12.443 | 10.851 | 9.591 | 8.260 | 7.434 |
| 21 | 41.401 | 38.932 | 35.479 | 32.671 | 29.615 | 13.240 | 11.591 | 10.283 | 8.897 | 8.034 |
| 22 | 42.796 | 40.289 | 36.781 | 33.924 | 30.813 | 14.041 | 12.338 | 10.982 | 9.542 | 8.643 |
| 23 | 44.181 | 41.638 | 38.076 | 35.172 | 32.007 | 14.848 | 13.091 | 11.689 | 10.196 | 9.260 |
| 24 | 45.559 | 42.980 | 39.364 | 36.415 | 33.196 | 15.659 | 13.848 | 12.401 | 10.856 | 9.886 |
| 25 | 46.928 | 44.314 | 40.646 | 37.652 | 34.382 | 16.473 | 14.611 | 13.120 | 11.524 | 10.520 |
| 26 | 48.290 | 45.642 | 41.923 | 38.885 | 35.563 | 17.292 | 15.379 | 13.844 | 12.198 | 11.160 |
| 27 | 49.645 | 46.963 | 43.195 | 40.113 | 36.741 | 18.114 | 16.151 | 14.573 | 12.879 | 11.808 |
| 28 | 50.993 | 48.278 | 44.461 | 41.337 | 37.916 | 18.939 | 16.928 | 15.308 | 13.565 | 12.461 |
| 29 | 52.336 | 49.588 | 45.722 | 42.557 | 39.087 | 19.768 | 17.708 | 16.047 | 14.256 | 13.121 |
| 30 | 53.672 | 50.892 | 46.979 | 43.773 | 40.256 | 20.599 | 18.493 | 16.791 | 14.953 | 13.787 |
| 40 | 66.766 | 63.691 | 59.342 | 55.758 | 51.805 | 29.051 | 26.509 | 24.433 | 22.164 | 20.707 |
| 50 | 79.490 | 76.154 | 71.420 | 67.505 | 63.167 | 37.689 | 34.764 | 32.357 | 29.707 | 27.991 |
| 60 | 91.952 | 88.379 | 83.298 | 79.082 | 74.397 | 46.459 | 43.188 | 40.482 | 37.485 | 35.534 |
| 70 | 104.215 | 100.425 | 95.023 | 90.531 | 85.527 | 55.329 | 51.739 | 48.758 | 45.442 | 43.275 |
| 80 | 116.321 | 112.329 | 106.629 | 101.879 | 96.578 | 64.278 | 60.391 | 57.153 | 53.540 | 51.172 |
| 90 | 128.299 | 124.116 | 118.136 | 113.145 | 107.565 | 73.291 | 69.126 | 65.647 | 61.754 | 59.196 |
| 100 | 140.169 | 135.807 | 129.561 | 124.342 | 118.498 | 82.358 | 77.929 | 74.222 | 70.065 | 67.328 |

# Index

Addition rule, 121, 132
Alternative hypothesis, 236
Appendix, 305
Area under the standard normal curve, 174
Average, 28

Bar charts, 3, 11–12, 107–109
Bell shaped, 52, 171
Bernoulli experiment, 154
  trials, 144, 153–154
Binomial distribution, 144, 154–156
  experiment, 154
  table, 305
Box plots, 69–72

Calculators, 17
Categorical frequency distributions, 6
  variables, 103
Causation, 87–88
Census, 5
Central Limit Theorem, 189
  difference between two sample means, 201
  difference between two sample proportions, 199
  sample means, 195
  sample proportions, 199
Charts, 11
  bar, 3, 11–12, 107–109

Charts (*Cont.*):
  Pareto, 3, 17
  pie, 1, 16–17
Chi-square distribution, 289–291
  table, 314
  test for goodness-of-fit, 289, 291–293
  test for independence, 289, 293–295
Class width, 10
Classical probability, 123
Coefficient of determination, 90–91
  correlation, 86
  variation, 51
Conditional distributions, 103, 105
  probability, 121, 133
Confidence intervals, 215
  difference between two means (large samples), 215, 221–222
  difference between two means (small samples), 265, 270–272
  difference between two proportions, 215, 220–221
  means (large samples), 215, 218–219
  means (small samples), 265, 267–268
  means (dependent samples), 265, 274–275
  proportions, 215–217,
  repeated sample interpretation, 217
Contingency tables, 107

Continuous variables, 4, 5
Complement of an event, 131
  rule, 132
Compound events, 128
Computer software, 17
Correlation, 86–87
Critical region, 237
  value, 217, 237
Cumulative frequency, 8
  relative frequency, 8

Data, 4
Deciles, 68
Decision rule, 237
Degrees of freedom, 266
Dependent samples, 274
Descriptive statistics, 4
Deviations, 29
Discrete variables, 4–5
Distributions, 6
  binomial, 144, 154–156, 306–311
  chi-square, 289–291, 315
  conditional, 103, 105
  continuous, 171
  discrete, 144, 147
  frequency, 6
  marginal, 103, 104–105
  normal, 166, 167–170, 312
  probability, 121, 133
  sampling, 189, 190–203
  skewed, 33–34
  $t$, 266–267, 313
Dot plots, 3, 10–11

Empirical probability, 125
  rule, 51–53, 172–173
Error, 236
  Type I, 236
  Type II, 236
Events, 123
  complement, 131
  compound, 128
  independent, 134
  intersection, 129
  mutually exclusive, 130
  union, 128
Expectation, 144
Expected values, 144, 148–151

Factorials, 154
Frequency, 6
  cumulative, 8

Frequency (*Cont.*):
  distribution, 3, 6, 9
  polygons, 3, 13–14
  relative, 8

Grouped frequency distribution, 9
Goodness-of-fit test, 290–293

Histograms, 3, 12–13
Hypothesis tests, 235
  alternative, 236
  critical value, 237
  difference between two means (large samples),
    235, 246–248
  difference between two means (small samples),
    265, 272–274
  difference between two proportions (large samples), 235, 243–246
  five step procedure, 237
  mean (large samples), 235, 241–243
  mean (small samples), 265, 268–270
  mean (dependent samples), 265, 275–276
  null, 237
  one-tailed, 237
  proportion (large samples), 235, 237–241
  two-tailed, 137

Independence, 103
  categorical variables, 109–110
  events, 133–134
  probability, 121, 133–134
  samples, 265
Inferential statistics, 4
Influential points, 92
Interquartile range, 45–46
Intersection of events, 129–130

Law of Large Numbers, 121, 125–127
Laws of probability, 127
Least squares regression, 82
  line, 88
Level of significance, 236
Line of best fit, 88–90

Marginal distributions, 103, 104–105
Mean, 28–30
  absolute deviations, 46–47
Measures of central tendency, 27
  position, 63
  variability, 43
Median, 30–32

Minitab, 18, 35, 72, 93, 157, 179–180, 194, 197, 200, 223, 250, 277–278, 295
Mode, 32–33
Multiplication rule, 121
Mutually exclusive events, 130–131

Negatively skewed distribution, 34
Noncritical region, 237
Nonrejection region, 237
Normal probability distribution, 166, 167–170
  applications of, 177–179
  properties of, 170–172
Null hypothesis, 237

One-tailed test, 237
Outliers, 68–69, 92

Parameter, 5
Pareto charts, 3, 17
Percentiles, 65–68
Pie charts, 3, 16–17
Plots, 10
  box, 69–72
  dot, 3, 10–11
  scatter, 82–85
  stem-and-leaf, 3, 14–15
Population, 5
  coefficient of variation, 51
  mean, 28
  standard deviation, 50
  variance, 50
Positively skewed distribution, 33
Probability, 122
  for discrete random variables, 146–147
  distributions, 144
  relative frequency, 121, 125
*P* value, 248–249

Qualitative data, 5
  variables, 4
Quantitative data, 5
  variables, 4
Quartiles, 68

Random experiment, 122
  sample, 6
  variables, 145
Randomness, 122
Range, 44–45
Regression, 88
  analysis, 89
  least squares, 82, 88
  line, 88

Rejection region, 237
Relative frequency, 8
Residual plots, 91–92

Sample, 5
  coefficient of variation, 51
  mean, 28
  standard deviation, 50
  size for mean, 219–220
  size for proportion, 218
  space, 122–123
  variance, 48–49
Sampling distribution, 189
  difference between two sample means, 189, 200–203
  difference between two sample proportions, 189, 197–200
  sample mean, 189, 193–196
  sample proportion, 189, 190–193
Scatter plots, 82–85
Significance level, 236
Simpson's paradox, 103, 111–113
Skewness, 33
  negative, 34
  positive, 33
Slope of the line of best fit, 89
Standard deviation, 153
  binomial distribution, 156
  difference between two sample means, 201, 202
  difference between two sample means (dependent samples), 274
  difference between two sample proportions, 198, 199
  normal distribution, 170
  population, 50
  random variable, 153
  sample, 50
  sample means, 194, 195
  sample proportions, 191,192
  standard normal distribution, 173
Standard normal distribution, 173
Statistic, 5
Statistics, 4
Statistical hypothesis, 235
  test, 236
Stem-and-leaf plots, 3, 14–15
Symmetrical distribution, 34

*t* distribution, 265, 266–267
  table, 313
Technology, 17, 35, 53, 72, 92, 113, 135, 157, 179, 203, 223, 249, 277, 295
Test statistic 236, 237

TI-83, 35, 53–54, 72, 93, 157, 179–180, 223, 250, 277–278, 295
Time series graphs, 15–16
Two-tailed test, 137
Two-way tables, 103

Uncertainty, 122
Ungrouped frequency distribution, 7
Union of events, 128–129

Variables, 4
    categorical, 103
    continuous, 4, 5

Variables (*Cont.*):
    discrete, 4, 5
    qualitative, 4
    quantitative, 4
Variance, 47, 144
    binomial distribution, 156
    discrete random variable, 151–153
    population, 50
    sample, 48
    *t*-distribution, 266

*z* score, 64–65, 166, 217

# About the Author

**Lloyd R. Jaisingh, Ph.D.,** is a Professor of Mathematics at Morehead State University, Morehead, Kentucky. He is the author of both articles and books on elementary statistics and in the use of technology in the classroom. He has taught statistics at the university level for fifteen years and is the recipient of a number of grants and fellowships and an award as Outstanding Researcher of the Year at Morehead State.

When it comes to statistics, Dr. Jaisingh has been there and done that. He knows what it takes to eliminate the confusion, and he's written it all down for you in a manner that will take the stress out of studying statistics.